Super-radiance

Multiatomic Coherent Emission

Optics and Optoelectronics Series

Series Editors: **E R Pike** FRS, **B E A Saleh** and **S Lowenthal**

Optics and Optoelectronics Series

Super-radiance
Multiatomic Coherent Emission

M G Benedict

Attila József University, Szeged, Hungary

A M Ermolaev

University of Durham, UK
Université Libre de Bruxelles, Belgium

V A Malyshev

Vavilov State Optical Institute, St Petersburg, Russia

I V Sokolov

St Petersburg State University, St Petersburg, Russia

E D Trifonov

Russian Herzen University, St Petersburg, Russia

CRC Press
Taylor & Francis Group
Boca Raton London New York

CRC Press is an imprint of the
Taylor & Francis Group, an **informa** business

CRC Press
Taylor & Francis Group
6000 Broken Sound Parkway NW, Suite 300
Boca Raton, FL 33487-2742

First issued in paperback 2019

© 1996 by Taylor & Francis Group. LLC

CRC Press is an imprint of Taylor & Francis Group, an Informa business

No claim to original U.S. Government works

ISBN-13: 978-0-7503-0283-8 (hbk)
ISBN-13: 978-0-367-40144-3 (pbk)

Library of Congress catalog number· 96-31132

Library of Congress Cataloging-in-Publication Data

Catalog record is available from the Library of Congress

Visit the Taylor & Francis Web site at
http://www.taylorandfrancis.com

and the CRC Press Web site at
http://www.crcpress.com

To the memory of

Maria Ivanovna Petrashen

an outstanding scientist, teacher and personality

Contents

Preface

The plan for this book on super-radiance was motivated by the original work done on the subject by the authors over the last two decades. It is devoted mainly to the theoretical methods of investigating this effect. During the preparation of the manuscript, however, we had in mind that the book must be of value to a wide readership, the large community of researchers and graduate students in quantum electronics and quantum optics. To this end we have included an introductory chapter to tell the reader about the main features of super-radiance in a simple way. We have also devoted a chapter to reviewing some important experiments on the effect, and at the end, a brief chapter on some new applications of super-radiance.

The task of writing the chapters was divided amongst the authors as follows: Chapter 1 was written by Trifonov, Ermolaev and Benedict, Chapter 2 by Trifonov and Ermolaev, Chapter 3 by Sokolov, Chapter 4 by Trifonov, Chapter 5 by Benedict, Malyshev and Trifonov, Chapters 6, 7 and 8 by Malyshev, Chapter 9 by Benedict and Trifonov, Chapter 10 by Benedict and Malyshev and Chapter 11 by Benedict, Ermolaev, Sokolov and Trifonov, with most of the material being jointly discussed and collated during our meetings at Durham, Szeged and St Petersburg. Many of the authors' original works had been written with other collaborators. Without their work and assistance the book could hardly have been made possible. We would like to thank especially A Troshin, A Zaitsev, Yu Avetisyan, R Malikov, N Shamrov, V Pirogov, N Kaneva and D Bulyanitsa, and acknowledge the important contribution of the late D Smirnov. We thank Vl V and V V Kocharovsky for supplementing the material of section 11.3. Part of the work reported in the book has been done in collaboration with experimental groups. In particular, the experiments on coherent amplification and induced super-radiance were analysed jointly with Professor A Leontovich and his group at the Lebedev Physics Institute of the Russian Academy of Sciences, Moscow. The experimental work on super-radiance in nuclear-spin systems was examined together with the group of A I Kovalev at the Institute of Nuclear Physics, St Petersburg.

We would like to express our thanks to Professor V I Perel and Professor E E Fradkin for many fruitful discussions on the problems treated in this volume. We are grateful to Professor E R Pike of Kings College, London for the opportunity of discussing the original plan of this book, and for his valuable comments and suggestions.

Many thanks are due to Dr L Schwan of the University of Düsseldorf for providing us with his extensive bibliography on super-radiance, which we have used in our book, together with other sources. Because of the enormous number of papers on the subject it was impossible to include a fully comprehensive reference list in the book. We apologize to those authors whose work is not mentioned in our bibliography.

The authors thank the University of Durham, where work on the book started. M Benedict extends his thanks also to his home university at Szeged, and to the Hungarian Science Foundation (OTKA) for partial support under Grant No 1977; A M Ermolaev acknowledges support of the Université Libre de Bruxelles under the EC Grants ERBCHRXCT-940470 and ERBCBGCT-940552 during the final preparation of the manuscript; V Malyshev is grateful for the hospitality of the University of Salamanca where part of his work on the book was done; E D Trifonov acknowledges the support of the International Soros Science Program in 1994–96. E D Trifonov and M Benedict also express their thanks to the University of Durham for the hospitality shown to them during their visits to the UK. Many of the figures are taken from research papers and other publications, and we here acknowledge our debt of gratitude to the authors and publishers for their permission to use them.

Special thanks are due to Michael Cole who undertook the difficult task of typesetting the mathematical text, as well as for his invaluable help in improving the language of the manuscript. Our thanks go to Pauline Russell of the Department of Physics, University of Durham for the precision drawing of the illustrations for the book, and to Kati Kis Kovács of Szeged University for her help in compiling the bibliography.

Introduction

"For want of a better term, a gas which is radiating strongly because of coherence will be called 'super-radiant'."

R H Dicke 1954

If an atom is initially in an excited state, the transition to its ground state (i.e. spontaneous decay) will occur as a result of interacting with the vacuum fluctuations of the electromagnetic field. The theory of this phenomenon was first proposed by Dirac [D27] and then by Wigner and Weisskopf [WW30] at an early point in the development of quantum electrodynamics. A direct application of this theory to a system of several atoms is possible only if the spontaneous decays of the atoms are not correlated, and if they proceed independently. One then obtains the well known exponential law of luminescence damping with the natural radiation time of a single atom. For allowed optical transitions this time is of the order of 10^{-8} s.

Dicke [D54] was the first to recognize that within the framework of the problem of the interaction of atoms with the vacuum state of the electromagnetic field, the independence of the spontaneous decay of several identical atoms is an assumption only, and that a more exact consideration of this problem leads to quite different results. Dicke showed that the radiation decay time of the multi-atomic system depends, in fact, upon the number of atoms. It is proportional to the inverse of that number, and thus can be extremely short for sufficiently dense media. Furthermore, as has been shown by subsequent investigations, the decay of the electromagnetic pulse is no longer exponential, but has a peak (or some ringing) after a certain delay. This type of spontaneous radiation has become known as *super-radiance* (sometimes also referred to as *super-fluorescence*). The interaction of the atoms with each other through the common radiation electromagnetic field results in correlation between the atomic dipole moments, which leads to the creation of macroscopic optical polarization. The latter is proportional to the number of atoms, N. So the radiation intensity is proportional to the square of that number, and the duration of the radiation is

1

inversely proportional to N.

Super-radiance (SR) belongs to the class of cooperative coherent optical phenomena. Cooperation is easily seen to be happening in this effect—the radiation from atoms influences each other. The term 'coherent', in this case, refers not only to the electromagnetic field, but also to the atomic system itself. Indeed, super-radiance is effective only when other interactions, such as collisions, thermal noise etc, which are always present besides the interaction with the radiation electromagnetic field, do not disturb the phase of the atomic wavefunctions during emission. In other words, super-radiance exhibits itself only under conditions in which the phase of the wavefunction has a 'long memory'. As these additional interactions determine the width, Γ, of the spectral line (in addition to the natural width), the phase memory time is thus of the order of Γ^{-1}. When Γ^{-1} is shorter than the spontaneous decay time one has ordinary spontaneous emission, or amplification of the latter: *amplified spontaneous emission* (ASE), an effect which is sometimes also called *super-luminescence*. ASE can be described as a sequence of the real elementary acts of spontaneous and induced emission by the individual atoms. Against this, super-radiance proceeds so fast, that it is impossible, in principle, to subdivide the process into a sequence of separate events: the whole atomic system (if its dimensions do not exceed a certain cooperative length) radiates as a single complex. In one of his papers, Dicke [D64] referred to super-radiance as a *coherence-brightened laser*. The feedback in this laser is produced not by mirrors, but by the phase memory of the system which is 'burned' by the photons that have escaped the sample.

The main condition for super-radiance, as we have already said above, is the preservation of the phase memory of the atomic system during the process of spontaneous emission. One uses the term super-radiance, in a broader sense, also for any emission processes by such a system (e.g. under different initial conditions and ways of excitation) if it has relatively long phase memory. The partial and very important case of the super-radiance phenomenon—cooperative spontaneous emission from uncorrelated fully excited states—has gradually gained the special name of *super-fluorescence* (SF), which was introduced by Bonifacio and Lugiato [BL75a, BL75b]. Super-radiance is sometimes understood (in the narrow sense) to mean that emission occurs from an initially correlated state. Such a state can be prepared, for example, by coherent pulse pumping. A more detailed classification of super-radiance phenomena can be found in [Se86a]. However, since this terminology is not commonly adopted elsewhere, we prefer to use throughout the book the original name *super-radiance* in the broad sense, pointing out the initial condition or the character of excitation as and when needed. Nevertheless, in Chapter 2, which is a review of experimental work on super-radiance, we have kept to the name SF wherever it was used by the authors.

In 1973 the pioneering work of Skribanowitz *et al* [SHMF73] turned super-radiance from hypothesis into reality. Their experiment was the first to demonstrate super-radiance in the laboratory in HF gas. Since then super-radiance has been observed, in a series of experiments, in other gases as well as in solids.

Significant interest in the phenomenon of SR has been sustained to date, particularly as a result of the widening of the conditions where it can be observed. A considerable body of both theoretical and experimental work has been devoted to various aspects of SR. These have already been discussed and reviewed in several publications [F72, A74, AE75, SVP81, GH82, VG82, PY85, LPV86, AEI88, GSK88, ZKK89a]. None of these contributions is fully comprehensive and each is mainly based on the specific results achieved by their authors. This is a characteristic of the present book, as well.

Chapter 1 gives an introduction to the problem, in which we discuss the main features of super-radiance using some simplified models. This justifies its title: *The elementary theory*. This chapter is designed for the reader who is approaching this problem for the first time.

Chapter 2 presents a review of experimental work on super-radiance. We first describe experiments in gases, where super-radiance was observed in the nanosecond range. Recent experiments in solids have shown that super-radiant pulses can also be observed in the picosecond range. The second part of this chapter discusses observations of super-radiance in solids.

The quantum electrodynamical theory of super-radiance is considered in Chapter 3. It presents a generalization of the Wigner–Weisskopf approach to the N-atom system using a diagram method of non-stationary perturbation theory. We would like to draw particular attention to two results of this chapter: to the spectrum and time evolution of super-radiation of a two-atomic system, and to the photon angular correlation function for an N-atom system.

The results in Chapter 4 are based on the theorem of Glauber, Haake and co-workers [GH78, HKSHG79], which gives a foundation for the semiclassical approach. We discuss here several correlation functions for the field and the polarization, which characterize the coherence properties of super-radiation. Computer simulation of the process with stochastic initial conditions exhibits the self-organization of dipole moments in many-atomic systems—one of the main features of super-radiance.

In Chapter 5 we use the semiclassical approach to investigate the influence of de-phasing processes on radiation efficiency, or in other words, the influence of homogeneous and inhomogeneous broadening of the spectral line on the dynamics and spectrum of super-radiance. This problem is of practical interest because de-phasing quenches the coherent interaction of the atoms. Its solution permits one to obtain the threshold conditions for the observation of super-radiance. With the increase of homogeneous broadening, super-radiance transforms into amplified spontaneous emission. Though it is not super-radiance, some features of ASE are considered in detail. The semiclassical approach is also applied to describe coherent Raman scattering, which is a three-level super-radiant effect. We also discuss here briefly the possibility of lasing without inversion, a related question of recent interest.

Dipole–dipole atomic coupling is taken into account in Chapter 6. It is particularly significant for small systems whose dimensions are less than the wavelength of the radiation. The simplest model of a finite regular atomic chain is considered in detail. The results may be interesting for clarifying the conditions for the observation of super-radiance in a high-density exciton system.

Dipole–dipole coupling destroys super-radiance in small irregular systems. However, super-radiance can be observed in such a system if it is inside a cavity, as was first pointed out by Bloembergen and Pound [BP54]. This situation arises in a multi-spin system placed in a magnetic field. This problem is considered in Chapter 7. Attention is mainly paid to the interpretation of the experimental results [BBZKMT90] on radiational relaxation of proton spin polarization.

In Chapter 8 we investigate propagation effects in super-radiance in more detail than in Chapter 5. Two- and three-dimensional diffraction problems are considered, taking into account the stochastic initial condition of atomic polarization. We compare our theoretical results with the diffraction patterns of super-radiance obtained from $KCl:O_2^-$ in the experiments of Schiller *et al* [SSS87].

Chapter 9 treats reflection from the surface of a resonant medium. To this end we solve the corresponding boundary value problem without using the slowly varying amplitude approximation in space. It is shown, in particular, that if the thin-layer super-radiance condition is satisfied, then strong reflection of an incident short pulse occurs. Other results which we would like to point out are the retarded reflection of such a pulse, and the correlation of counter-propagating super-radiance pulses caused by internal resonant reflection.

In Chapter 10 we consider the boundary value problem for the transmission and reflection of short pulses, taking into account the local field correction. Two effects are of special interest: the population-dependent frequency shift and the possibility of mirror-less bistability of ultra-short electromagnetic pulses.

In the last chapter we give a brief discussion of some questions of present and future interest that are not included in the rest of the book. These are super-radiance in a high-gain, free-electron laser, gamma-ray super-radiance and recombination super-radiance in semiconductors. The analysis of the effect called sub-radiance (SBR), a description of a recent experiment that shows the realization of the original concepts of Dicke and the problems of squeezing and non-classical light in super-radiance close the chapter.

At the end of the book we also present a relatively comprehensive bibliography of super-radiance, divided into years of publication. References to other works, not dealing directly with super-radiance, are separated into another section called 'Other References' and cited numerically.

Chapter 1

The elementary theory of super-radiance

In this chapter we shall consider super-radiance on an elementary level in the sense that the main notions and concepts of the phenomenon will be treated as they evolved in the course of the early studies starting with the classical works by Dicke [D54]. The attention of the reader will be directed mainly to the physical aspects of the problem. In doing so the treatment will be restricted to simple models where results of some general significance can be obtained without using complicated mathematical tools. By making the exposition more accessible we have intended to lay down the groundwork for subsequent chapters of the book where some results obtained here will be refined and generalized.

1.1 Cooperative spontaneous emission of two two-level atoms separated by a distance less than the wavelength

We shall start with a system of two identical atoms separated by a distance less than the emission wavelength, but at the same time where the direct interaction between the atoms is neglected. This is the simplest model of an atomic system for which the spontaneous decay, as we shall see, is no longer that of a single atom.

We shall assume that each of the atoms may be either in the ground state $|g\rangle$ or in the excited stationary quantum state $|e\rangle$, $\langle g|g\rangle = \langle e|e\rangle = 1$, $\langle g|e\rangle = 0$ with respective non-degenerate energy levels E_g and E_e. The existence of other (excited) states of the atom will be neglected. The two-level model of the atom [AE75] is an idealization that is useful in the situations where coherent resonant interactions involve only two states of the atom. Radiative transitions in the atom will be considered in the dipole approximation. The transition matrix element d is given by

$$d = d_{eg} = \langle e|\widehat{d}|g\rangle \tag{1.1.1}$$

where $\widehat{d}$ is the operator of the electric dipole moment of the atom. The diagonal matrix elements are assumed to be zero, $\langle e|\widehat{d}|e\rangle = \langle g|\widehat{d}|g\rangle = 0$.

Let us consider the case when both of the atoms are initially in the excited state. We shall investigate the evolution of this two-atom system interacting with the vacuum state of the electromagnetic field.

The spontaneous decay rate (i.e. probability per unit time) for a single atom is given (see e.g. Loudon [50]) by the Dirac formula

$$\gamma = \frac{4}{3}\frac{d^2\omega_0^3}{\hbar c^3} \tag{1.1.2}$$

where $\hbar$ is Planck's constant ($\hbar = h/2\pi$), c the velocity of light, $\omega_0 = (E_e - E_g)/\hbar$, and d is the modulus of the vector d. Expression (1.1.2) is valid not only for an atom, but also for an arbitrary quantum system with an allowed dipole transition, provided the dimensions of such a system are less than that of the emission wavelength. So we can use equation (1.1.2) for evaluating the radiation decay rate of our two-atom system as well.

Under the assumptions made above, the system of two identical two-level atoms has the following three equidistant energy levels (see figure 1.1)

$$2E_g \qquad E_g + E_e \qquad 2E_e \tag{1.1.3}$$

the second level being two-fold degenerate. The corresponding symmetrized and normalized states of the total system are

$$|gg\rangle = |1, g\rangle|2, g\rangle$$
$$|s\rangle = \frac{1}{\sqrt{2}}(|1, g\rangle|2, e\rangle + |2, g\rangle|1, e\rangle)$$
$$|a\rangle = \frac{1}{\sqrt{2}}(|1, g\rangle|2, e\rangle - |2, g\rangle|1, e\rangle)$$
$$|ee\rangle = |1, e\rangle|2, e\rangle \tag{1.1.4}$$

where the numbers '1' and '2' label the atoms.

The states corresponding to the ground and the upper level are symmetrical, and there are one symmetrical and one antisymmetrical states corresponding to the intermediate level. Now, it is easy to obtain the dipole matrix elements of the total dipole moment $\widehat{d}_1 + \widehat{d}_2$ for transitions between all of these states

$$\langle ee|\widehat{d}_1 + \widehat{d}_2|s\rangle = \langle s|\widehat{d}_1 + \widehat{d}_2|gg\rangle = d\sqrt{2}$$
$$\langle ee|\widehat{d}_1 + \widehat{d}_2|a\rangle = \langle a|\widehat{d}_1 + \widehat{d}_2|gg\rangle = \langle ee|\widehat{d}_1 + \widehat{d}_2|gg\rangle = 0. \tag{1.1.5}$$

From (1.1.2) and (1.1.5) it follows that the corresponding transition rates are

$$\gamma_{ee,s} = \gamma_{s,gg} = 2\gamma \tag{1.1.6}$$

$$\gamma_{ee,a} = \gamma_{a,gg} = \gamma_{ee,gg} = 0.$$

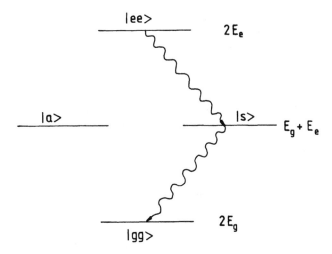

Figure 1.1. A schematic diagram of energy levels and allowed dipole transitions for two two-level atoms.

We find that a cascade of transitions 'ee' → 's' → 'gg' is permitted, but the direct transition 'ee' → 'gg' as well as a transition into the state 'a' are forbidden. This is a consequence of the symmetry properties of the dipole moment operator of the two-atom system which is invariant with respect to transposition of the atoms, so transitions are allowed only between states of the same symmetry. Another important result is that the transition rate at each of the steps of the cascade is twice the spontaneous decay rate for a single atom. This is a manifestation of the cooperative effect in the spontaneous radiation decay of a two-atom system.

For the spontaneous radiation decay of a single atom, the probability $P_e(t)$ of finding the atom in the excited state at time t satisfies the rate equation

$$\frac{dP_e(t)}{dt} = -\gamma P_e(t). \tag{1.1.7}$$

Its solution is the well known exponential decay law

$$P_e(t) = \exp(-\gamma t) \tag{1.1.8}$$

with radiation decay time γ^{-1}.

Let us generalize equation (1.1.7) to the case of a two-atom system. We can write, with the help of equations (1.1.6), the following system of rate equations

$$\frac{dP_{ee}(t)}{dt} = -2\gamma P_{ee}(t)$$

$$\frac{dP_s(t)}{dt} = 2\gamma P_{ee}(t) - 2\gamma P_s(t)$$

$$\tag{1.1.9}$$

$$\frac{dP_a(t)}{dt} = 0$$

$$\frac{dP_{gg}(t)}{dt} = 2\gamma P_s(t)$$

where $P_k(t)$ is the probability of finding the system in the state 'k' at time t. Integrating this system with the initial conditions $P_{ee}(0) = 1$, $P_s(0) = P_a(0) = P_{gg}(0) = 0$, which correspond to the initial excitation of both of the atoms, we obtain

$$P_{ee}(t) = \exp(-2\gamma t)$$

$$P_s(t) = 2\gamma t \exp(-2\gamma t)$$

$$P_a(t) = 0 \tag{1.1.10}$$

$$P_{gg}(t) = 1 - (1 + 2\gamma t)\exp(-2\gamma t).$$

We can now find the average photon emission rate as a function of time

$$W(t) = \gamma_{ee,s}P_{ee} + \gamma_{s,gg}P_s$$

$$= 2\gamma P_{ee} + 2\gamma P_s \tag{1.1.11}$$

$$= 2\gamma(1 + 2\gamma t)\exp(-2\gamma t).$$

This result differs from that for two isolated atoms, $2\gamma \exp(-\gamma t)$, by a factor of $(1 + 2\gamma t)\exp(-\gamma t)$. As will be shown in the next section, this difference becomes even more radical as the number of atoms in the system increases.

The refined quantum electrodynamical theory of a system of two two-level atoms will be given in section 3.5 of Chapter 3. At the end of the book, in section 11.5, we discuss a recent experiment [DVB96], where the ideal situation described above has been almost realized.

Let us now consider how a de-phasing process may reduce the cooperative spontaneous decay of the system to the ordinary one, as in fact usually happens. The atomic ground and excited states evolve in time with the phase factors $\exp(-iE_g t/\hbar)$ and $\exp(-iE_e t/\hbar)$ respectively. Any external perturbation which is additional to the interaction with the radiation electromagnetic field disturbs the states, leading to some energy shifts of the levels, $\hbar \Delta_g$ and $\hbar \Delta_e$, and to the appearance of the corresponding additional phase factors in the states. For the (primed) perturbed states we shall write

$$|g'\rangle = |g\rangle \exp(-i\Delta_g t)$$

$$|e'\rangle = |e\rangle \exp(-i\Delta_e t). \tag{1.1.12}$$

For a sufficiently complicated interaction the energy shifts are arbitrary, and the corresponding phase factors destroy the symmetry of the intermediate states (see equations (1.1.4)). We shall examine the consequences of this for the square modulus of the matrix element of the dipole moment

$$|\langle ee|\widehat{d}_1 + \widehat{d}_2|s\rangle|^2 \equiv w(ee, s) \qquad (1.1.13)$$

which determines the rate of the transition 'ee'→'s'. For simplicity, let us consider a particular case where the distribution $g(\Delta)$ of the transition frequency shifts $\Delta_j = \Delta_{ej} - \Delta_{gj}$, $j = 1, 2$, is given by the Lorentz formula

$$g(\Delta) = \frac{\Gamma}{\pi} \frac{1}{\Delta^2 + \Gamma^2} \qquad (1.1.14)$$

where Γ is the half-width of the distribution, assumed to be the same for both atoms. We use the perturbed states (1.1.12) in formula (1.1.13), and average it over the shifts with the help of (1.1.14)

$$w(e'e', s')_{\text{AV}} = \int_{-\infty}^{\infty} d\Delta_1 \int_{-\infty}^{\infty} d\Delta_2 \, g(\Delta_1) g(\Delta_2) w(e'e', s'). \quad (1.1.15)$$

Making use of the integrals

$$\int_{-\infty}^{\infty} g(x) \, dx = 1 \qquad (1.1.16)$$

and

$$\int_{-\infty}^{\infty} g(x) \exp(\pm ixt) \, dx = \exp(-\Gamma t) \qquad (1.1.17)$$

we obtain the following final result

$$w(e'e', s')_{\text{AV}} = |d|^2 (1 + \exp(-2\Gamma t)). \qquad (1.1.18)$$

The right-hand side of expression (1.1.18) coincides with the corresponding result without de-phasing, obtained from equation (1.1.5), but only for a sufficiently small time. When the time exceeds Γ^{-1} the magnitude of the matrix element (1.1.18) approaches that for the ordinary spontaneous decay. Therefore Γ^{-1} can be taken as a time scale of the phase memory of the quantum state (i.e. *phase memory time*). If the phase memory time is less than the radiation decay time, cooperative spontaneous decay is impossible. Hence, the necessary condition for observation of the cooperative spontaneous emission is that the phase memory time be sufficiently long. As seen from the discussion above, the loss of phase memory is in close correspondence with a broadening of the corresponding spectral line, which in the present case is described by (1.1.14). This type of broadening is called homogeneous, because the disturbances leading to the perturbation of the phase factors, as given by (1.1.12), are the same on average for each atom. Γ^{-1} is often denoted by T_2, and it is called the time constant of homogeneous broadening. We note that there exists another effect which may disrupt cooperativity, namely inhomogeneous broadening, resulting from a spread in the transition frequencies, $(E_e - E_g)/\hbar$ of the different atoms in the system. A detailed discussion of these broadening mechanisms can be found in [AE75]. In order to reduce the problem to its bare essentials we shall assume in this chapter that the phase memory time of the atomic system is infinite. The discussion of the influences of the homogeneous and inhomogeneous broadenings on super-radiance will be given in Chapter 5.

1.2 Super-radiance of a system of N two-level atoms in a small volume: the Dicke model

The spontaneous emission of an ensemble of N two-level atoms confined to a volume with dimensions small compared with the wavelength of the emitted radiation can be treated as an extension of the two-atom theory developed in the previous section. The problem was first formulated and solved by Dicke in his original paper [D54], therefore this model of super-radiance is usually referred to as the *Dicke model*. Here we shall generally follow his presentation which was based on spin-$\frac{1}{2}$ formalism.

Since any state $|\psi\rangle$ of a two-level atom is a superposition of two states $|e\rangle$ and $|g\rangle$

$$|\psi\rangle = c_e|e\rangle + c_g|g\rangle \tag{1.2.1}$$

it can be represented by the column matrix

$$\psi = \begin{pmatrix} c_e \\ c_g \end{pmatrix} \tag{1.2.2}$$

where c_e and c_g are arbitrary complex coefficients subject to the condition $|c_e|^2 + |c_g|^2 = 1$. Accordingly, any quantum mechanical operator applied to such a state can be represented as a 2×2 square matrix. Then for the free atomic Hamiltonian we have

$$\widehat{H}_1 = \begin{pmatrix} \frac{1}{2}\hbar\omega_0 & 0 \\ 0 & -\frac{1}{2}\hbar\omega_0 \end{pmatrix} \tag{1.2.3}$$

where the zero energy has been chosen at the mid-point of the energy interval $\hbar\omega_0 = E_e - E_g$. The atomic dipole moment operator $\widehat{d}$ can be written as an off-diagonal matrix

$$\widehat{d} = \begin{pmatrix} 0 & d_{eg} \\ d_{ge} & 0 \end{pmatrix}. \tag{1.2.4}$$

All the two-level atomic Hermitian operators can be expressed in terms of the Pauli matrices and the unit 2×2 matrix

$$\widehat{\sigma}_1 = \begin{pmatrix} 0 & 1 \\ 1 & 0 \end{pmatrix} \qquad \widehat{\sigma}_2 = \begin{pmatrix} 0 & -i \\ i & 0 \end{pmatrix} \qquad \widehat{\sigma}_3 = \begin{pmatrix} 1 & 0 \\ 0 & -1 \end{pmatrix}$$

$$\widehat{\sigma}_0 = \begin{pmatrix} 1 & 0 \\ 0 & 1 \end{pmatrix}.$$

(1.2.5)

In particular, in accordance with (1.2.3) and (1.2.4)

$$\widehat{H}_1 = \tfrac{1}{2}\hbar\omega_0\widehat{\sigma}_3 \qquad \widehat{d} = \operatorname{Re} d\,\widehat{\sigma}_1 + \operatorname{Im} d\,\widehat{\sigma}_2 \qquad (1.2.6)$$

where $\operatorname{Re} d$ and $\operatorname{Im} d$ are the corresponding real and imaginary parts of d_{ge}. Owing to the arbitrariness of phase of the state vector we can always assume that the dipole matrix element d is real.

Let us introduce the quasi-spin operators for the ith atom

$$\widehat{R}_1^{(i)} = \tfrac{1}{2}\widehat{\sigma}_1^{(i)} \qquad \widehat{R}_2^{(i)} = \tfrac{1}{2}\widehat{\sigma}_2^{(i)} \qquad \widehat{R}_3^{(i)} = \tfrac{1}{2}\widehat{\sigma}_3^{(i)} \qquad (1.2.7)$$

and the total quasi-spin operators of the N-atom system

$$\widehat{R}_\alpha = \sum_{i=1}^{N} \widehat{R}_\alpha^{(i)} \qquad \alpha = 1, 2, 3 \qquad (1.2.8)$$

which obey the usual commutation rules of angular momentum operators

$$[\widehat{R}_\alpha^{(k)}, \widehat{R}_\beta^{(j)}] = i\delta_{kj}\widehat{R}_\gamma^{(k)} \qquad [\widehat{R}_\alpha, \widehat{R}_\beta] = i\widehat{R}_\gamma \qquad (1.2.9)$$

where α, β, γ is any cyclic permutation of the numbers 1, 2, 3. Then the energy operator of an ensemble of N free identical two-level atoms can be expressed in the form

$$\widehat{H}_N = \sum_{i=1}^{N} \hbar\omega_0 \widehat{R}_3^{(i)} = \hbar\omega_0 \widehat{R}_3 \qquad (1.2.10)$$

and, consequently, its eigenvalues are equal to $\hbar\omega_0 M$, where $M = \tfrac{1}{2}N, \tfrac{1}{2}N - 1, \ldots, -\tfrac{1}{2}N$ is the eigenvalue of $\widehat{R}_3$.

Hence, a system of N two-level atoms has $N+1$ equidistant energy levels. The bottom level corresponds to the ground state of the system, whilst at the top level all atoms are in the excited state. All the levels, save these two, are degenerate.

The Hamiltonian for the interaction of such an N-atom system with the electromagnetic field can be chosen to be in the form

$$\widehat{H}_{\text{int}} = -\widehat{\mathcal{E}} \sum_{i=1}^{N} \widehat{d}^{(i)} = -\widehat{\mathcal{E}}d \sum_{i=1}^{N} 2\widehat{R}_1^{(i)} \qquad (1.2.11)$$

where $\widehat{\mathcal{E}}$ is the operator of the electric field, $\widehat{d}^{(i)}$ is the electric dipole moment operator of the ith atom, $\widehat{d}^{(i)} = 2d\widehat{R}_1^{(i)}$.

Let us now suppose that all atoms are initially in their excited states (i.e. the system is at the top level). Let us investigate the subsequent time evolution of this system.

The initial state is the eigenstate of $\widehat{R}_3$, with the maximum eigenvalue $\frac{1}{2}N$. This state, as is known from the theory of angular momentum [47], is the eigenstate of the square of total angular momentum $\widehat{R}^2 = \widehat{R}_1^2 + \widehat{R}_2^2 + \widehat{R}_3^2$, with the eigenvalue $\frac{1}{2}N(\frac{1}{2}N + 1)$.

It is easily verified that the operator $\widehat{R}^2$ commutes with the total Hamiltonian $\widehat{H}_N + \widehat{H}_{\text{int}}$. This follows from the symmetry of the Hamiltonian with respect to permutation of any pair of the pseudo-spin operators for the individual atoms. So $\widehat{R}^2$ is the integral of motion, and its eigenvalues must be conserved in time. The states corresponding to the maximal eigenvalue of $\widehat{R}^2$ will be denoted as $|\frac{1}{2}N, M\rangle$. They satisfy the relations

$$\widehat{R}^2|\tfrac{1}{2}N, M\rangle = \tfrac{1}{2}N(\tfrac{1}{2}N + 1)|\tfrac{1}{2}N, M\rangle$$
$$\widehat{R}_3|\tfrac{1}{2}N, M\rangle = M|\tfrac{1}{2}N, M\rangle. \tag{1.2.12}$$

$\widehat{H}_{\text{int}}$ can be written in the form

$$\widehat{H}_{\text{int}} = -\widehat{\mathcal{E}}d(\widehat{R}_+ + \widehat{R}_-) \tag{1.2.13}$$

where $\widehat{R}_+ = \widehat{R}_1 + i\widehat{R}_2$ and $\widehat{R}_- = \widehat{R}_1 - i\widehat{R}_2$.

Owing to the property

$$\widehat{R}_+|\tfrac{1}{2}N, M\rangle = \sqrt{(\tfrac{1}{2}N - M)(\tfrac{1}{2}N + M + 1)}\,|\tfrac{1}{2}N, M + 1\rangle$$
$$\widehat{R}_-|\tfrac{1}{2}N, M\rangle = \sqrt{(\tfrac{1}{2}N - M + 1)(\tfrac{1}{2}N + M)}\,|\tfrac{1}{2}N, M - 1\rangle \tag{1.2.14}$$

the operators $\widehat{R}_+$, $\widehat{R}_-$ have non-vanishing matrix elements only between the states with neighbouring values of M. Consequently, the radiative decay of an ensemble of the initially excited atoms is a cascade of transitions between the adjacent states with the same eigenvalue of $\widehat{R}^2$ equal to $\frac{1}{2}N(\frac{1}{2}N + 1)$

$$|\tfrac{1}{2}N, \tfrac{1}{2}N\rangle \to |\tfrac{1}{2}N, \tfrac{1}{2}N - 1\rangle \to \cdots \to |\tfrac{1}{2}N, -\tfrac{1}{2}N\rangle \tag{1.2.15}$$

The corresponding diagram of states is shown in figure 1.2.

Let us find the probability per unit time, $\gamma_{M,M-1}$ for a transition $M \to M - 1$. For this purpose we can use the expression (1.1.2) with a transition

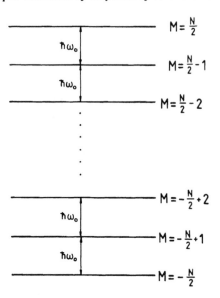

$M = \frac{N}{2}$

$\hbar\omega_0$

$M = \frac{N}{2} - 1$

$\hbar\omega_0$

$M = \frac{N}{2} - 2$

$M = -\frac{N}{2} + 2$

$\hbar\omega_0$

$M = -\frac{N}{2} + 1$

$\hbar\omega_0$

$M = -\frac{N}{2}$

Figure 1.2. An energy-level diagram for the initially fully excited system of N two-level atoms. Energy levels with $|M| < \frac{1}{2}N$ are degenerate, but transitions may take place only between states with a fixed eigenvalue of $\widehat{R}^2$. The figure shows the transitions between the states with the eigenvalue $\frac{1}{2}N(\frac{1}{2}N + 1)$ of $\widehat{R}^2$.

matrix element $d_{M,M-1}$, which can be obtained with the help of (1.2.14)

$$d_{M,M-1} = 2d\langle \tfrac{1}{2}N, M|\widehat{R}_1|\tfrac{1}{2}N, M - 1\rangle$$
$$= d\sqrt{(\tfrac{1}{2}N + M)(\tfrac{1}{2}N - M + 1)}. \qquad (1.2.16)$$

Putting this expression into (1.1.2), we obtain

$$\gamma_{M,M-1} = \gamma(\tfrac{1}{2}N + M)(\tfrac{1}{2}N - M + 1). \qquad (1.2.17)$$

It is seen that $\gamma_{M,M-1}$ attains the maximum value for $M = 0$ if the number of atoms is even, or for $M = \frac{1}{2}$ if the number of atoms is odd, i.e. for the states in which the numbers of atoms in the upper and ground states are equal or differ by one. In both these cases the decay rate is proportional to N^2.

Now we can write the rate equations which generalize equations (1.1.9), for the case of an N atom ensemble

$$\frac{dP_M(t)}{dt} = \gamma_{M+1,M}P_{M+1}(t) - \gamma_{M,M-1}P_M(t) \qquad (1.2.18)$$

$$(M = \tfrac{1}{2}N, \tfrac{1}{2}N - 1, \ldots, -\tfrac{1}{2}N)$$

where $P_M(t)$ is the probability of finding the system in the state $|\frac{1}{2}N, M\rangle$ at time t. The equations (1.2.18) are to be solved with the initial conditions $P_{N/2}(0) = 1$ and $P_M(0) = 0$ for $M \neq \frac{1}{2}N$.

In principle, the solution of equations (1.2.18) can be obtained in an analytical form similar to (1.1.10) for a two-atom system, by successive integration starting from $M = \frac{1}{2}N$. However, for a large number of atoms in the ensemble such a procedure leads to cumbersome analytical expressions for $P_M(t)$ and it is more convenient to solve the system numerically [BSH71b]. Some results obtained in [BSH71b] for $N = 200$ are shown in figure 1.3. It is seen that close to the initial moment the probability distribution P_M exhibits a sharp peak about the point $M = \frac{1}{2}N$. The distribution then becomes broader, and its variance reaches a maximum value of the order of N^2. This indicates a large uncertainty in the number of photons emitted up to that time (which corresponds to the maximum of the radiation intensity). At the end of the process the distribution P_M narrows again, peaking at $M = -\frac{1}{2}N$.

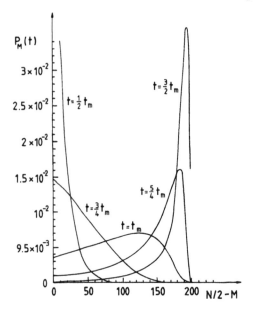

Figure 1.3. The probability distribution $P_M(t)$ at various times. t_m is the instant of the pulse's maximum, the number of atoms is $N = 200$, and the system starts from the state where all atoms are in their excited states: $P_M(0) = \delta_{N/2,M}$ [BSH71b].

Once the distribution P_M is known, we can calculate the mean radiation intensity (the average number of photons emitted per unit time) as the decreasing rate of the mean atomic energy

$$\bar{I}(t) = -\frac{d}{dt}\langle M \rangle = -\frac{d}{dt} \sum_{M=-N/2}^{N/2} M P_M(t) = -\sum_{M=-N/2}^{N/2} M \frac{dP_M(t)}{dt}. \qquad (1.2.19)$$

With the aid of equations (1.2.18) we can readily obtain that $\bar{I}(t)$ can be represented as a mean probability emission rate

$$\bar{I}(t) = \sum_{M=-\frac{1}{2}N+1}^{N/2} \gamma_{M,M-1} P_M(t). \tag{1.2.19a}$$

Let us now find the average time for the emission of n photons, $\bar{t}(n)$. The quantity n can be also treated as the number of atoms decayed to the ground level, so we can set $n = \frac{1}{2}N - M$. The time $\bar{t}(n)$ can be evaluated as a sum of the average times $\gamma_{M,M-1}^{-1}$ of adjacent transitions (1.2.15)

$$\bar{t}(n) = \sum_{M=\frac{1}{2}N-n+1}^{N/2} \gamma_{M,M-1}^{-1}$$

$$= \gamma^{-1} \sum_{M=\frac{1}{2}N-n+1}^{N/2} [(\tfrac{1}{2}N+M)(\tfrac{1}{2}N-M+1)]^{-1}. \tag{1.2.20}$$

If N is sufficiently large we can approximate the summation in (1.2.20) by an integration

$$\bar{t}(n) = \gamma^{-1} \int_{\frac{1}{2}N-n}^{N/2} \frac{\mathrm{d}M}{(\frac{1}{2}N+M)(\frac{1}{2}N-M+1)}$$

$$= \frac{2\gamma^{-1}}{N+1} \tanh^{-1} \frac{2M+1}{N+1} \Big|_{M=\frac{1}{2}N-n}^{M=N/2}. \tag{1.2.21}$$

Solving equation (1.2.21) with respect to n, we obtain for $N \gg 1$

$$n = \tfrac{1}{2}N + \tfrac{1}{2}N \tanh[\gamma \tfrac{1}{2}N(\bar{t} - t_{\mathrm{D}})]$$

$$t_{\mathrm{D}} = (\gamma N)^{-1} \ln N. \tag{1.2.22}$$

This gives the following expression for the radiation intensity

$$\tilde{I}(t) = \frac{\mathrm{d}n}{\mathrm{d}t} = \gamma \tfrac{1}{4}N^2 \operatorname{sech}^2[\gamma \tfrac{1}{2}N(t - t_{\mathrm{D}})] \tag{1.2.23}$$

where we have replaced $\bar{t}$ by t.

We note that the intensity (1.2.23) is not precisely equivalent to the average intensity determined by equation (1.2.19), in view of the different averaging procedures. The intensity (1.2.23), and the intensity (1.2.19), are plotted in figure 1.4. These curves are generally close to each other, although they have different peak values and slightly different peak times.

From equation (1.2.23) the main features of the super-radiant pulse are readily seen.

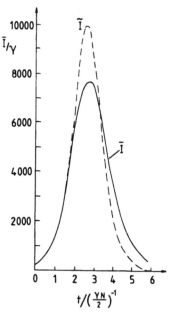

Figure 1.4. Mean radiation intensities $\tilde{I}(t)$ (1.2.19) and $\bar{I}(t)$ (1.2.23) for an ensemble of $N = 200$ atoms.

(i)   The initial intensity of the super-radiant pulse is the same as that of ordinary spontaneous emission of N atoms, $I(0) = \gamma N$.

(ii) Its duration measured at half maximum is of the order $(\gamma N)^{-1}$, i.e. N times *shorter* than the radiation decay time of a single atom.

(iii) The maximum of the intensity is proportional to the square of the total number of atoms N, and appears after a delay time of $t_D = (\gamma N)^{-1} \ln N$.

The reason for (ii) and (iii) is the creation of the macroscopic polarization of the atomic ensemble in the course of spontaneous emission. In order to show this, let us consider the expectation value of the product of the operators $\widehat{D}_\pm = d\widehat{R}_\pm$ for the total electric dipole moment. We have

$$\langle \widehat{D}_+ \widehat{D}_- \rangle = \sum_{M=-N/2}^{N/2} \langle \tfrac{1}{2}N, M | \widehat{D}_+ \widehat{D}_- | \tfrac{1}{2}N, M \rangle P_M(t). \qquad (1.2.24)$$

Using the relation

$$\langle \tfrac{1}{2}N, M | \widehat{R}_+ \widehat{R}_- | \tfrac{1}{2}N, M \rangle = (\tfrac{1}{2}N + M)(\tfrac{1}{2}N - M + 1) \qquad (1.2.25)$$

which follows from (1.2.14), we can express the expectation value (1.2.24) in terms of the mean intensity (1.2.19a)

$$\langle \widehat{D}_+ \widehat{D}_- \rangle = d^2 \sum_{M=-N/2}^{N/2} \langle \tfrac{1}{2}N, M | \widehat{R}_+ \widehat{R}_- | \tfrac{1}{2}N, M \rangle P_M(t)$$

$$= \frac{1}{\gamma} d^2 \bar{I}(t). \tag{1.2.26}$$

So the magnitude of $\bar{I}(t)$ directly characterizes the coherence of the total electric dipole moment. We see that the latter does attain the macroscopic value of order N near the maximum of intensity.

Let us show that this is related to the dipole–dipole correlation of different atoms in the ensemble. Indeed, we have

$$\langle \tfrac{1}{2}N, M | \widehat{R}_+ \widehat{R}_- | \tfrac{1}{2}N, M \rangle$$
$$= \langle \tfrac{1}{2}N, M | \sum_i \widehat{R}_+^{(i)} \sum_j \widehat{R}_-^{(j)} | \tfrac{1}{2}N, M \rangle$$
$$= \langle \tfrac{1}{2}N, M | \sum_{i \neq j} \widehat{R}_+^{(i)} \widehat{R}_-^{(j)} | \tfrac{1}{2}N, M \rangle + \langle \tfrac{1}{2}N, M | \sum_i \widehat{R}_+^{(i)} \widehat{R}_-^{(i)} | \tfrac{1}{2}N, M \rangle.$$

$$\tag{1.2.27}$$

Since all states $|\tfrac{1}{2}N, M\rangle$ are symmetrical with respect to any transposition of the atoms [47] it then follows that

$$\langle \tfrac{1}{2}N, M | \sum_{i \neq j} \widehat{R}_+^{(i)} \widehat{R}_-^{(j)} | \tfrac{1}{2}N, M \rangle$$
$$= N(N-1)\langle \tfrac{1}{2}N, M | \widehat{R}_+^{(i)} \widehat{R}_-^{(j)} | \tfrac{1}{2}N, M \rangle. \tag{1.2.28}$$

Now, using the relation [47]

$$\widehat{R}_+^{(i)} \widehat{R}_-^{(i)} = \widehat{R}^{(i)2} - \widehat{R}_3^{(i)2} + \widehat{R}_3^{(i)} \tag{1.2.29}$$

we have

$$\langle \tfrac{1}{2}N, M | \sum_i \widehat{R}_+^{(i)} \widehat{R}_-^{(i)} | \tfrac{1}{2}N, M \rangle = \tfrac{1}{2}N + M. \tag{1.2.30}$$

Then with the aid of (1.2.25) and (1.2.30) we obtain

$$(\tfrac{1}{2}N + M)(\tfrac{1}{2}N - M + 1)$$
$$= N(N-1)\langle \tfrac{1}{2}N, M | \widehat{R}_+^{(i)} \widehat{R}_-^{(j)} | \tfrac{1}{2}N, M \rangle + \tfrac{1}{2}N + M \tag{1.2.31}$$

whence

$$\langle \tfrac{1}{2}N, M | \widehat{R}_+^{(i)} \widehat{R}_-^{(j)} | \tfrac{1}{2}N, M \rangle = \frac{\tfrac{1}{4}N^2 - M^2}{N(N-1)}. \tag{1.2.32}$$

Expression (1.2.32) has the meaning of a pair-correlation function of the dipole moments of different atoms in the ensemble. It can be seen that for a fixed N this function changes from zero to a maximum value of $\frac{1}{4}$ when M goes from $\frac{1}{2}N$ to zero, and that it then decreases down to zero when M goes from zero to $-\frac{1}{2}N$.

In concluding this section, we point out that the theory of super-radiance presented above assumed that all the atoms in the system are strictly identical. In reality, unless the atoms are arranged in a very symmetrical way, so that they have an equivalent environment, the near-field dipole–dipole coupling between them will lead to a site-dependent perturbation of the relevant atomic levels. This effect causes a de-phasing, as was first pointed out by Friedberg *et al* [FHM72, FHM73]. A theory which includes the dipole–dipole coupling will be developed in Chapter 6. Nonetheless, the present simplified model can be applied to 'pencil-shaped' extended systems where the influence of near-field dipole–dipole coupling is negligible (see also Chapters 3 and 6). In this latter case, however (as will be shown in Chapter 3), only those photon states give a contribution to the probability of the cooperative spontaneous decay, for which the wavevectors are confined in the diffraction solid angle λ^2/A, where A is the cross-section of the pencil-shaped sample (for the spontaneous decay of a small system, considered above, one takes into account all possible directions of photon wavevectors). Accordingly, the time scale of super-radiance will be of the order

$$T_R \approx \left(\gamma N \frac{\lambda^2}{A}\right)^{-1} = (\gamma N_0 \lambda^2 L)^{-1} \qquad (1.2.33)$$

where N_0 is the number of atoms per unit volume of the sample of length L. Thus we see that for an extended system the duration of the super-radiant pulse decreases by a factor equal to the number of atoms in the volume $\lambda^2 L$, rather than by a factor equal to the total number of atoms N as in the case of a small sample. We shall return to the extended system in section 1.4 where the semiclassical theory of super-radiation will be formulated on an elementary level.

1.3   Photon statistics of super-radiance in the Dicke model

Here we consider, as above, super-radiance from a system with dimensions small compared with the wavelength of the emission, i.e. the Dicke model. The results to be obtained are also applicable to the single-mode super-radiance of an extended system discussed briefly at the end of the previous section.

The temporal evolution of an ensemble of N two-level excited atoms is governed by equations (1.2.18). The quantity $P_M(t)$ in these equations can also be interpreted as the probability of $n = \frac{1}{2}N - M$ photons being emitted up to the instant t after the initial excitation. Let us consider the statistics of the total number of emitted photons. The result of photon counting will depend upon the measurement procedure, because of the reduction of the quantum mechanical state by an observation. If we suppose that at the initial instant all atoms are in upper states (total inversion) then the probability of the first observation can be obtained from equations (1.2.18) solved with the initial condition $P_M(0) = \delta_{M,N/2}$. Let us assume that the result of the first measurement is n_0 detected photons. Then the probability of the next observation will be obtained from the same equation with the new initial condition $P_M(0) = \delta_{M,N/2-n_0}$ etc. As will be shown, the solution of (1.2.18) is very sensitive to the initial condition. Large fluctuations of the photon number will manifest themselves only at the initial and final stages of super-radiance. The main part of the process can be treated as non-fluctuating, to good accuracy.

In order to support this qualitative picture, let us investigate in more detail how the initial condition affects the distribution $P_M(t)$ in the early stage of super-radiance. We restrict ourselves to the case $n = \frac{1}{2}N - M \ll N$ when from (1.2.17) we have approximately: $\gamma_{M,M-1} = \gamma N(n+1)$ and $\gamma_{M+1,M} = \gamma Nn$. Then equation (1.2.18) can be written as

$$\frac{dP_M(t)}{dt} = -\gamma N(n+1)P_M(t) + \gamma Nn P_{M+1}(t). \qquad (1.3.1)$$

For the initial condition

$$P_M(0) = \delta_{\frac{1}{2}N-M,n_0} \qquad (1.3.2)$$

the solution of (1.3.1) is

$$P_{\frac{1}{2}N-n}(t) = C_n^{n_0} \exp[-\gamma N(n_0+1)t]\{1 - \exp[-\gamma Nt]\}^{n-n_0} \qquad (1.3.3)$$

where $C_n^{n_0}$ is a binomial coefficient. In particular, for $n_0 = 0$ we obtain from (1.3.3)

$$P_{\frac{1}{2}N-n}(t) = \exp[-\gamma Nt]\{1 - \exp[-\gamma Nt]\}^n. \qquad (1.3.4)$$

In this case the distribution decreases monotonically with increasing n, and, as time goes on, its slope decreases.

For the initial condition (1.3.2) with $n_0 > 0$, the distribution $P_{N/2-n}(t)$ has a different behaviour. Initially it is also a monotonically decreasing function of n. However, after a certain time interval, which can be determined from the equation

$$\exp[-\gamma Nt] = \frac{n_0}{n_0+1} \qquad (1.3.5)$$

the distribution peaks at around the mean photon number

$$\langle n(t) \rangle = \sum_n n P_{\frac{1}{2}N-n}(t) = (n_0+1)\exp(\gamma Nt) - 1 \qquad (1.3.6)$$

with variance

$$\langle(n(t) - \langle n(t)\rangle)^2\rangle = \langle n^2(t)\rangle - \langle n(t)\rangle^2$$
$$= (n_0 + 1)[\exp(2\gamma Nt) - \exp(\gamma Nt)]$$
$$\approx \frac{\langle n(t)\rangle^2}{n_0 + 1}. \tag{1.3.7}$$

Hence, for a fixed value of $\langle n\rangle$ the distribution $P_{N/2-n}$ narrows as n_0 increases, its width being proportional to $1/\sqrt{n_0}$. These results are valid only for the initial stage of super-radiance, when the number of emitted photons is much less than the number of atoms. It can be readily shown that the mean intensity (1.2.19) in this case is

$$\overline{I_{n_0}(t)} = \gamma N(n_0 + 1)\exp(\gamma Nt). \tag{1.3.8}$$

In particular, for the fully excited initial state we get

$$\overline{I_0(t)} = \gamma N \exp(\gamma Nt). \tag{1.3.9}$$

This result is in agreement with that determined by (1.2.23). It is interesting to compare the intensity (1.3.9) with that for ordinary spontaneous emission

$$I_{sp}(t) = \gamma N \exp(-\gamma t). \tag{1.3.10}$$

In both cases the intensity has the same initial value, γN, but for ordinary spontaneous emission there is an exponential decrease with rate γ, whilst for super-radiance there is an exponential increase with the anomalous rate γN.

The system (1.2.18) can also be simplified in another limit, $\frac{1}{2}N + M \ll \frac{1}{2}N$, i.e. at the end of the cascade transition (1.2.12). Then, instead of (1.3.1), we obtain

$$\frac{dP_M(t)}{dt} = -\gamma Nm P_M(t) + \gamma N(m + 1)P_{M+1}(t) \tag{1.3.11}$$

where $m = \frac{1}{2}N + M$ is the number of excited atoms. The solution of (1.3.11) for the initial condition

$$P_M(0) = \delta_{\frac{1}{2}N+M,m_0} \tag{1.3.12}$$

is the binomial distribution

$$P_{m-\frac{1}{2}N}(t) = C_{m_0}^m \exp(-\gamma Nmt)[1 - \exp(-\gamma Nt)]^{m_0-m}. \tag{1.3.13}$$

Consequently, the mean value $\langle m\rangle$ and the variance $\langle(m - \langle m\rangle)^2\rangle$ in that case are

$$\langle m(t)\rangle = m_0 \exp(-\gamma Nt) \tag{1.3.14}$$
$$\langle(m(t) - \langle m(t)\rangle)^2\rangle = m_0 \exp(-\gamma Nt)[1 - \exp(-\gamma Nt)]. \tag{1.3.15}$$

Up to now we have discussed statistical properties of super-radiance but have not considered how these properties can be measured. In principle this can be done by an experiment registering the intensity fluctuations of the radiation emitted. In such experiments one can determine a probability distribution, $W_k(\Delta t)$, the probability of observing k photons in a time interval Δt during the emission process.

The theoretical and experimental determination of photocount distributions, W_k, for several types of light sources are discussed in a number of reviews [5, 25, 52, 54] and monographs [24, 50, 71]. It is well known, for instance, that for a coherent (laser) field W_k follows the Poisson distribution $W_k^{\mathrm{P}} = (\bar{k})^k \exp(-\bar{k})/k!$. On the other hand, a source in thermal equilibrium emits photons that obey the Bose–Einstein distribution $W_k^{\mathrm{BE}} = (\bar{k})^k/(1+\bar{k})^{1+k}$. In these expressions $\bar{k}$ denotes the mean value of k with respect of the distribution W_k. The parameter which determines the size of the fluctuations of k about the mean value is the variance $D = \overline{k^2} - (\bar{k})^2$ of the photon number emitted in the time interval Δt. A useful parameter that characterizes the type of the photocount distribution is the Mandel parameter [53]

$$Q = \frac{D - \bar{k}}{\bar{k}}. \tag{1.3.16}$$

For photon statistics with Poissonian distribution, i.e. for a coherent field, the variance is equal to the mean value: $D^{\mathrm{P}} = \bar{k}$, and hence the Mandel parameter is equal to zero: $Q^{\mathrm{P}} = 0$. Conversely, a zero value of the Mandel parameter indicates coherence of the radiation. For the Bose–Einstein distribution $D^{\mathrm{BE}} = \bar{k}(1 + \bar{k})$, the Mandel parameter for thermal radiation is equal to the mean value: $Q^{\mathrm{BE}} = \bar{k}$. In the case where Q is negative, the photon flux is more regular than for the coherent radiation [50]. This effect is called antibunching, to emphasize the difference from Bose–Einstein statistics, where photons tend to come in groups, and photon bunching takes place.

Now it will be interesting to reveal the type of statistics displayed by a super-radiant pulse.

It is clear that the probability $W_k(t, t + \Delta t)$ of emission of k photons in the time interval $(t, t + \Delta t)$ can be expressed through the probabilities $P_M(t)$ as

$$W_k(t, t + \Delta t) = \sum_M P_M(t) P_{M-k}^M(\Delta t) \tag{1.3.17}$$

where $P_{M-k}^M(\Delta t)$ is the probability that k photons are emitted in the interval Δt, provided the system reached the state with quantum number M in the moment t. Therefore $P_{M-k}^M(\Delta t)$ is the solution of (1.2.18) with the initial condition: $P_{M'}^M(0) = \delta_{M,M'}$.

Restricting our considerations to the symmetrical states of the atomic system (i.e. to eigenstates of $\widehat{R}^2$ with maximum eigenvalue $\frac{1}{2}N(\frac{1}{2}N + 1)$) we first consider two particular solutions of equation (1.2.18). If the initial state is close to total inversion, then by the help of (1.3.6) and (1.3.7) we get (at $t = 0$)

$$Q = \frac{\bar{k}}{\frac{1}{2}N - M_0 + 1} \qquad \text{if } \tfrac{1}{2}N - M_0 \ll \tfrac{1}{2}N \quad \tfrac{1}{2}N - M \ll \tfrac{1}{2}N. \qquad (1.3.18)$$

Here M_0 is the initial value of the quantum number M. It can be seen from (1.3.18) that for an initially totally inverted system ($M_0 = \frac{1}{2}N$) the photon statistics follows a Bose–Einstein distribution: $Q = \bar{k}$. If, to the contrary, the initial state is close to the ground state, we have to use (1.3.14) and (1.3.15), and obtain

$$Q = -\frac{\bar{k}}{\frac{1}{2}M + M_0} \qquad \text{if } \tfrac{1}{2}N + M_0 \ll \tfrac{1}{2}N \quad \tfrac{1}{2}N + M \ll \tfrac{1}{2}N. \qquad (1.3.19)$$

In this case Q is negative, and this type of photon statistics is usually called sub-Poissonian.

We will show now that when during its evolution the atomic system goes through the state $M = 0$, the photon statistics is nearly Poissonian and consequently the radiation is almost coherent. We have to choose the time interval Δt to be small enough compared with the characteristic time $(\gamma N)^{-1}$ of the process. Then, in accordance with (1.2.18), probabilities $P_{M'}^M(\Delta t)$ in (1.3.17) can be represented in the form

$$P_M^M(\Delta t) = 1 - \gamma_{M,M-1}\Delta t$$
$$P_{M-1}^M(\Delta t) = \gamma_{M,M-1}\Delta t \qquad (1.3.20)$$
$$P_{M-2}^M(\Delta t) = \gamma_{M,M-1}\gamma_{M-1,M-2}\tfrac{1}{2}(\Delta t)^2.$$

All other probabilities $P_{M'}^M(\Delta t)$, with $M' < M - 2$, are of higher order in Δt than $(\Delta t)^2$, and can be neglected.

Now we have

$$\bar{k} = \sum_M P_M(t)[\gamma_{M,M-1}\Delta t + \gamma_{M,M-1}\gamma_{M-1,M-2}(\Delta t)^2]$$
$$= \langle \gamma_{M,M-1}\rangle \Delta t + \langle \gamma_{M,M-1}\gamma_{M-1,M-2}\rangle (\Delta t)^2 \qquad (1.3.21)$$

$$\overline{k^2} = \sum_M P_M(t)[\gamma_{M,M-1}\Delta t + 2\gamma_{M,M-1}\gamma_{M-1,M-2}(\Delta t)^2]$$
$$= \langle \gamma_{M,M-1}\Delta t + 2\gamma_{M,M-1}\gamma_{M-1,M-2}(\Delta t)^2\rangle \qquad (1.3.22)$$

where angular brackets, as before, denote averaging with the probabilities $P_M(t)$.

Then for the variance D and the Mandel parameter (1.3.16) we obtain, using matrix elements (1.2.14) of the operators $\widehat{R}_+$ and $\widehat{R}_-$

$$D = \langle \widehat{R}_+\widehat{R}_-\rangle \gamma \Delta t + 2\langle \widehat{R}_+\widehat{R}_+\widehat{R}_-\widehat{R}_-\rangle (\gamma \Delta t)^2 - (\langle \widehat{R}_+\widehat{R}_-\rangle \gamma \Delta t)^2 \quad (1.3.23)$$

$$Q = \frac{\langle \widehat{R}_+\widehat{R}_+\widehat{R}_-\widehat{R}_-\rangle - \langle \widehat{R}_+\widehat{R}_-\rangle^2}{\langle \widehat{R}_+\widehat{R}_-\rangle}\gamma\Delta t. \qquad (1.3.24)$$

With the help of (1.3.23) and (1.3.24), we can draw some conclusions about the photon statistics of the Dicke model. Here we restrict ourselves only to a qualitative estimation of the Mandel parameter. We assume that the distribution $P_M(t)$ is sufficiently narrow and can be approximately replaced by a δ-type distribution. In this case Q has a simplified form

$$\begin{aligned}
Q &= \frac{\langle M|\widehat{R}_+\widehat{R}_+\widehat{R}_-\widehat{R}_-|M\rangle - (\langle M|\widehat{R}_+\widehat{R}_-|M\rangle)^2}{\langle M|\widehat{R}_+\widehat{R}_-|M\rangle}\gamma\Delta t \\
&= \frac{\gamma_{M,M-1}\gamma_{M-1,M-2} - \gamma_{M,M-1}^2}{\gamma_{M,M-1}}\Delta t \\
&= 2(M-1)\gamma\Delta t. \qquad (1.3.25)
\end{aligned}$$

This equation shows that while $M > 1$, the Mandel parameter is positive, but when M is less than 1, Q becomes negative. As mentioned above, the time interval Δt must be less than $(\gamma N)^{-1}$. Therefore Q lies inside the interval $(-1, 1)$, and in the neighbourhood of $M = 0$ the Mandel parameter Q is close to 0, and super-radiance has a coherent character.

We emphasize that the observed single super-radiant pulse is a macroscopic realization of the stochastic quantum mechanical process, reduced by observation and not the quantum mechanical average. We can obtain such a realization by a Monte Carlo simulation of the cascade transitions (1.2.15). Indeed, let us consider some ideal (*gedanken*) experiment on the measurement of the super-radiant intensity, assuming the resolving time of the detection of photons to be sufficiently small. If initially our atomic system is in the state 'M', then after a time interval Δt there is the probability

$$w_{M,M-1} = \gamma_{M,M-1}\Delta t \ll 1 \qquad (1.3.26)$$

of finding the system in the state '$M - 1$' (with detection of a photon), and the probability

$$w_{M,M} = 1 - w_{M,M-1} \qquad (1.3.27)$$

of finding the system in the same initial state 'M'.

The suggested Monte Carlo procedure is as follows. Assume that initially all atoms are in the excited states, then for the first observation there is the probability $w_{N/2,(N-1)/2}$ of a photon being emitted by a transition from the state '$\frac{1}{2}N$' to the state '$\frac{1}{2}N - 1$', and there is a probability $w_{N/2,N/2}$ of it remaining in the initial state '$M = \frac{1}{2}N$'. To recognize what will happen, let us choose an arbitrary number r_m from the range [0, 1] using a random-number generator. If the current number r_m satisfies the inequality

$$r_m < w_{M,M-1} \qquad (1.3.28)$$

for a given M, then the photon emission 'M'$\rightarrow$'$M-1$' takes place, otherwise the system remains in the initial state 'M'. By repeating these tests for as long as is required to satisfy (1.3.28), we obtain a stochastic realization of the super-radiant pulse as a sequence of photon transitions starting from $M = \frac{1}{2}N$ and ending at $M = -\frac{1}{2}N + 1$. For the practical realization of this process it is convenient to choose the time intervals Δt_M in equation (1.3.26) in such a way that the transition probability $w_{M,M-1}$ remains the same at each step. Figure 1.5 presents two randomly chosen stochastic realizations of an super-radiance pulse generated by an ensemble of 10 000 atoms. The figure illustrates the statistical properties of the super-radiance pulses discussed above. The fluctuations are relatively large in the initial and final stages of the process (these are shown in the histograms with the vertical scale enlarged by a factor of 100). In the central part, the super-radiance pulses are deterministic enough. Nevertheless, the fluctuations at the early stage lead to fluctuations in the delay time of the stochastic pulses. The shape of these pulses, as can be seen in figure 1.5, is close to that of the secant hyperbolic pulse given by equation (1.2.23). The latter can be also obtained from the stochastic process described if we put $w_{M,M-1} = 1$ in the test condition (1.3.16). The secant hyperbolic pulse is shown in figure 1.5 as a broken curve. For comparison, in the same figure we display the mean stochastic pulse obtained by averaging 100 individual pulses (the full curve). It can be seen that as a result of fluctuations of the delay time the mean stochastic pulse is broader and lower in height than the secant hyperbolic pulse. This conclusion is in correspondence with the curves shown earlier in figure 1.4.

1.4 Semiclassical theory of super-radiance of an extended system

Up to now we have considered mainly the cooperative spontaneous emission of a collection of N identical atoms confined within a small volume. If we wish to investigate the super-radiance of an extended system, the dimensions of which exceed the emission wavelength, the semiclassical method is more convenient. Here we give an introduction to the method and delay more substantial discussion until Chapters 4 and 5.

1.4.1 Maxwell–Bloch equations

In essence, the semiclassical approach discussed here combines the quantum mechanical treatment of the atomic system that uses the density matrix formalism with the classical treatment of the radiation field.

Let us consider a system of identical two-level atoms interacting with the electromagnetic field. In the single-atom approximation the wavefunction obeys the Schrödinger equation

$$i\hbar\frac{\partial\psi}{\partial t} = (\widehat{H}_1 + \widehat{H}_{int})\psi \tag{1.4.1}$$

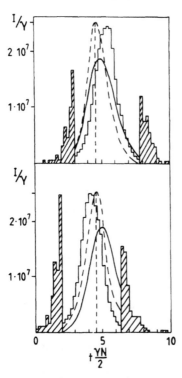

Figure 1.5. Monte Carlo simulation of the super-radiance pulse generated by an ensemble of 10 000 two-level atoms. Two randomly chosen stochastic realizations are shown, the initial and final stages (hatched areas) are enlarged by a factor of 100. The broken curve is the intensity $\tilde{I}(t)$ of the secant hyperbolic pulse calculated from (1.2.23). The mean stochastic pulse $\bar{I}(t)$ of equation (1.2.19) is also displayed (full curve).

where $\widehat{H}_1$ is, as above, the free atomic Hamiltonian, and $\widehat{H}_{\text{int}}$ is the operator of the interaction between the atom and the electromagnetic field generated by other atoms of the ensemble, plus the field, if present, of an external source.

In the dipole approximation $\widehat{H}_{\text{int}}$ can be represented in the form

$$\widehat{H}_{\text{int}} = -\widehat{d}\mathcal{E}(r, t) \tag{1.4.2}$$

where $\mathcal{E}(r, t)$ is the classical electric field. The solution of equation (1.4.1) is written as

$$\psi = \begin{pmatrix} c_e(t) \\ c_g(t) \end{pmatrix} \tag{1.4.3}$$

with the coefficients c_e and c_g to be determined as functions of time.

Let us introduce a single-atom density matrix

$$\hat{\rho}(\boldsymbol{r}, t) = \begin{pmatrix} \rho_{ee}(\boldsymbol{r}, t) & \rho_{eg}(\boldsymbol{r}, t) \\ \rho_{ge}(\boldsymbol{r}, t) & \rho_{gg}(\boldsymbol{r}, t) \end{pmatrix} \equiv \begin{pmatrix} c_e c_e^* & c_e c_g^* \\ c_g c_e^* & c_g c_g^* \end{pmatrix} \tag{1.4.4}$$

where the radius vector $\boldsymbol{r}$ marks the position of the atom. From the Schrödinger equation (1.4.1) there follows the von Neumann equation for the density matrix

$$i\hbar \frac{\partial \hat{\rho}}{\partial t} = [\widehat{H}_1 + \widehat{H}_{int}, \hat{\rho}]. \tag{1.4.5}$$

We further assume the local averaging of the atomic density matrix (1.4.4), which permits us to consider ρ as a continuous function of $\boldsymbol{r}$.

As follows from Maxwell's equations, the electric field $\mathcal{E}$ obeys the inhomogeneous wave equation [46]

$$\nabla \times \nabla \times \mathcal{E}(\boldsymbol{r}, t) + \frac{1}{c^2} \frac{\partial^2}{\partial t^2} \mathcal{E}(\boldsymbol{r}, t) = -\frac{4\pi}{c^2} \frac{\partial^2}{\partial t^2} \mathcal{P}(\boldsymbol{r}, t) \tag{1.4.6}$$

where $\mathcal{P}(\boldsymbol{r}, t)$ is the electric polarization density (the polarization) of the atomic medium, which can be treated as the quantum mechanical expectation value of the electric dipole moment per unit volume

$$\mathcal{P}(\boldsymbol{r}, t) = N_0 \operatorname{Tr}(\hat{\rho} d) = N_0 (d\rho_{eg} + d^* \rho_{ge}) \tag{1.4.7}$$

where, we recall, N_0 is the number of atoms per unit volume.

The self-consistent system of equations (1.4.5) and (1.4.6) will be considered in this section only in a one-dimensional version. We take $\hat{\rho}$, $\mathcal{P}$, and $\mathcal{E}$ to be dependent only upon one spatial coordinate, say x, along the direction of radiation, and homogeneous in the transverse dimensions. The direction of the dipole moments of the transition of all the atoms will be assumed to be the same, and to be perpendicular to the x axis. Then the vectors $\mathcal{P}$ and $\mathcal{E}$ have the same direction and we may omit the vector designations all together. Hence the system of equations (1.4.5) and (1.4.6) can be rewritten as follows

$$\frac{\partial \rho_{ee}}{\partial t} = \frac{id}{\hbar} \mathcal{E}(\rho_{ge} - \rho_{eg})$$

$$\frac{\partial \rho_{gg}}{\partial t} = -\frac{id}{\hbar} \mathcal{E}(\rho_{ge} - \rho_{eg})$$

$$\frac{\partial \rho_{eg}}{\partial t} = -i\omega_0 \rho_{eg} + \frac{id}{\hbar} \mathcal{E}(\rho_{gg} - \rho_{ee}) \qquad \rho_{ge} = \rho_{eg}^* \tag{1.4.8}$$

$$\frac{\partial^2 \mathcal{E}}{\partial x^2} - \frac{1}{c^2} \frac{\partial^2 \mathcal{E}}{\partial t^2} = \frac{4\pi}{c^2} \frac{\partial^2 \mathcal{P}}{\partial t^2}.$$

A solution of equations (1.4.8) is sought in the form of plane waves propagating in the positive direction of the x axis

$$\rho_{ge}(x, t) = \tfrac{1}{2} R(x, t) e^{-i(\omega_0 t - k_0 x)}$$

$$(1.4.9)$$

$$\mathcal{E}(x,t) = \tfrac{1}{2}E(x,t)e^{-i(\omega_0 t - k_0 x)} + \text{c.c.}$$

where $k_0 = \omega_0/c$, and $R(x,t)$ and $E(x,t)$ are complex amplitudes. On substituting (1.4.9) into (1.4.8) the density matrix equations will be simplified by neglecting the fast oscillating terms containing exponential factors $\exp(\pm 2i\omega_0 t)$. This is known as the *rotating wave approximation* (RWA). The last equation of (1.4.8) will be simplified by applying the slowly varying envelope approximation (SVEA). According to this, the amplitudes $R(x,t)$ and $E(x,t)$ are assumed to be slowly varying functions of their arguments, in the sense that

$$\left| \frac{\partial E}{\partial t} \right| \ll \omega_0 |E| \qquad \left| \frac{\partial E}{\partial x} \right| \ll k_0 |E|$$

$$\left| \frac{\partial R}{\partial t} \right| \ll \omega_0 |R| \qquad \left| \frac{\partial R}{\partial x} \right| \ll k_0 |R| \qquad (1.4.10)$$

so that the spatial and temporal modulations of the waves are slow over distances of the order of the wavelength $2\pi/k_0$, and over the oscillation period $2\pi/\omega_0$. Then the second derivatives in the field equation can be neglected. The resulting reduced form of the equations is

$$\frac{\partial R}{\partial t} = -\frac{id}{\hbar} EZ$$

$$\frac{\partial Z}{\partial t} = \frac{id}{2\hbar}(ER^* - E^*R) \qquad (1.4.11)$$

$$\frac{\partial E}{\partial x} + \frac{1}{c}\frac{\partial E}{\partial t} = i\frac{2\pi\omega_0}{c} dN_0 R$$

where $Z = \rho_{ee} - \rho_{gg}$ is the population difference. The first two equations of this system are similar to those introduced by Bloch in his work on magnetic resonance. This is why the system of equations (1.4.11) is often referred to as the *Maxwell–Bloch system*. We shall follow this convention.

It is convenient to write the Maxwell–Bloch system (1.4.11) using the dimensionless variables τ and ξ

$$\tau = t\Omega_0 \qquad \xi = x\frac{\Omega_0}{c} \qquad (1.4.12)$$

and the dimensionless field amplitude ε

$$\varepsilon = -i\frac{E}{\sqrt{2\pi N_0 \hbar \omega_0}} = -i\frac{dE}{\hbar\Omega_0} \qquad (1.4.13)$$

where the physical meaning of the frequency

$$\Omega_0 = (2\pi N_0 d^2 \omega_0/\hbar)^{1/2} \qquad (1.4.14)$$

will be clarified in the next section. Then we obtain

$$\frac{\partial R}{\partial \tau} = Z\varepsilon$$

$$\frac{\partial Z}{\partial \tau} = -\tfrac{1}{2}(\varepsilon^* R + \varepsilon R^*) \qquad (1.4.15)$$

$$\frac{\partial \varepsilon}{\partial \tau} + \frac{\partial \varepsilon}{\partial \xi} = R.$$

The solution of (1.4.15) with the appropriate initial and boundary conditions describes the temporal and spatial evolution of super-radiance. We shall assume that the intensity of super-radiance is given as the number of photons emitted from the end cross-section of the sample per unit time and per atom. This can be expressed in terms of the Poynting vector, averaged over an oscillation period

$$I = \frac{c}{8\pi}|E|^2 \frac{A}{N\hbar\omega_0} = \frac{c}{4L}|\varepsilon|^2. \qquad (1.4.16)$$

Here A is the cross-sectional area of the sample and L is its length.

1.4.2   Spatially homogeneous approximation

Let us consider first the spatially homogeneous solution of the reduced Maxwell–Bloch system (1.4.11). This can be obtained if we assume the amplitudes of the field, the polarization, and the population difference to be functions only of the time and not of the spatial coordinate, or, equivalently, that the values of such quantities at each point can be replaced by the corresponding mean values in space (the *mean-field theory* of super-radiance developed by Bonifacio and Lugiato [BL75a, BL75b]). In this case, we can always regard the values of ε and R as being real, so that $R^* = R$, and $\varepsilon^* = \varepsilon$. Then equations (1.4.15) can be written as

$$\frac{dR}{d\tau} = \varepsilon Z$$

$$\frac{\partial Z}{\partial \tau} = -\varepsilon R \qquad (1.4.17)$$

$$\frac{d\varepsilon}{d\tau} = R.$$

It follows readily from the first two equations of (1.4.17) that $B^2 = R^2 + Z^2$ is the integral of motion, $B^2 = $ constant. Therefore we can search for a solution of (1.4.17) in the form

$$R(\tau) = B\sin\theta(\tau) \qquad Z(\tau) = B\cos\theta(\tau) \qquad (1.4.18)$$

where the quantities B and θ are usually referred to as the modulus and the polar angle of the so-called Bloch vector B with components R and Z.

Substituting (1.4.18) into the first or second equation of (1.4.17), we obtain

$$\varepsilon(\tau) = \frac{d\theta(\tau)}{d\tau}. \tag{1.4.19}$$

Note that according to (1.4.19), θ (up to its initial value) has the meaning of the area under the field amplitude ε

$$\theta(\tau) = \theta_0 + \int_0^\tau \varepsilon(\tau') \, d\tau'.$$

Using (1.4.18) and (1.4.19), the third equation of (1.4.17) then becomes the pendulum equation for θ

$$\frac{d^2\theta(\tau)}{d\tau^2} = B \sin \theta(\tau) \tag{1.4.20}$$

i.e. the time dependence of the solution of (1.4.17) can be modelled by the motion of a pendulum. We see that for small oscillations the Bloch vector 'swings' about the position $\theta = \pi$ with characteristic frequency $\Omega = \sqrt{B}$ (in the usual units $\Omega = \Omega_0\sqrt{B}$).

Let us seek the solution of equation (1.4.20) with the initial condition $\theta(0) = \theta_0$. By virtue of (1.4.18), this initial condition corresponds to $Z(0) = B \cos \theta_0$ and $R(0) = B \sin \theta_0$. If, for instance, all atoms are initially in the excited state, then it follows that $Z(0) = 1$ (see the definition of Z in (1.4.11)), and consequently that $\theta_0 = 0$ and $B = 1$. In this case, the pendulum initially is in the position of unstable equilibrium and equation (1.4.20) has only a zero solution. However, in order to take into account the quantum fluctuations of polarization to be discussed in section 1.5 and Chapter 4, we assume θ_0 to be of the order $N^{-1/2}$. Then the pendulum does swing, and as is well known [45], the period of its oscillations is

$$T = 4\Omega^{-1} K(\cos^2 \tfrac{1}{2}\theta_0) \tag{1.4.21}$$

where $K(x)$ is the elliptic integral of the first kind. For small values of θ_0 the asymptotic value of K is [2]

$$K(\cos^2 \tfrac{1}{2}\theta_0) \simeq \ln \frac{8}{\theta_0}. \tag{1.4.22}$$

The spatially homogeneous solution considered above satisfies a cyclic boundary condition: the input field is equal to the output field, as in a single-mode ring laser. Since the influence of the input field upon the output field manifests itself only in the time it takes for the light to propagate, L/c, the spatially homogeneous solution of (1.4.17) is correct at the output end of the sample during this time interval, independently of the boundary conditions imposed at the input end. Such a solution describes a flow of energy oscillating back and forth between atoms and the field mode. It was first obtained by Bonifacio and Preparata [BP70]. We note that the oscillation of the population difference and polarization cannot be explained in this case as the usual Rabi oscillation, since the intensity of the electric field is not constant. In accordance with (1.4.21), the frequency of this oscillation, T^{-1}, is proportional to Ω_0. It also depends upon the value of the initial tipping angle, θ_0, and, via B, upon the initial population difference.

The pendulum-like solution can be used to describe super-radiance if a damping term representing the release of radiation from the sample is added to the third equation of (1.4.17) [BSH71a]

$$\frac{d\varepsilon}{d\tau} + \kappa\varepsilon = R \tag{1.4.23}$$

where κ is the damping constant. It is reasonable to put κ equal to the reciprocal of the time a photon needs to travel the length L of a sample, in our units $\kappa = c/(L\Omega_0)$. Then equation (1.4.20) for the tipping angle of the Bloch vector is modified to the form

$$\frac{d^2\theta(\tau)}{d\tau^2} + \kappa\frac{d\theta}{d\tau} = B\sin\theta(\tau). \tag{1.4.24}$$

This is the equation for a pendulum with friction. We shall consider here only the special case of heavily over-damped oscillations, i.e. $\kappa \gg 1$. We may then neglect $d^2\theta/d\tau^2$, in comparison with $\kappa\, d\theta/d\tau$, in equation (1.4.24)

$$\frac{d\theta}{d\tau} = \frac{B}{\kappa}\sin\theta(\tau). \tag{1.4.25}$$

The exact solution of (1.4.25)

$$\theta(\tau) = 2\tan^{-1}[\exp(B\tau/\kappa)\tan\tfrac{1}{2}\theta_0] \tag{1.4.26}$$

with the help of (1.4.19) gives a secant hyperbolic pulse

$$\varepsilon(\tau) = \tau_R^{-1}\operatorname{sech}\frac{\tau - \tau_0}{\tau_R} \tag{1.4.27}$$

where $\tau_R = \kappa/B$ and $\tau_0 = \tau_R\ln(\theta_0/2)$. The duration of this pulse at half its maximum is about $2.64\tau_R$. Recall that we chose Ω_0^{-1} as the unit of time. In arbitrary units the characteristic time T_R of the pulse becomes

$$T_R = \Omega_0^{-1}\frac{\kappa}{B}. \tag{1.4.28}$$

In particular, for the case of total initial inversion ($B = 1$) we obtain, using the definition (1.4.14) of Ω_0 and (1.1.2)

$$T_R = \Omega_0^{-2}\frac{c}{L} = \frac{\hbar c}{2\pi\omega_0 d^2 N_0 L} = \left(\frac{3}{8\pi}\gamma N_0\lambda^2 L\right)^{-1} \tag{1.4.29}$$

which is in agreement with the estimate of the super-radiant time scale for the extended system, given at the end of section 1.2. We note that some authors use a different definition for T_R, namely without the factor 3 in (1.4.29). Referring to these works, if necessary, we shall use the notation $T_R' = 3T_R$.

1.4.3 Auto-modelling solution of the Maxwell–Bloch equations

Now we consider the auto-modelling solution of the Maxwell–Bloch equations, which will be applied to the super-radiance problem in the next section. Originally this type of solution was obtained by Burnham and Chiao [BC69] for modelling the coherent resonant fluorescence excited by a short light pulse (see also [L69, AC70, Z80]).

Assuming, as above, that R and ε are real, we can rewrite equations (1.4.15) as

$$\frac{\partial R}{\partial \tau'} = Z\varepsilon$$

$$\frac{\partial Z}{\partial \tau'} = -R\varepsilon \qquad (1.4.30)$$

$$\frac{\partial \varepsilon}{\partial \xi} = R$$

where we have introduced the retarded time τ'

$$\tau' = \tau - \xi. \qquad (1.4.31)$$

Since by virtue of these the Bloch vector's length does not change, we can look for a solution in the form

$$R = B \sin \theta(\tau', \xi) \qquad Z = B \cos \theta(\tau', \xi). \qquad (1.4.32)$$

Substituting (1.4.32) into (1.4.30) we obtain

$$\frac{\partial R}{\partial \tau'} = \frac{\partial \theta}{\partial \tau'} B \cos \theta = \varepsilon B \cos \theta \qquad (1.4.33)$$

whence it follows that

$$\frac{\partial \theta}{\partial \tau'} = \varepsilon. \qquad (1.4.34)$$

By differentiating ε with respect to ξ and using the last equation of (1.4.30), we obtain the sine–Gordon equation

$$\frac{\partial^2 \theta}{\partial \tau' \partial \xi} = B \sin \theta. \qquad (1.4.35)$$

We note that Arecchi and Bonifacio [AB65] were the first to apply this equation to a problem of nonlinear optics.

It is easy to check that equation (1.4.35) is invariant under the scale transformation

$$\tau' \to \beta\tau' \qquad \xi \to \beta^{-1}\xi \qquad\qquad (1.4.36)$$

where β is an arbitrary parameter. Therefore there exists an auto-modelling solution of this equation in the form of a function of the invariant argument $\tau'\xi$. For such a solution, where the dependence on τ' and ξ is expressed by a single variable, the partial differential equation (1.4.35) becomes an ordinary differential equation, and thus the problem is simplified a great deal. It is useful to consider the case when the solution $\theta(w)$ is a function of the variable $w = 2\sqrt{\tau'\xi}$. Expressing the partial derivatives with respect to τ' and ξ through the derivative with respect to w, we obtain from (1.4.35)

$$\frac{d^2\theta}{dw^2} + \frac{1}{w}\frac{d\theta}{dw} = B\sin\theta. \qquad\qquad (1.4.37)$$

Once equation (1.4.37) is solved, we can find the population difference, and the polarization, as well as the field as follows

$$Z(\tau', \xi) = B\cos\theta(w)$$
$$R(\tau', \xi) = B\sin\theta(w) \qquad\qquad (1.4.38)$$
$$\varepsilon(\tau', \xi) = \frac{\partial}{\partial\tau}\theta(w) = \frac{\xi}{w}\frac{d\theta(w)}{dw}.$$

Burnham and Chiao [BC69] obtained the numerical solution of equation (1.4.37) with the initial conditions

$$\theta(0) = \theta_0 \qquad \frac{d\theta(0)}{dw} = 0. \qquad\qquad (1.4.39)$$

Their results are shown in figure 1.6. They suggest the following interpretation of the solution. The two-level atomic system, started in the ground state, was excited to a state with the angle θ_0 by a short coherent pulse of area $\pi - \theta_0$. The same initial situation can be produced by starting with all atoms in the excited state and de-exciting them by means of an incident pulse of area θ_0. The response to such retarded excitation is radiation in the same direction as the incident pulse. The ringing produced in the radiation has a simple interpretation. Atoms located at position ξ are prepared in state 'θ_0' at $\tau' = 0$. As they radiate away their stored energy θ decreases to zero. They then begin to absorb energy radiated by atoms at positions $\xi' < \xi$ and become excited. When the incident radiation has been attenuated to zero, the atoms again radiate, but with the opposite phase. This process repeats as the angle θ rings down to zero.

1.5 Solution of the Maxwell–Bloch equations for super-radiance

Let us first formulate the initial conditions corresponding to the semiclassical treatment of super-radiance. Postponing the rigorous consideration until Chapter 4, we give here only a preliminary discussion.

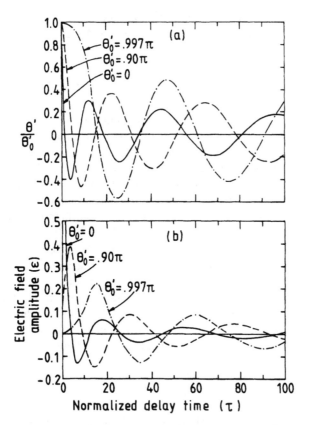

Figure 1.6. Time evolution of the polar angle of the Bloch vector $\theta' = \pi - \theta$ and of the field ε for three values of the initial excitation angle: $\theta'_0 = \pi - \theta_0$. The field has been normalized to give $\int \varepsilon^2(\tau)\,d\tau = 1$. The curve for $\theta'_0 \approx 0$ corresponds to $\varepsilon(\tau = 0) = 1$ [BC69].

We shall consider all atoms to be initially in the upper state. So the initial value of the population difference, $Z(0) = 1$, is spatially homogeneous. For the sake of simplicity, we shall assume the initial value of the polarization amplitude $R(0)$ also to be spatially homogeneous. If a two-level atom is in a stationary state, the mean value of its electric dipole is zero (see the matrix for the dipole moment (1.2.4)). But the variance of the atomic dipole moment does not vanish. Squaring the matrix (1.2.4), we obtain

$$\hat{d}^2 = \begin{pmatrix} |d|^2 & 0 \\ 0 & |d|^2 \end{pmatrix}. \tag{1.5.1}$$

So the variance of the dipole moment in both stationary states is $|d|^2$. Since there is no correlation between the atoms at the initial instant, the variance of the total dipole moment of N atoms is Nd^2. We take $\sqrt{Nd^2}$ as the initial value for the total dipole moment NdR. Then $R(0) = 1/\sqrt{N}$, and the initial conditions for Z and R are

$$Z(0, \xi) = 1 \qquad R(0, \xi) = \frac{1}{\sqrt{N}} \qquad (N \gg 1, \ B = 1). \quad (1.5.2)$$

To determine the initial condition for the field, let us estimate the vacuum field fluctuations. For the resonant field mode, which is localized inside the sample volume V, the energy of the vacuum fluctuations is $\frac{1}{2}\hbar\omega_0$, so that the corresponding electric field intensity E can be estimated from the relation $|E|^2 V \sim \hbar\omega_0$. Then for the dimensionless amplitude ε, with the aid of (1.4.12), we obtain $\varepsilon \sim 1/\sqrt{N}$. If the temperature T of a sample is relatively large, $k_B T > \hbar\omega_0$, where k_B is the Boltzmann constant, the field fluctuations will be determined by the thermal noise and $\varepsilon \sim \sqrt{kT/N\hbar\omega_0}$. This magnitude of the field fluctuations can be taken as the initial value of the field.

The Burnham–Chiao solution gives a good description of super-radiance of an extended system if the transient time L/c is less than the characteristic time of the process. In the latter case the retarded time in equations (1.4.30) can be treated as the ordinary time. If we take the initial conditions for Z and R to be (1.5.2), then with the aid of the third equation of (1.4.30) we obtain the following expression for the initial magnitude of the field

$$\varepsilon(0, \xi) = \int_0^\xi R(0, \xi')\, d\xi' = \frac{1}{\sqrt{N}}\xi. \quad (1.5.3)$$

This is of the same order of magnitude as the vacuum fluctuations of the field, that is $1/\sqrt{N}$. The linear dependence of (1.5.3) upon ξ has no particular physical significance, and it may be considered the price to be paid for having a simple solution. In fact, we may accept (1.5.3) because, as is shown by numerical integration, the results depend weakly upon the details of the initial conditions. The Burnham–Chiao solution of the sine–Gordon equation describes super-radiance if we take the initial value of the tipping angle $\theta(0)$ to be of the order $\sqrt{N}$.

Taking into account the scaling invariance of the sine–Gordon equation, it is reasonable to choose a new unit of length equal to that of the sample, L. If we want to keep the old units for w, i.e. $(c/\Omega_0^{-2})^{1/2}$, then the new unit of time becomes

$$T_R = c\Omega_0^{-2}L^{-1} = \frac{c\hbar}{2\pi N_0 d^2 \omega_0 L} \quad (1.5.4)$$

which is equal to the characteristic time of super-radiance (1.4.29) obtained for the spatially homogeneous model.

Figure 1.7 demonstrates a Burnham–Chiao-type solution for the intensity per atom (1.4.16) of the super-radiant pulse. The solution is given by a universal curve that weakly depends upon $\theta(0)$. The dependence of the solution on all other parameters is given only via the time scale, i.e. T_R, determined by (1.5.4). Since the intensity is normalized to one photon per atom, the area under the curve is equal to 1. Therefore the intensity amplitude is inversely proportional to T_R, and, consequently, in accordance with (1.5.4), it is proportional to the concentration of inverted atoms N_0 and to the length of the sample, L. Hence we obtain two main features of super-radiance:

(i)   its characteristic time scale, T_R, is inversely proportional to N_0;
(ii)   the intensity of radiation per atom is proportional to N and therefore the total intensity is proportional to N^2.

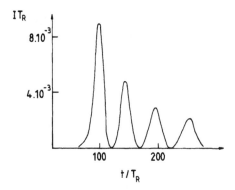

Figure 1.7. The auto-modelling solution for the intensity of the super-radiant pulse. The initial polarization here corresponds to $\theta_0 = 10^{-8}$.

As has been shown in section 1.2, these features of super-radiance result from the build-up of the macroscopic dipole moment during the emission, in proportion to the number of atoms in the system.

The curve in figure 1.7 depicts a burst of radiation with ringing, preceded by some delay T_D. The reason for the ringing has already been discussed in the previous section. The relatively long delay is a result of the small initial value of θ_0; the system needs time to create the macroscopic dipole moment from the small quantum fluctuation of the polarization. MacGillivray and Feld [MGF76] give an estimate for the delay time T_D defined as the time of the appearance of the first maximum of the pulse. They consider the linear limit of equation (1.4.37), when $\sin\theta$ can be replaced by θ. It then reduces to the equation of the modified Bessel function

$$\theta'' + \frac{1}{w}\theta' - \theta = 0. \tag{1.5.5}$$

For our initial condition, $\theta(0) = \theta_0$, $\theta'(0) = 0$, the solution is

$$\theta(w) = \theta_0 I_0(w) \tag{1.5.6}$$

where I_0 is the modified Bessel function of order zero. Assuming $T_D > T_R$ we can use the asymptotic expression for I_0 for the large values of the argument, $I_0(w) = e^w/\sqrt{2\pi w}$ [2]. Setting $\theta = \pi$ and $w = 2\sqrt{T_D/T_R}$ we obtain

$$T_D = \tfrac{1}{4} T_R \left| \ln \frac{\theta_0}{2\pi} \right|^2 \tag{1.5.7}$$

where the numerical factor $1/2\pi$ was introduced in [MGF76] to improve the fit to the results of the numerical integration of equation (1.4.37). We see that the delay time depends logarithmically upon θ_0. Since $\theta_0 = 1/\sqrt{N}$, and in a realistic situation it can be of the order of 10^{-5}–10^{-10}, the delay time can exceed the characteristic time T_R by two orders of magnitude (see figure 1.7).

Let us consider the scaling properties of the solution for the field, polarization, and population difference. Since, by virtue of (1.4.22), ε/ξ, R and Z are functions of w only, i.e. of $\tau\xi$, the dependences of these quantities on τ and ξ are closely related. This permits us to draw the following conclusions.

(i)   The dependence of R, Z and ε/ξ on τ at a fixed value of ξ coincides with the dependence of these quantities on ξ at a fixed value of τ.
(ii) The distribution of R, Z and ε/ξ over the length of the sample at the instant τ_1 is obtained by scaling the distribution of these quantities at the instant τ_2 with a scaling coefficient equal to τ_2/τ_1 (this feature is illustrated in figure 1.8).
(iii) The time-evolution curves of R, Z and ε/ξ at the spatial point ξ_1 are obtained by scaling the same curves of these quantities at the spatial point ξ_2 with a scaling coefficient of ξ_2/ξ_1.

The present treatment neglects retardation, i.e. it is valid only for $T_D > L/c$, in other words, for relatively short samples. In this case the system radiates in a certain correlated régime. But otherwise, for a sufficiently long sample, the system breaks into a number of independently super-radiating segments, as was first described by Arecchi and Courtens [AC70]. They introduced the notion of cooperative length L_c, which gives a limit to the size of a system radiating as one single system, via the relation

$$T_R(L_c) = \frac{L_c}{c} \tag{1.5.8}$$

whence, with the aid of (1.4.23), we obtain

$$L_c = \frac{c}{\sqrt{2\pi N_0 d^2 \omega_0}}. \tag{1.5.9}$$

Therefore by virtue of (1.4.23)

$$L_c = \frac{c}{\Omega_0}. \tag{1.5.10}$$

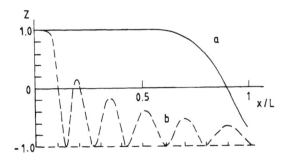

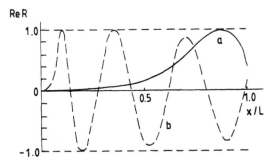

Figure 1.8. Distributions of the inversion Z (upper graph) and the real part of the polarization Re R (lower graph), along the medium at the time moments corresponding to (a) $\tau = 60$, (b) $\tau = 400$. The initial polarization was $R_0 = 4 \times 10^{-4}$. Curves (b) can be obtained by scaling curves (a) by a factor of 400/60.

It is readily seen that, for $L = L_c$

$$T_R(L_c) = \frac{1}{\Omega_0} = \frac{L}{c} \qquad (1.5.11)$$

so that $1/\Omega_0$ is the lower limit for the characteristic radiation time when the system radiates as a whole.

1.6 Concluding remarks

In this chapter, we have treated super-radiance as the cooperative spontaneous radiation of an ensemble of identical quantum emitters (atoms or molecules). The models we have used have completely neglected any possible de-phasing process. However, in real situations the relatively strong de-phasing that usually takes place in the ensemble suppresses super-radiance. A more realistic theory that allows us to estimate the influence of relaxation processes upon the generation of super-radiance will be introduced in Chapters 5 and 6.

Chapter 2

The observation of super-radiance

For nearly two decades, starting from the prediction of super-radiance by Dicke in 1954, the subject was generally considered as being mainly of theoretical significance. The situation changed in 1973 when Skribanowitz, Herman, MacGillivray and Feld [SHMF73] performed experiments with low-pressure HF gas, and realized conditions where the phase memory time of the radiation centres was longer than the cooperative spontaneous emission time. This enabled them to carry out the first observation of super-radiance. Their work created a new interest in coherent optics, and other observations of super-radiance followed in the subsequent years.

The experiments can be divided into two classes with respect to the time scale of the relaxation processes. In gases, the relaxation times are typically in the nanosecond range, whereas in solids these processes are faster, by at least one or two orders of magnitude. In this chapter we shall give a description of some of these experiments both in gases and in solids, following the original papers.

2.1 Super-radiance in gases

2.1.1 The first observation of super-radiance in optically pumped HF gas

The first experimental study of super-radiance was carried out in 1973 at the MIT Spectroscopy Laboratory by Skribanowitz et al [SHMF73]. Their experimental set up consisted of a hydrogen fluoride pump laser, a long (30–100 cm) stainless steel cell containing the low-pressure HF gas at room temperature, and the detection system with a helium cooled InGe detector and fast pulse amplifiers, schematically shown in figure 2.1. The laser operating on the $R_1(J)$ and $P_1(J)$ branch transitions of the vibrational ground state ($\lambda \approx 2.5~\mu$m) produced intense short pulses of 200–400 ns at peak powers of a few kilowatts per square centimetre. These pulses were pumped into the HF gas of the cell selectively, putting the HF molecules into a particular rotational level in the first excited vibrational state. This resulted in nearly complete population inversion between two adjacent rotational levels. In this way, it was possible to study the $J+1 \rightarrow J$ rotation transitions in the $v = 1$ band, with J ranging from 0 to 4, which corresponded to the infrared region with wavelengths of respectively 252, 126, 84, 63 and 50 μm.

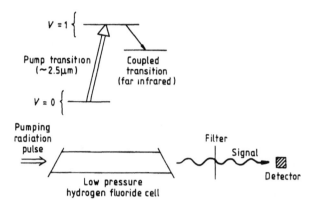

Figure 2.1. Level scheme for hydrogen fluoride, HF, and the principle of the experimental arrangement of Skribanowitz *et al*. The windows of the sample cell are tilted to prevent multiple reflections [MGF81a].

The sample cell was optically pumped in a single pass, which led, after a considerable delay ($\approx$ microseconds), to a burst of super-radiant output pulses with peak intensities estimated to be in the 100 μW cm^{-2} range, and pulse widths in the range of several hundred nanoseconds, depending upon the gas pressure and power of the pump laser (see figures 2.2 and 2.3).

At pressures below 5 mTorr the observed pulses appeared with delays of 500–2000 ns after the beginning of the pumping pulse. With the pressure decreasing the delays increased, the pulses broadened and their magnitude decreased. Pulses often showed ringing, with as many as four after-lobes. Also pulses were seen to be emitted backwards, i.e. propagating anti-parallel to the travelling wave radiation of the pump. The radiation pattern was highly directional. Almost all radiation was emitted within a very small solid angle along the axis of the pump's beam.

If the radiation emitted by this system had been ordinary spontaneous emission, it would have had a long exponential decay (as the radiative lifetime of these transitions is in the 1–10 s range), and the radiation would have been emitted isotropically. Furthermore, the peak intensity observed was ten orders of magnitude greater than would have been expected for ordinary spontaneous radiation. Neither was the signal amplified spontaneous emission, because the transit time through the cell was over 100 times shorter than the time it took for the pulse to evolve. The radiation observed was also distinct from

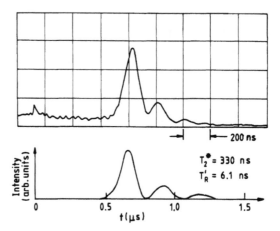

Figure 2.2. Oscilloscope trace of the super-radiance pulse observed by Skribanowitz *et al* [SHMF73] in HF gas at 84 μm ($J = 3 \rightarrow 2$), pumped by the $R_1(2)$ laser line, and the theoretical fit. The parameters are: pump intensity $I = 1$ kW cm^{-2}, $p = 1.3$ mTorr, $L = 100$ cm. The small peak on the oscilloscope trace at $t = 0$ is the 3 μm pump pulse, highly attenuated.

that emitted by the laser, of which the peak output intensity was directly proportional to the total population difference between the levels of the laser's transition. Therefore, when the length or pressure was increased, the peak intensity increased proportionally. In contrast to this, the peak intensity of the HF output pulse was proportional to the pressure squared (see figure 2.4). These findings led Skribanowitz *et al* to the conclusion that the phenomenon observed was super-radiance. This was supported by estimates for the de-phasing homogeneous and inhomogeneous relaxation times, T_2 and T_2^* (see Chapter 5). At the millitorr pressures used in the experiments, T_2, determined by collisions, was in the microsecond range, and enormously exceeded the pulse delay. The inhomogeneous time T_2^* (the result of Doppler broadening) was about 200–300 ns, i.e. of the same order as the pulse's duration.

The experimental results were compared with the calculated super-radiant pulses computed by the same authors with the help of the semiclassical theory (the coupled Maxwell–Bloch equations) outlined in Chapter 1 (see also [MGF76, MGF81a]) and to be discussed in more detail in Chapter 5. Using parameters corresponding to their experimental arrangement, Skribanowitz *et al* were able to calculate the delays, peak intensities and shapes of the super-radiant pulses in good agreement with experiment, as seen in figures 2.2 and 2.3.

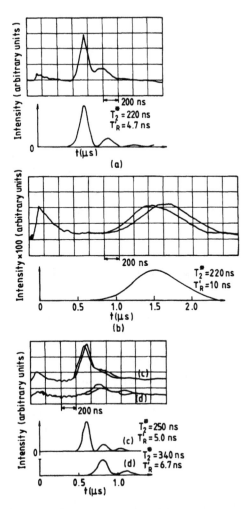

Figure 2.3. Oscilloscope traces (upper panels) of super-radiant pulses from HF and computer fits (lower panels) [SHMF73]. (a) $J = 3 \rightarrow 2$ transition at 84 μm pumped by the $P_1(4)$ laser line. Pump intensity: $I = 2.2$ kW cm^{-2}, $p = 4.5$ mTorr, $(L = 100$ cm). (b) Same as (a) but $p = 2.1$ mTorr. Note increased delay and broadening of pulse. (c) Same transition as in (a), but pumped by the $R_1(2)$ laser line. $I = 1.7$ kW cm^{-2}, $p = 1.2$ mTorr, $(L = 100$ cm). (d) Same as (c) except $I = 0.95$ kW cm^{-2}. The same intensity scale is used in fitting curves (a) and (b), and (c) and (d). The oscilloscope traces are well reproduced in double exposure. The authors of this experiment used $T_R' = 3T_R$.

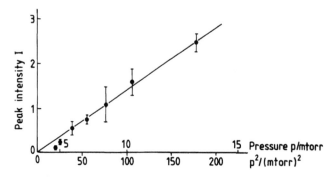

Figure 2.4. The peak intensity of a super-radiant pulse at 84 μm ($J = 3 \to 2$), pumped by the $P_1(4)$ laser line, as a function of the square of the pressure p of HF in the sample cell [FM80, MGF81b].

2.1.2 Near-infrared super-radiance in cascading transitions in atomic sodium

Super-radiance was subsequently observed by many authors. Gross *et al* [GFPH76] used a short-pulse tunable dye laser to optically pump sodium vapour in the $5S_{1/2}$ state. They observed super-radiance in the near-infrared in transitions cascading down from this level.

Atoms were prepared in the $5S_{1/2}$ level by stepwise excitation from the ground state via the intermediate $3P_{3/2}$ level (see figure 2.5(a)). All transitions of the cascade from the $5S_{1/2}$ level (except those leading to the ground and $3P_{3/2}$ states) are totally inverted at a given stage in the atomic decay, and so are capable of exhibiting super-radiance. These transitions are listed in table 2.1 with the relevant parameters T_2^*, $T_R' = 3T_R = 8\pi(\gamma N_0 L\lambda^2)^{-1}$, excited-atom density N_0, and the threshold density $N_t = N_0 T_R'/T_2^*$ (i.e. the density for which $T_R' = T_2^*$). In a pressure range corresponding to $N_0 = 10^9 - 10^{10}$ cm^{-3} the three transitions at 3.41 μm, 2.21 μm and 9.10 μm (which are drawn as solid arrows in figure 2.5(a)) appear to be above the threshold. As a result, the system first goes super-radiant on the 5S–4P transition, and then, after almost all of the population has been transferred to the 4P levels, super-radiates again on one—or both—of the competing 4P–4S and 4P–3D transitions.

The set-up used for detecting the super-radiant emission is sketched in figure 2.5(b). Two optical pulses B_1 and B_2 provided excitation at $\lambda_1 = 0.5890$ μm (3S–3P$_{3/2}$) and $\lambda_2 = 0.6160$ μm (3P$_{3/2}$–5S$_{1/2}$). These pulses, which were produced by two dye lasers simultaneously pumped by a 1 MW N$_2$ laser, were of about 10 kW peak power. They were able to saturate both λ_1 and λ_2 transitions, so that about a quarter of the total number of atoms in the active volume could be prepared in the $5S_{1/2}$ state. The duration of the pumping pulses was about 2 ns, and their spectral width of about 10 GHz was large enough to excite the whole Doppler profile of the pumping transitions. The pulses propagated

Table 2.1. Transitions of the cascade from the $5S_{1/2}$ level with their relevant parameters for super-radiance [GFPH76].

Transition	λ (μm)	T_2^* (ns)	$N_0 T_R'$ (s cm^{-3})	N_t (cm^{-3})
5S–4P$_{3/2}$	3.41	1.7	4	3×10^9
5S–4P$_{1/2}$	3.41	1.7	8	5×10^9
4P–4S	2.21	1.1	5	5×10^9
4P–3D$_{5/2}$	9.10	4.6	23	5×10^9
4P–3D$_{3/2}$	9.10	4.6	34	7×10^9
4S–3P$_{1/2}$	1.14	0.6	16	3×10^{10}
3D–3P$_{1/2}$	0.82	0.4	16	4×10^{10}
5S–3P$_{1/2}$	0.615	0.3	200	6×10^{11}

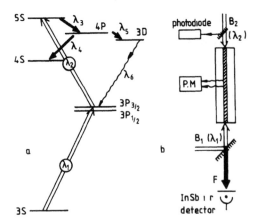

Figure 2.5. (a) Diagram of Na energy levels relevant to the super-radiance experiment of Gross *et al* [GFPH76]. Double-line arrows show pumping transitions at $\lambda_1 = 0.5890$ μm and $\lambda_2 = 0.6160$ μm; solid arrows—super-radiant transitions $\lambda_3 = 3.41$ μm, $\lambda_4 = 2.21$ μm and $\lambda_5 = 9.10$ μm; wavy line—transition at $\lambda_6 = 0.8191$ μm detected off-axis by the photomultiplier. (b) Sketch of the experimental set-up showing collinear pumping beams B_1 and B_2, on-axis InSb detector and off-axis photomultiplier.

along the same axis in order to prepare a pencil-shaped active volume in a temperature-regulated cylindrical cell containing Na vapour. A semi-reflecting mirror allowed the exciting pulses to be separated from the super-radiant pulses emitted along the same axis. A fast infrared InSb detector was used to monitor the infrared signals, the wavelengths of which were selected by suitable filters.

Above a threshold of $N_t = 6 \times 10^9$ cm^{-3} (Na pressure 10^{-6} Torr) two directionally emitted infrared signals of a few nanoseconds' duration were detected, with wavelengths of 3.41 μm and 2.21 μm. The 2.21 μm pulse followed the laser excitation after a few nanoseconds' delay (the longest observed delay was 7 ns). The 2.21 μm pulse was delayed by several nanoseconds with respect to the 3.41 μm one. Figure 2.6 shows a recording of the 3.41 μm pulses (trace (b)) and the 2.21 μm pulses (trace (c)), which appear clearly delayed with respect to the exciting pulse B_2 (trace (a)). The 3.41 μm pulse was recorded for different excitation densities in order to show variations of pulse heights and delays versus excitation intensities. The authors noted the ringing in the wings of the largest 3.41 μm pulse. For a given Na pressure above the threshold, and for a non-saturating excitation by light, the height h and delay T_D of the pulses were seen to be a function of the intensity I of the exciting pulse B_2. The variation of h and T_D versus I just reflected the way these quantities varied with N_0. Over a range of excitation densities $N_t < N_0 < 4N_t$ the signal increased as N_0^2 and was delayed as N_0^{-1}, which was good evidence of its super-radiant character. When N_0 was increased above $4N_t$ the delay of the pulse became so short that it fell into the wings of the exciting pulse ($T_D < T_E \sim 2$ ns). The system then started operating under quasi-stationary conditions. The pulse amplitude was no longer proportional to N_0^2, but saturated as N_0. The system was no longer in the super-radiant régime but evolved continuously towards amplified spontaneous emission as N_0 was increased, analogously to the emission observed in a high-gain amplifying medium (mirror-less laser).

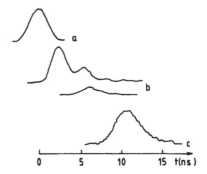

Figure 2.6. Time variation of pulse signals from Na monitored by an InSb detector [GFPH76]. Trace (a): exciting pulse B_2; trace (b): 3.41 μm pulse for two different excitation intensities; trace (c): 2.21 μm pulse for the same excitation as the second 3.41 μm pulse of trace (b).

Above a threshold of about $4N_t$ (a pressure of 4×10^{-6} Torr), when the 3.41 μm and 2.21 μm pulses were not delayed any longer, there was a very sharp burst of fluorescence at 0.82 μm, typically delayed by several tens of nanoseconds with respect to the pumping excitation. This burst was caused by a delayed and very fast transfer of population from the 4P to the 3P level, and was indirect evidence of the third super-radiance emission at 9.10 μm, which could not be directly detected, because the corresponding pulse was absorbed in the Pyrex wall of the cell. The delay corresponding to the maximum of the burst decreased as the excitation increased.

Increasing the pressure above 4×10^{-5} Torr ($N_0 > 40 N_t$), two directly emitted infrared pulses at 1.14 μm and 0.82 μm were observed, which corresponded to 4S–3P and 3D–3P transitions. These pulses occurred in the wings of the exciting pulse, and no clear evidence of delayed super-radiance emission was obtained at these wavelengths. Difficulties in observing super-radiance on these transitions were a result of their large Doppler effect, yielding very small T_2^* values (see table 2.1).

2.1.3 Single-pulse super-fluorescence in caesium

This experiment was carried out by Gibbs *et al* [GVH77] to study the super-fluorescence[1] (SF) output in caesium under a wide range of experimental conditions (see table 2.2).

Table 2.2. Conditions of the caesium experiment of Gibbs *et al* [GVH77]. The amplitude gain at the centre of the atomic line is $\alpha L/2 = T_2/T_R$. All times are in nanoseconds. The full width of the pump beam at I_{max}/e is d, $T_E = L/c$.

	L (cm)	d (μm)	T_2^*	T_E	T_R	T_D	$\alpha L/2$
Cell	5.0	432	5	0.17	0.15–1	6–20	35–5
Beam	3.6	366	18	0.12	0.10–1.8	5–35	180–10
Beam	2.0	273	32	0.07	0.15–1.3	6–25	215–25
Cell	1.0	193	5	0.035	0.12–0.5	5–12	45–10

The sample consisted of Cs atoms in a cell or an atomic beam of variable length; a Fresnel number, F, close to 1 for the SF wavelength, was realized by adjusting the diameter of the pump beam. The $7P_{3/2}$ level was excited from the ground state $6S_{1/2}$ with a dye-laser pulse 2 ns long with a 500 MHz bandwidth at 0.455 μm. The pump had a peak intensity of about 10 kW cm^{-2} on its axis. The transverse intensity was studied by projecting the beam onto a television camera tube, and scanning a narrow slit through the beam. The atomic density in the cell or beam was measured carefully, and the excited-state density was then calculated assuming complete saturation of the pump transition, with a precision of $(-40, +60)\%$. In a transverse magnetic field of about 2.8 kOe, and with σ polarization of the pump, the sublevel $7P_{3/2}$ ($m_J = -\frac{3}{2}$, $m_I = \frac{5}{2}$) emits σ polarization radiation (see figure 2.7).

---

[1] Here, and throughout the chapter, we keep the terminology used by Gibbs *et al*.

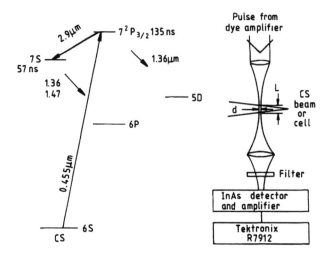

Figure 2.7. Schematic diagram of the atomic energy levels and of the experimental set-up for observation of SF in caesium [GVH77].

The output pulses (of comparable energy) were observed simultaneously in the forward and backward directions, with equal time delays. Quantitative estimates indicated that at least 20% of the stored energy was emitted in each direction. For the experiments with $F = 1$, the energy was found to be emitted into a solid angle close to the diffraction-limited value. Normalized output pulse shapes and delay times are shown in figure 2.8. A single-pulse SF was always observed for delay times beyond 7 ns. For shorter delay times, multiple pulses occurred with shapes that fluctuated greatly from pulse to pulse. time was the same as that of the first pulse. The observed value of T_R at which the transition from single pulses to multiple pulses took place was approximately $2T_E = 2L/c$. The authors noticed that the occurrence of a single-pulse SF could not be explained by relaxation processes destroying coherent ringing, since, for instance, in a 2 cm atomic beam single pulses were observed for a delay time of 8 ns, four times smaller than T_2^* and nearly ten times smaller than T_1 and T_2. Therefore a possible reason for the appearance of a single SF pulse could be random initial polarization and diffraction effects (see Chapter 6).

2.1.4 Super-radiance at 1.3 μm in atomic thallium vapour

Super-radiance in atomic thallium vapour at 1.30 μm was observed by Flusberg *et al* [FMH76]. The pulse was generated by a single dye laser working at 379.1 nm, which produced stimulated Raman scattering at 538 nm and first populated the $6\,^2P_{3/2}$ level and then the $7\,^2P_{1/2}$ state, by virtue of the exact coincidence of the laser frequency with that of the E2-allowed $6\,^2P_{3/2}$–$7\,^2P_{1/2}$ transition (see figure 2.9). Super-radiance was observed by monitoring the 1.3

μm radiation from the $7\,^2P_{1/2}-7\,^2S_{1/2}$ transition. This radiation was collimated and intense—at high laser powers it followed the laser intensity, whilst at low laser powers it showed itself as a weaker, broadened pulse, considerably delayed after the laser excitation.

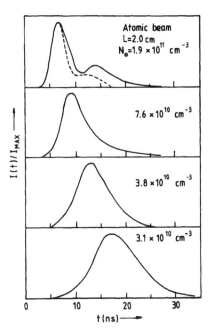

Figure 2.8. Normalized single-shot pulses for several densities in the caesium experiment of [GVH77]. Uncertainties in the values of the atomic densities are estimated to be $(+60, -40)\%$.

The experimental apparatus consisted of a nitrogen-laser-pumped dye laser, an oven-heated thallium vapour cell and a fast Ge detector. The laser line width was 0.3 cm^{-1}, the pulse width was 5–6 ms, and the pulse energy was approximately 25 μJ. A lens of 30 cm focal length focused the linearly polarized laser beam into the centre of the 15 cm long thallium vapour cell of 2.5 cm diameter, which was heated to temperatures in the range 770–850 °C. The infrared 1.3 μm pulse, of nanosecond duration, appeared with a delay of up to 12 ns with respect to the pump Stokes pulse. This long pulse delay was interpreted as a major characteristic of super-radiance. The authors also observed a correlation between the pulse height and the delay. Meanwhile, since the Doppler de-phasing time for the $7\,^2P_{1/2}-7\,^2S_{1/2}$ transition in thallium (at 800 °C) is $T_2^* = 1$ ns, so that T_D is several times T_2^*, this experiment has to be considered as a demonstration of super-radiance suppressed by the inhomogeneous broadening.

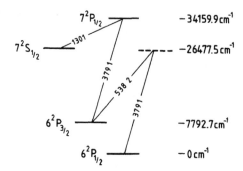

Figure 2.9. Energy-level diagram of thallium relevant to the super-radiance experiment of Flusberg *et al* [FMH76]. The indicated transition wavelengths are in nm. The laser at 379.1 nm is at exact resonance with the $7\,^2P_{1/2}$–$6\,^2P_{3/2}$ transition frequency. The broken line indicates the virtual level 99.2 cm^{-1} below the $7\,^2S_{1/2}$ level.

2.1.5 Doppler beats in super-radiance

Gross *et al* [GRH78] observed Doppler-shifted beats of super-radiance. In their experiment they prepared two groups of caesium atoms, with different velocities, in the same $7P_{1/2}$ state. A modulation in the super-radiant emission was observed on the $7P_{1/2}$–$7S_{1/2}$ transition. The frequency of modulation depended upon the Doppler shift between the two selected velocities and differed for emissions in opposite directions.

The Cs energy levels relevant to the experiment are shown in figure 2.10. As in the earlier super-radiance beat experiment of Vrehen *et al* [VHG77], the atoms were prepared by a short laser pulse at 4594 Å, which excited them from one of the hyperfine sublevels, with $F = 3$ or 4, of the $6S_{1/2}$ ground state (*g*) to the $7P_{1/2}$ level. The subsequent super-radiant emission at 3.1 μm brought the

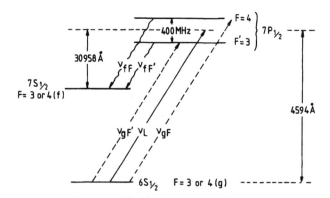

Figure 2.10. Scheme of caesium energy levels relevant to the super-radiance Doppler beat experiment of Gross *et al* [GRH78].

atoms to the two $7S_{1/2}$ final states, with $F = 3$ and 4. Since hyperfine splitting of 2175 MHz between these states was too large to give rise to observable beats, this structure could be neglected, and only one final state (f) was considered. The $7P_{1/2}$ state split into two hyperfine substates, with $F = 4$ and $F' = 3$, separated by $\nu_{FF'} = 400$ MHz. This structure lay in the background of the Doppler profile of the pump transition ($\Delta\nu_{\text{Doppler}} = 500$ MHz for each hyperfine component). In order to select two groups in the Maxwellian distribution of the excited state, a pump laser with a spectral width $\Delta\nu_L$ much smaller than $\Delta\nu_{\text{Doppler}}$ was used.

The atoms were prepared in the $7P_{1/2}F$ ($7P_{1/2}F'$) state provided their velocity along the laser beam, v_F ($v_{F'}$), satisfied, respectively, the following conditions

$$\nu_{gF} = \nu_L - k_L v_F \qquad \nu_{gF'} = \nu_L - k_L v_{F'}. \qquad (2.1.1)$$

In this equation, ν_L and k_L are the mean frequency and wavevector of the laser field, and ν_{gF} ($\nu_{gF'}$) are the optical frequencies of the two pump transitions, in the rest frame of the atom. Two groups with different velocities were thus excited and were able to super-radiate at frequencies $\nu_{fF} + k_{SR}v_F$ and $\nu_{fF'} + k_{SR}v_{F'}$ respectively (in the last expression, ν_{fF} and $\nu_{fF'}$ were the rest frame optical frequencies of the super-radiant transitions, and k_{SR} was their mean wavevector). Correspondingly, the super-radiant signal was expected to be modulated at the frequency

$$\nu_{\text{beat}} = \nu_{FF'}(1 - k_{SR}k_L^{-1}). \qquad (2.1.2)$$

This frequency was red-shifted with respect to $\nu_{FF'}$ for an emission occurring in the same direction as the laser pulse (forward emission, $k_{SR}k_L^{-1} > 0$) and it was blue-shifted for super-radiance in the opposite direction (for backward emission, $k_{SR}k_L^{-1} < 0$). The frequency shift, $\pm|k_{SR}k_L^{-1}| = \pm 59$ MHz, corresponded to the Doppler shift between the two velocity groups observed for the infrared transition.

Typical modulated signals obtained in coincidence in both forward and backward directions are shown in figure 2.11. The delay of the backward pulse was subtracted, in order to reproduce the exact timing of events ($t = 0$ corresponded to the laser pulse's maximum). The beat contrast of these single-shot signals was good, and several beats could be counted. The forward signal appears to have had, as expected, a smaller frequency than the backward one. The histogram of figure 2.11 represents the results of beat frequency measurements for 100 good laser shots. The forward and backward pulses clearly exhibited different frequency components ($\nu_{\text{fore}} = 344$ MHz and $\nu_{\text{back}} = 454$ MHz). The average frequency, $(\nu_{\text{back}} + \nu_{\text{fore}})/2 = 399$ MHz, was in good agreement with the known value of $\nu_{FF'}$. The half-frequency difference $(\nu_{\text{back}} - \nu_{\text{fore}})/2 = 55$ MHz was also in good agreement with the Doppler shift evaluated above.

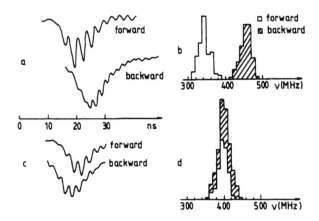

Figure 2.11.   Super-radiant Doppler beats observed in caesium [GRH78].   (a) Doppler-shifted beats recorded in forward and backward directions following narrow-band excitation. (b) Histogram of beat frequencies for narrow-band pumping. White and hatched boxes represent forward and backward signals respectively. (c) Forward and backward signals following broad-band excitation. (d) Histogram of beat frequencies for broad-band excitation.

This experiment gave direct evidence of the temporal coherence of the super-radiant pulse. It demonstrated the possibility of detecting beats coming from different atoms.

2.1.6   Direct measurement of the effective initial tipping angle

A direct measurement of the effective initial tipping angle θ_0 for SF in the $7P_{3/2}(m_j = -\frac{3}{2}, m_I = -\frac{5}{2}) \rightarrow 7S_{3/2}(m_j = -\frac{1}{2}, m_I = -\frac{5}{2})$ transition in Cs vapour, was carried out by Vrehen and Schuurmans [VS79] as an extension of their previous work [GVH77] described in subsection 2.1.3.

In the semiclassical Maxwell–Bloch treatment of SF (see Chapter 1), the initial tipping angle θ_0 represents a coherent pulse of small area propagating along the axis of the cell, which triggers the individual Bloch vectors. It models quantum fluctuations that cause the beginning of the SF pulse. Vrehen and Schuurmans suggested injecting such a pulse at the SF wavelength into the sample immediately after the sample had been completely inverted by a short pump pulse. As long as the area θ of the tipping pulse was smaller than θ_0 the delay time of the SF output pulse would not be affected. However, when $\theta > \theta_0$ the delay time would be reduced. Therefore, by measuring the delay time as a function of the area θ of the injected pulse the magnitude of θ_0 could be found.

The experimental set-up shown in figure 2.12 was as follows: two caesium cells 1 and 2, at a mutual distance D of 50 cm, were successively pumped by the same pump laser beam. This beam had a diameter d_1 of 270 μm at the first cell, and a diameter d_2 of 450 μm at the second cell, which corresponded to the

Fresnel number $F = 2$ for cell 1 ($L_1 = 1$ cm) and to $F = 1$ for cell 2 ($L_2 = 5$ cm). The pump pulse at 455 nm had a duration of about 2 ns full width at half its maximum, a bandwidth of 400 MHz and peak power of 35 W. The vapour density in cell 1 was adjusted so that the SF pulse was emitted with a delay time of about 1.5 ns; its width was found to be nearly 2 ns. This SF pulse was the infrared injection pulse for cell 2. The spatial coherence of the injection pulse over the cross-section of the pumped volume in cell 2 was guaranteed by the geometry: $d_1 d_2 \ll \lambda D$.

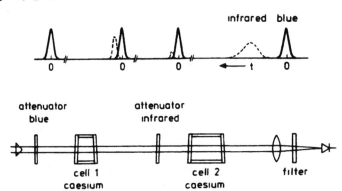

Figure 2.12. Set-up of the experiment to measure the effective initial tipping angle, θ_0, in SF [VS79].

The evidence for the temporal coherence came from the observation of 'classical' beats in SF on two different uncoupled transitions observed by Vrehen *et al* [VHG77], and from the SF in different groups of atoms observed by Gross *et al* [GRH78] (see subsection 2.1.5). The area of the infrared pulse emitted from cell 1 was estimated roughly as π. It was reduced with the aid of calibrated attenuators, consisting of Perspex plates of 1 mm thickness. At the SF wavelength of 2931 nm their transmittance was 0.04, whereas at the pump pulse wavelength of 450 μm transmittance was full, apart from small reflection losses. The density in cell 2 was adjusted so that, without injection, the delay time of the SF pulse was approximately 13 ns. For injected pulses with areas above 5×10^{-4} the delay was reduced considerably. To be able to compare their results with the theory of Burnham and Chiao [BC69] (see subsection 1.4.3) the authors plotted the relation between the average delay time T_D and $|\ln(\theta/2\pi)|^2$ for one particular experiment, i.e. at constant vapour density in both cells while varying attenuation of the injection pulse. This plot is shown in figure 2.13. As expected, T_D increased linearly for large injection pulses (see equation (1.5.7)). For small injection pulses the initiation of SF was dominated by quantum fluctuations, and T_D was constant. At the cross-over $\theta = \theta_0$, by definition. From similar plots for several experiments with a total number of excited atoms $N = 2 \times 10^8$, the authors calculated the most probable value of θ_0. It was found to be $\theta_0 = 5 \times 10^{-4}$ radians. Taking into account the shot-

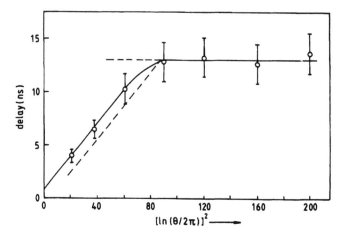

Figure 2.13. Delay time T_D of output pulse versus $[\ln(\theta/2\pi)]^2$. The dashed line is used to correct for the delay of the injection pulse with respect to the pump pulse [VS79].

to-shot fluctuations and the uncertainty connected with the measurement of θ_0, they estimated that $10^{-4} < \theta_0 < 2.5 \times 10^{-3}$ radians.

An injection pulse of area 5×10^{-4} and 2 ns duration carries a total energy of one single photon through the cross-section of the sample. So the authors concluded that the SF evolution was triggered by the first photon emitted spontaneously along the axis of the sample. The authors compared their result with the quantum theory of SF initiation developed by Glauber and Haake [GH78], and by Schuurmans *et al* [SPV78] (see Chapter 4 and section 1.5), which gave the result $\theta_0 \simeq 1/\sqrt{N}$ for the effective tipping angle, in good agreement with the measured value.

2.1.7 Triggered super-radiance

Carlson *et al* [CJSGH80] investigated the effect of triggered super-radiance on the $7P_{3/2}$–$7S_{1/2}$ transition in caesium, by using the output of a tunable infrared colour-centre laser.

The changes induced in super-radiance by external triggering could be detected in several ways, as shown in figure 2.14. At first, (a), the injection field (represented by a black arrow) caused the emission to occur somewhat faster. It was thus possible to observe the shortening of the time delay T_D. This was the same technique as that used by Vrehen and Schuurmans [VS79]. Since T_D depends logarithmically upon the injection power, it was difficult to obtain a precise measurement of the noise-equivalent power by this method. Much more sensitive signal changes could be observed by taking advantage of the property that super-radiance is never a one-dimensional effect, but necessarily has a multi-mode character. In the case of a cylindrical sample, the non-triggered

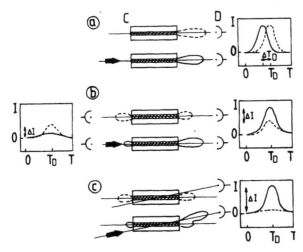

Figure 2.14. Sketches illustrating various ways in which super-radiance can be modified by external triggering [CJSGH80]: (a) change in super-radiance delay; (b) change in right/left intensity ratio; (c) change in super-radiance direction. In each case the upper and lower drawings correspond to the unperturbed and modified super-radiant emission respectively. C is the emitting cell and D the super-radiant detector. The injection beam is represented by a black arrow. The corresponding signal, as it would appear on an oscilloscope connected to the detector, is shown in the box frames. The broken-line trace represents the unperturbed signal and the full-line trace the modified signal.

super-radiance was able to occur symmetrically in two opposite directions (upper part of figure 2.14(b)). For each of these directions the emission would have a transverse pattern and occur in several modes with k wavevectors having slightly different orientations (upper part of figure 2.14(c)). The injection field favoured some of the modes by triggering emission into these modes more rapidly than into others. This resulted in a strong change of the unperturbed emission pattern. If the triggering light beam propagated along the sample axis towards the right (lower part of figure 2.14(b)) the signal observed in this direction would become systematically larger than the one detected in the opposite direction. The observed change in this case is the average relative difference $(I_{\text{right}} - I_{\text{left}})/(I_{\text{right}} + I_{\text{left}})$ between the right and left super-radiance signals. If the injected beam was slightly off-axis (lower part of figure 2.14(c)) the emission was enhanced in the direction of this beam, and the signal detected along the direction of the injection field strongly increased. The last method of super-radiance triggering was the simplest, since it was possible, by conveniently adjusting the position of the signal detector, to obtain a very large signal for this change.

Figure 2.15 shows a sketch of the experimental set-up with a diagram of the relevant Cs energy levels. The super-radiance cylindrical sample (length $L = 10$

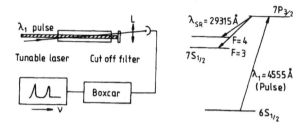

Figure 2.15. Sketch of the caesium super-radiance triggering spectroscopy experiment of Carlson *et al* [CJSGH80]. On the right is shown a diagram of the relevant energy levels of Cs.

cm, diameter $d = 4$ mm) was prepared by a dye-laser light pulse exciting the $6S_{1/2}$–$7P_{3/2}$ transition at $\lambda = 455.5$ nm. Super-radiance then occurred on the totally inverted $7P_{3/2}$–$7S_{1/2}$ transition. Both levels involved in this transition have a hyperfine structure. The pump laser intensity and atomic vapour pressure were set so that super-radiance occurred with a typical delay of 10–20 ns, corresponding to $T_R \simeq 0.5$ ns. The injection signal was a 4 mm diameter light beam emitted at 2.93 μm by a colour-centre laser pumped by a krypton-ion laser. The colour-centre laser had an air-spaced intra-cavity etalon for single-mode operation. The injection intensity was controlled by using calibrated neutral filters. The laser frequency was scanned across the atomic transition lines by varying the gas pressure in the arm of the laser cavity containing the etalon.

Carlson *et al* [CJSGH80] observed the qualitative changes in super-radiance using all three methods described above. They also made a quantitative study of the triggered super-radiance by using an off-axis beam. The injection beam made an angle of 1.5° with the axis of super-radiance, which was small enough to ensure that the pumping and triggering beams overlapped throughout the whole length of the cell. An InAs detector was aligned along the direction of the injection beam. As a result, when the injection beam was off resonance only a very small signal was detected. When the injecting field was tuned across resonance, the signal increased significantly. Figure 2.16 shows typical spectra obtained for the injection flux of several intensities. The observed doublet corresponded to the hyperfine splitting of the $7S_{1/2}$ level ($\Delta\nu_{hfs} = 2100$ (±150) MHz), in good agreement with previous determinations of this structure. The much smaller $7P_{3/2}$ structure was not resolved. The lowest triggering flux recorded in this experiment, was 8×10^{-9} W mm^{-2} (the lower trace in figure 2.16). The determination of this threshold intensity was made with an uncertainty factor of two to three. This corresponded to 10–20 resonant photons per T_R in an area $a^2 = L$, and to a tipping angle of $\theta_0 \simeq 10^{-3}$ radians, in good agreement with the earlier experiment by Vrehen and Schuurmans [VS79].

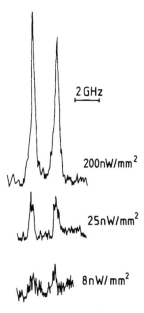

Figure 2.16. Super-radiance triggering spectrum of the $7P_{3/2}$–$7S_{1/2}$ caesium transition for three power fluxes of the injection field [CJSGH80]. The doublets correspond to the hyperfine splitting of the $7S_{1/2}$ level.

2.1.8 Super-radiance in a system of Rydberg atoms

Rydberg atoms are bound atomic systems in which a valence electron has been excited to a state with very high principal quantum number n. From the theory of the hydrogen atom, which is applicable to Rydberg atoms, it is known that the separation between the energy levels with adjacent n values is of the order of n^{-3} (in atomic units), and that it corresponds to the millimetre-wave region for $n \approx 20$–50. The electric dipole matrix elements between neighbouring Rydberg levels are of the order of n^2, and for $n \approx 20$–50 they turn out to be about three orders of magnitude larger than for low-lying atomic states. As a result, Rydberg atoms interact resonantly, and strongly, with millimetre-wave radiation. Ensembles consisting of relatively small numbers of such atoms, suitably excited by a short laser pulse, can be used as the active medium in maser systems which exhibit super-radiance characteristics.

The first realization of these conditions was reported by Moi *et al* [MGGRFH83]. In their experiments a beam of atomic Na, originating from an oven heated to about 400 °C, was made to cross a millimetre-wave open Fabry–Pérot cavity (with a finesse ≈ 100) which had a semi-confocal structure (see figure 2.17). The atoms were excited to the upper Rydberg level of the maser transition by two collinear N_2 laser-pumped pulsed dye lasers crossing the atomic beam at right angles inside the cavity. The lasers were tuned, respectively, to the

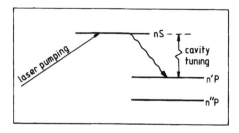

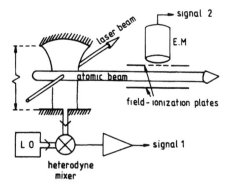

Figure 2.17.   General scheme of the Rydberg atom experiment of Moi *et al* [MGGRFH83]. Signal 1 corresponds to the direct detection of the microwave emitted by the atoms. Signal 2 is an indirect detection consisting of measuring the radiative atomic population transfer through field ionization of the Rydberg atoms. (In the inset: energy levels relevant to the emission. The final $n'P$ state of the emission is selected by cavity tuning.)

transitions $3S$–$3P_{1/2}$ ($\lambda_1 = 5896$ Å) and $3P_{1/2}$–$nS_{1/2}$ ($n = 20$–40, $\lambda_2 = 4100$ Å). The laser pulse lasted for about 5 ns. The size of the active volume in the cavity was controlled by adjusting the focus of the pumping laser beams using lenses of various focal lengths. The projection of the beam could be reduced, in this way, to a spot less than 0.2 mm in diameter or expanded to over 0.5 cm (i.e. made much less than or much more than λ for the maser transition).

Immediately after the laser pulse the atomic system was totally inverted in all the transitions connecting the Rydberg level to states which were more tightly bound. If the cavity was tuned to the frequency of one of these transitions, and if the population inversion was larger than a certain threshold (10^4 atoms for an $nS_{1/2} \rightarrow (n-1)P_{1/2}$ transition, with $n \approx 30$), the inverted medium emitted a short burst of radiation and decayed within a few hundred nanoseconds to the lower state of the transition. The emission process was detected either directly, by recording the microwave signal, or indirectly, by monitoring the fast transfer of the atomic population. The first method used a millimetre-wave heterodyne receiver coupled to the cavity through a waveguide. In the second method, the

Rydberg atoms were ionized after they had left the cavity by an electric field pulse produced by two parallel condenser plates. The ejected electrons were detected with the help of an electron multiplier. The threshold electric field for ionizing the atoms depended upon their excitation. This enabled one to distinguish between the contributions to the ionization current from the upper and lower states of the maser transition, and hence to measure the population transfer. Only the first method allowed a real-time analysis of the pulse shapes, fluctuations and delays. The second method offered the advantages of higher sensitivity (emission from a few atoms can be detected) and simplicity of operation. Figure 2.18 shows the time-resolved ion signals corresponding to the upper and lower states of the transition.

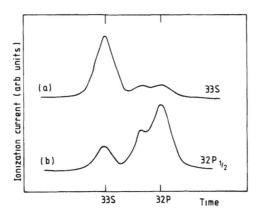

Figure 2.18. Examples of time-resolved ion signals from Na ions averaged over 100 laser pulses as observed in the Rydberg atom experiment of Moi *et al* [MGGRFH83]. Trace (a) is the ion signal associated with field ionization of the 33S level, directly prepared by laser excitation. Trace (b) is the ion signal corresponding to the $32P_{1/2}$ level. This level is prepared by laser excitation of the 32S level (which ionizes in a field larger than the one applied to the atom) followed by a microwave-induced $32S \rightarrow 32P_{1/2}$ transition. The 33S and 32P markers on the time axis indicate the respective maxima of these ion pulses. Note the partial time overlapping of these two signals.

The bursts of radiation were so short that any atomic relaxation processes (collision and Doppler broadening, Stark de-phasing) would have been negligible for the experimental conditions considered. The bursts could be described by a damped-pendulum-like solution (see section 1.4) with the characteristic time $T_R^c = [(8/\pi)f\gamma N\mu]^{-1}$, where μ is the diffraction factor of the cavity, f is the cavity's finesse and γ is the radiation decay constant for a single atom. T_R^c is inversely proportional to the number of inverted atoms, and in a cavity this quantity can be considered as the super-radiation time of the collective system. In the case of this experiment the appropriate data were $\mu \approx 10^{-2}$, $f \approx 100$ and γ was of the order of 20 s^{-1} for the principal quantum number $n \approx 30$. For

$N = 5 \times 10^5$, $T_R^f \approx 30$ ns, and the delay time $T_D \cong T_R^c \ln N$ is of the order of 400 ns, in good agreement with typical experimental results (see figure 2.19).

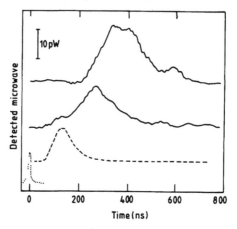

Figure 2.19. Two typical maser pulses detected at 107 892 GHz (33S → 32P$_{1/2}$ transition in Na) [MGGRFH83]. The dotted curve around time $t = 0$ represents the pumping laser pulse. The broken curve is the percussive response of the microwave receiver. The upper trace is a 20 pW microwave burst, corresponding to an actual ∼ 200 pW emission by the atoms (taking into account the 9.5 dB output coupling cavity loss). The estimated number of atoms is ∼ 5×10^5.

For a cavity, the characteristic time T_R^c shortens by a factor proportional to the cavity's finesse. This is particularly important if the medium's dimension is small compared with the wavelength of the radiation. The super-radiance time for free space is then of the same order of magnitude as the phase disruptive time resulting from the dipole–dipole interaction. The cavity enhancement factor speeds up the evolution of super-radiance without increasing the dipole–dipole couplings, which would have destroyed super-radiance in a sample of the same size in free space. (For a more detailed discussion, see Chapter 8.)

Moi *et al* [MGGRFH83] gave a full and very clear theoretical explanation of the observed phenomenon by interpreting it as super-radiance in a cavity. The theory of the transient maser effect is similar to that of super-radiance in free space, for the mean-field model (however, some important modifications for the extended system—standing-wave modulation of the field and the atomic characteristics—have to be considered). Moi *et al* obtained a theoretical estimate of the threshold value for the number of atoms for super-radiance in a cavity. This value essentially depended upon the transit time of the atomic beam through the waist of the cavity. For the parameters of the experiment the theoretical threshold was found to be $N \sim 13\,000$ for 300 K, being in fair agreement with the experimental threshold of $20\,000 \pm 800$ atoms. This is six to seven orders of magnitude smaller than for an ordinary maser operating in similar cavities at comparable wavelengths.

For further details and experiments with fewer Rydberg atoms in the active sample, we refer the reader to a review by Haroche and Raimond [HR85].

2.2   Observation of super-radiance in solid-state materials

2.2.1   Super-fluorescence of O_2^- centres in KCl

Schmid and co-workers [FSS82, FSS84a, SSS87, SSS88, SSS89b] reported a series of experiments that used O_2^- centres in KCl, the first solid-state system in which SF was observed.

The structure and optical spectra of the O_2^- ion which may substitute itself for a halide ion are well known. At low temperatures its molecular axis was oriented along a 110 direction. Optical excitation at about 250 nm yielded a characteristic fluorescence spectrum which was dominated by a progression of narrow zero-phonon lines resulting from transitions from the relaxed excited state to the excited vibrational states of the electronic ground state [60] as shown in figure 2.20. The width of each line at low temperatures was about 1 cm^{-1}, as a

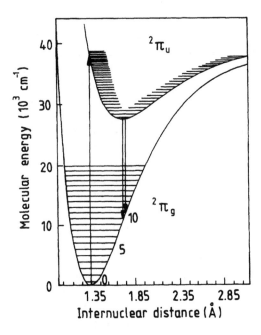

Figure 2.20.   Configurational–coordinate diagram for the O_2^- centre in KCl. Super-fluorescence was observed on the two transitions, 0–10 and 0–11, indicated in the figure [FSS82].

result of inhomogeneous broadening, i.e. $T_2^* \simeq 30$ ps. The phase relaxation time T_2 depended upon the temperature. Above 25 K the rotational motion of the

O_2^- ions was initiated, and the coherence of the SF state was destroyed (see the next subsection). For 10 K, T_2 was estimated as 100 ps. After optical excitation by an ultraviolet light pulse the system transferred to a non-equilibrium state, and relaxed within about 20 ps towards a new equilibrium state. The ensemble of the relaxed excited O_2^- centres was then totally inverted with respect to the ground-state vibrational levels to which optical transitions were able to occur. The most intense line in the spontaneous fluorescence spectrum resulted from the transition into the tenth vibrational level at 592.78 nm.

Sample and experimental method. In the experiments described, crystals containing O_2^- (with concentrations between 8×10^{16} cm^{-3}and 7×10^{17} cm^{-3}) were cleaved to produce samples of about $5 \times 5 \times 10$ mm^3. They were excited at temperatures below 30 K using the fourth harmonic of single pulses from a mode-locked Nd:YAG laser (with $\lambda_p = 266$ nm, a pulse duration of 30 ps and maximum pulse energy of 100 μJ). The pulses were focused to a spot of about 0.1 mm diameter at the sample surface. The resulting excitation volume was pencil shaped with a Fresnel number close to unity. The fluorescence light was monitored with a fast vacuum photodiode. The overall response time of the entire set-up was about 500 ps.

Experimental results. The upper part of figure 2.21 illustrates the principle of the experiment. At energies of the excitation pulse of less than 20 μJ (peak intensity 10 GW cm^{-2}) the authors observed the standard spontaneous fluorescence of the O_2^- ion to have a decay time of about 90 ns. At peak excitation intensities of more than 15 GW cm^{-2} the fluorescence at 625.04 nm (the 0–11 transition) became highly anisotropic, being collinear with the sample axis. Simultaneously the fluorescence intensity in the forward and backward directions increased by a factor of more than 10^4.

The lower panel of figure 2.21 represents the typical time dependence of this radiation for three different shots. In spite of the identical excitation conditions, the parameters of the observed signals fluctuated randomly. The pulse intensities varied by more than a factor of ten, whereas pulse widths were observed to vary between 0.5 ns and 0.6 ns, and pulse delay times between 0.5 ns and 10 ns. The occurrence of the signals did not depend upon the orientation of the sample. This ruled out any accidental laser activity caused by unintended specular reflections, for instance from the sample's surfaces. Moreover, this kind of laser activity was observed simultaneously with SF by adjusting the excitation beam to be just a few degrees away from being normal to the sample's surfaces. SF then propagated within the sample exactly along the excitation channel, and was observed outside the sample in the direction given by the law of refraction, whilst the laser radiation was observed perpendicular to the sample surfaces.

Two-colour SF. A simultaneous SF emission was observed for two transitions, 0–11 and 0–10 at 629.04 nm (red) and 592.78 nm (yellow), respectively, provided that the peak excitation intensity was raised above a threshold of about 30 GW cm^{-2}. The statistics of the pulse intensities and delay

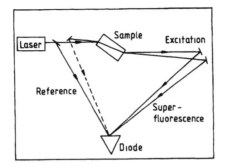

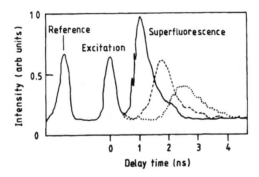

Figure 2.21. Upper panel: experimental set-up for measuring intensities and delay times of the SF pulse [FSS84a]. Lower panel: experimental results for the forward SF in three independent shots at identical excitation conditions. The backward SF can be monitored using the arrangement indicated by the broken line in the upper panel.

times confirmed that the red and yellow pulses originated at the same instant, to within the limits of error (±100 ps), whilst the relative pulse intensities fluctuated statistically, with a qualitative trend towards a preference for the yellow pulse at very short delay times. For relatively long delay times the red pulse usually dominated. In their one-dimensional experiments the authors succeeded in observing simultaneous SF at up to four different wavelengths: 592.78 nm (0–11), 629.04 nm (0–11), 669.38 nm (0–12) and 714.4 nm (0–13).

The theoretical interpretations of these observations were given by Haake and Reibold [HR82, HR84] and Schwendimann [Sn84].

'Two-dimensional' SF: diffraction pattern. To realize a two-dimensional excitation volume, the pump beam was expanded, using a cylindrical lens, as indicated in figure 2.22. In this way an active volume of about $8 \times 0.05 \times 1$ mm^3 was obtained, where the last dimension reflects the absorption length of the pump beam. When the pump intensity was increased above the threshold of about 500 MW cm^{-2} a SF emission was observed simultaneously in both directions of the excitation channel. The pulses emitted in opposite directions

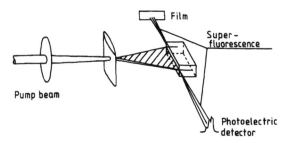

Figure 2.22. Experimental set-up for the observation of transverse two-dimensional SF [SSS87].

had identical properties (e.g. delay time, pulse, width, intensity and polarization). Figure 2.23 illustrates several examples. On the left-hand side of figure 2.23 photographic recordings are presented; on the right-hand side the pulses opposite the photographs were recorded photoelectrically in the same event. From top to bottom, the pump intensity was increased from about 500 to 800 MW cm^{-2}. The most striking feature was the observation of characteristic diffraction patterns with nodal planes perpendicular to the plane of the excitation volume. The number of nodal planes increased with increasing excitation intensity. Although the observed interference patterns resembled those of transverse laser modes, they did not result from such an effect. This was confirmed by performing experiments with simultaneous two-dimensional SF and laser activity, analogously to the one-dimensional experiment above. The theoretical interpretation of these observations will be made in Chapter 8, where we shall study the diffraction properties of the super-radiance emission. The reason for the simultaneous emission of counter-propagating pulses observed in these experiments is not fully understood at the moment. Section 9.5 will discuss one of the possible mechanisms for such a synchronization.

2.2.2 Transition from super-fluorescence to amplified spontaneous emission

In order to clarify the conditions for observing SF, Malcuit *et al* [MMSB87] studied the transition from the SF régime to amplified spontaneous emission (ASE) as temperature was increased.

The experiment used a KCl crystal, cleaved to dimensions of approximately $7 \times 7 \times 4$ mm^3, containing 2×10^{18} O$_2^-$ ions cm^{-3}. The crystal was mounted in a temperature-regulated cryostat, cooled by a closed-cycle helium refrigerator. The crystal was excited by a 30 ps pulse, of up to 60 μJ of energy, using a frequency-quadrupled Nd-doped YAG laser. The pulse was focused into the crystal with a cylindrical lens and created an interaction region which had the form of a cylinder of diameter 80 μm and length of 7 mm with a Fresnel number close to unity for the transition under consideration. For excitation energies above 10 μJ,

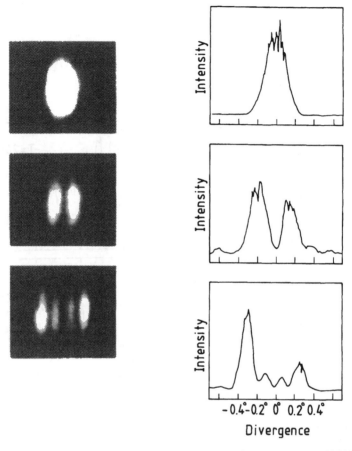

Divergence

Figure 2.23. Mode structure of two-dimensional SF observed by Schiller *et al* [SSS87].

highly directional emission of the 6294 Å vibronic transition in O_2^- was emitted from both ends of the crystal. The emission was detected by using a streak camera system, capable of providing a time resolution better than 2 ps. Figure 2.24 shows typical output pulses obtained under identical excitation conditions, for several different values of the crystal temperature from 10 K to 30 K. As the temperature was raised, the shape of the output pulse was seen to change continuously from that characteristic of SF to that characteristic of ASE, owing to the increase in the dipole de-phasing rate as the third power of the temperature. At the lowest temperature shown, the emission was typical SF with a pulse duration of approximately 60 ps and a time delay of approximately 160 ps. As the temperature of the crystal was increased slightly (cases (b)–(d)) the emitted pulse broadened and the time delay increased. As the temperature was increased still further (e) the pulse continued to broaden, but the time delay began to

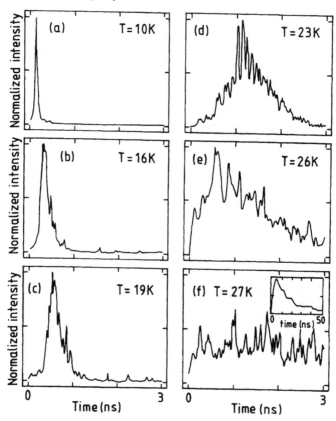

Figure 2.24. Typical experimental realizations of the temporal evolution of the emission from KCl : O_2^- for several different temperatures [MMSB87]. At the lowest temperature the emission is characteristic of SF, whereas at the highest temperature the emission is characteristic of amplified spontaneous emission. The inset in (f) shows the evolution of the emission on a greatly enlarged time scale.

decrease. At the highest temperature shown, the emission was characteristic of ASE: the time delay was almost too small to be measured, the output pulse was very noisy and the pulse duration had increased still further. This pulse duration was, however, considerably shorter than the 80 ns spontaneous emission time. In all the cases shown, the unsaturated line-centre single-pass gain was much greater than unity.

Figure 2.25 shows how the time delay and peak intensity depended upon the temperature. Because of the statistical nature of the emission process, many shots were collected at each temperature, so that the mean values and variations of the time delay and the peak intensity could be determined. The circles give the mean values, and the vertical bars indicate the range of values at each temperature. As the temperature of the crystal and, hence, the de-phasing rate were increased,

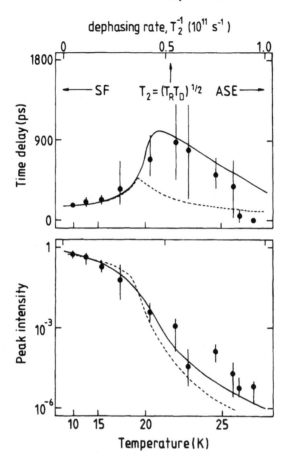

Figure 2.25. Time delay and peak intensity of the emission plotted versus temperature and versus de-phasing rate as recorded in [MMSB87]. Circles represent the mean values of the experimentally observed quantities, and the vertical lines represent the range of observed values. The emission is characteristic of sf for $T_2 \gg \sqrt{T_R T_D}$, and is characteristic of ASE for $T_2 \ll \sqrt{T_R T_D}$. The broken curve represents the predictions of a theory that treats the effects of noise only as an initial condition, whereas the full curve represents the predictions of a model that includes a temporarily fluctuating noise source.

the mean time delay was seen to change from 160 ps at the lowest temperature shown to a maximum of 800 ps, and then to decrease monotonically as the temperature was increased still further. The fluctuations in time delay were seen to increase in the transition region. From figure 2.25 we see that the peak intensity decreased monotonically over six orders of magnitude within the temperature range considered.

The authors compared their results with the prediction of the theories treating the initiation of SF in the presence of de-phasing [Ss80] and found a good qualitative agreement. The theoretical aspects of this problem will be considered in Chapter 5.

2.2.3   Coherent amplification of ultra-short pulses in YAG:Nd and ruby

Leontovich and co-workers [VKLMMT84, VGKMMBT86] studied the generation and amplification of ultra-short pulses in YAG:Nd and ruby at a temperature of about 100 K, when the pulse duration was less than the transverse relaxation time of the medium, i.e. under the super-radiance condition.

Investigated media and experimental method.   The samples of YAG:Nd used in this experiment were rods 8 mm in diameter and 80 mm long, with a Nd concentration of 0.6. The small-signal gain αL obtained for pumping with flash lamps amounted to ~ 3. The ruby rods were 12 mm in diameter and 120 mm long, and the concentration of Cr was 1.5×10^{19} cm^{-3}. Under flash lamp pumping the gain was $\alpha L \simeq 5.5$. The rods were cooled in a nitrogen vapour stream or in a stream of liquid nitrogen. The temperature of the rods could be varied within the range 80–100 K. The estimated phase memory time T_2 for both media for this temperature interval was within the range 100–300 ps. Under the conditions described, the line width was governed primarily by the inhomogeneous broadening, and that amounted to ~ 1 cm^{-1} in the case of YAG:Nd, and to ~ 0.3 cm^{-1} in the case of ruby. This corresponded to $T_2^* = 20$ and 60 ps respectively.

   In the experiment on amplification in YAG:Nd the source of the input pulses was a laser with an Nd-activated silicate glass active element, operating in the passive mode-locking régime. The laser emission wavelength was tuned by an intra-cavity Fabry–Pérot interferometer. This made it possible to vary the emission line wavelength within a band 50 cm^{-1} wide with a tuning error less than 0.1 cm^{-1}. The source of input pulses in the case of the ruby amplifier was a ruby laser operated at a low temperature in the passive mode-locking régime. The time parameters of radiation were determined by employing an image converter camera with a time resolution of ~ 20 ps.

Lethargic amplification and induced super-radiance.   The authors investigated the dependence of the output signal upon the area θ_0 of the input signal for various pulse durations. The value of θ_0 was estimated from the pulse duration and the energy, and it was varied using neutral optical filters as well as a dense saturable filter. Densitograms of the pulses representing amplification in ruby at 80 K are presented in figure 2.26. As can be seen, the duration of the input pulses decreases from the top to the bottom. In the topmost densitogram the pulse exhibits the so-called lethargic amplification, i.e. a gain which was low compared with that predicted by Beer's law. The peak radiation power increased only by a factor of ten. In the two lower densitograms the duration of the input pulses is shorter. The output pulses exhibited fragments corresponding to the

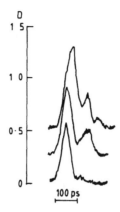

Figure 2.26. Densitometer traces of radiation from a self-mode-locked ruby amplifier at 100 K. The curves correspond to three successive steps of the attenuator mounted in front of the slit of the electro-optic camera. D: Photographic density [VKLMMT84].

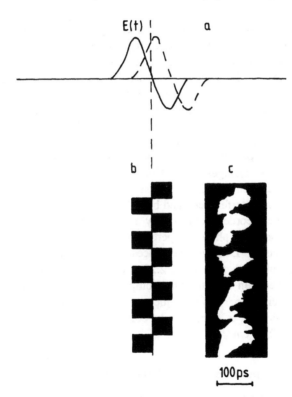

Figure 2.27. Interferograms of pulses with oscillating envelopes: (a) superposition of pulses (schematic); (b) position of interference bands (schematic); (c) time scan of the interference pattern recorded for the pulse from the ruby laser [VKLMMT84].

input pulses as well as the induced (i.e. triggered) super-radiance pulses. The latter were strongly suppressed owing to a relatively fast phase relaxation. These experimental results were in agreement with the predictions of the semiclassical theory (see Chapter 5).

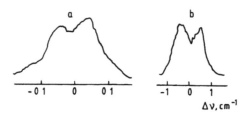

Figure 2.28. Densitometer tracings of the spectra of ultrashort pulses from (a) Nd:YAG and (b) ruby amplifiers [VKLMMT84].

Spectral and phase measurement. The change in sign of the field envelope can be determined from the evolution of the phase of the field along the pulse. This was achieved by time scanning of the interference patterns. Interference was produced between light beams reflected from the two surfaces of a plane-parallel glass plate. The far-field interference pattern was then swept across the electronic camera. The operation of the system is illustrated in figure 2.27. It shows interference between two pulses with oscillating envelopes when the delay between the pulses corresponds to half of the oscillation period. When the envelope of the first pulse passes through zero, the phase difference between the interfering pulses changes by π. Constructive interference is then replaced by destructive interference, and vice versa. A change of π in the phase is thus seen as a shift of the interference band by half the dispersion range of the etalon. These effects were recorded both in ruby and YAG:Nd.

The super-radiance phase properties manifested themselves also in the spectra of the pulses. Densitometer traces of the spectra of pulses amplified in YAG:Nd and ruby rods (see figure 2.28) clearly showed the doublet structure. The separation between the structure components was found to be consistent with the field oscillation. For ruby the separation between the components of the doublet was as large as 1 cm^{-1}, to be compared with a much smaller ground-state splitting of 0.38 cm^{-1}.

2.2.4 Super-radiance in a pyrene–diphenyl crystal

Zinoviev *et al* [ZLNSSS83, ZLNS84, ZLM85, ZMNS85], Naboikin *et al* [NSS83, NAZMSSS85] and Andrianov *et al* [AZMNSSS86] observed a super-radiance effect in a diphenyl crystal doped with pyrene molecules ($C_{12}H_{10}$). The latter is a plane molecule with symmetry group D_{2h}. A diphenyl crystal has two molecules per unit cell and the monoclinic spatial group symmetry $^5C_{2h}$ [40]. There are three kinds of pyrene centres corresponding to the three non-equivalent positions of pyrene molecules in a diphenyl crystal [30]. They can be identified with polarization spectroscopy at 4.2 K [10], and by the observed multiple structure of absorption and luminescence lines formed from the three components [AW76].

The main feature of the pyrene energy spectra in diphenyl crystals is the weak electron–phonon interaction; the intensities of phonon bands both in absorption spectra and in luminescence spectra are small. The line widths of the 0–0 transitions and fully symmetric electron vibrational transitions are of the order of 1 cm^{-1}, and their broadening is inhomogeneous.

The physical arrangement of the first super-radiance experiment with the pyrene–diphenyl crystal, together with the energy structure of the actual transition, are shown in figure 2.29. The experiment was carried out at a temperature of 4.2 K. Super-radiance was observed at the frequency

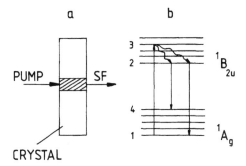

Figure 2.29. (a) Arrangement of the pyrene–diphenyl crystal experiment on super-radiance. (b) Scheme of actual levels and transitions in the system [ZLNSSS83].

corresponding to the transition between ground states of the electron vibrational terms $^1B_{2u}$ and ^1A_g of the pyrene molecule. The molecules were pumped by pulses of the third harmonic of a YAG:ND^{3+} laser, with a duration of 10 ns. The laser radiation was resonant with the transition between the ground-state level of the term $^1A_g(1, q = 0)$ and the third vibrational sublevel of the term $^1B_{2u}(2, q = 3)$, $\lambda = 353$ nm.

The absorption coefficient for the exciting wavelength was 10 cm^{-1}. The sample, cylindrical in form, had a length of 0.4 cm, and was excited by a laser beam of diameter 0.1 cm. Thus the excited volume also had a cylindrical form, and its length and cross-section were equal to 0.1 cm (because of the pump absorption) and 0.01 cm^{-2} correspondingly. The pyrene concentration was of the order of 10^{13}–10^{14} cm^{-3}.

Owing to fast non-radiative relaxation (a relaxation time of 10^{-11} s) the pyrene molecules initially settled in the ground state of the term $^1B_{2u}(2, q = 0)$. Emission corresponding to the transitions $^1B_{2u}(2, q = 0) \rightarrow {}^1A_g(1, q = 0)$

(the 0–0 transition, $\lambda = 373.9$ nm) and $^1B_{2u}(2, q = 0) \rightarrow {}^1A_g(1, q = 4)$ (the transition to the fourth electron vibrational sublevel of intra-molecular vibration having a frequency of 1408 cm^{-1}) was considered. Above the pump power thresholds of 5×10^5 W cm^{-2} for the transition $(2, q = 0) \rightarrow (1, q = 0)$ and 10^6 W cm^{-2} for the transition $(2, q = 0) \rightarrow (1, q = 4)$ a steep shortening of emission times of both transitions was seen: from 110 ns down to 5–6 ns for the 0–0 transition, and down to 10–12 ns for the other one.

Emission from the 0–0 transition had a delay of the order of 9 ns with respect to the pump pulse, and a sharp polar diagram (within a solid angle of the order of 0.1 sr). Its intensity exceeded the intensity of the usual spontaneous emission by a factor of the order of 1000. The measured dependence of this emission upon the concentration of pyrene proved to be quadratic (figure 2.30). Moreover, it had a higher degree of linear polarization. These features

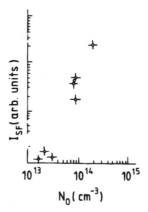

Figure 2.30. Concentration dependence of super-radiance intensity, I_{SF}, of pyrene molecules in diphenyl crystals at 4.2 K [ZLNSSS83].

of the radiation allowed Zinoviev *et al* [ZLNSSS83] to identify the emission corresponding to the 0–0 pyrene transition $^1B_{2u}(2, q = 0) \rightarrow {}^1A_g(1, q = 0)$ as super-radiance. The emission resulting from another transition $(2, q = 0) \rightarrow (1, q = 4)$ had an isotropic polar diagram, and the degree of the linear polarization was half that for the 0–0 transition. Its intensity exceed the intensity of spontaneous emission by a factor of 50. The emission was identified by the authors as super-luminescence (ASE).

2.3 Concluding remarks

In this section we have described only a few of the numerous super-radiance experiments, and more results can be found in other references given in the bibliography. Our intention has been to demonstrate that conditions for the experimental realization of super-radiance do exist, i.e. that under appropriate circumstances super-radiance is not suppressed by de-phasing or inhomogeneous relaxation.

Some of the features of super-radiance discovered through the experiments reviewed above will be the object of discussion in the next chapters. Of particular interest is triggered super-radiance observed in [VS79, CJSGH80]. In Chapter 5 we shall show that this effect can, in principle, be used for the generation of ultra-short pulses by relatively slow pumping, under conditions where spontaneous super-radiance is not possible, because of relaxation. Some other experiments will be reviewed in subsequent chapters. In Chapter 7 we shall analyse experiments of super-radiance in a proton spin system, where also the theory of this effect will be presented. In Chapter 8 we consider the diffraction pattern of super-radiance, observed in [SSS87]. The effect of correlations between counter-propagating pulses, which were registered in solids [FSS84a] but not observed in gases, will be discussed in Chapter 9. Some recent experimental studies of super-radiance will also be discussed in Chapter 11.

Chapter 3

Quantum electrodynamical approach

The quantum electrodynamical (QED) theory of the spontaneous decay of a single excited two-level atom (Weisskopf and Wigner [WW30], see also Heitler [29] and Pike and Sarkar [PS95]) was generalized and applied to the super-radiance problem by several authors [F66, ES68, DK70a, DK70b, L70a, CS71, PS71, BSH71a, BSH71b, M76, Ss74, RT77, AS81, R83, DS87, B83, Sl85]. A complete treatment of the spontaneous decay including a gauge-invariant formulation, renormalization and non-Markovian behaviour has been recently suggested by Seke [Se94].

The effective approach to the description of the evolution of a radiating polyatomic system is provided by master equation theory together with the use of the projection operator method. The state of the problem has been discussed in [Se91a]. For another widely used approach, based upon the Heisenberg equations, see the next chapter.

In the present chapter we shall consider the evolution of the system of atom plus field, applying an approach equivalent to the master equation method. To this end we shall use the diagram representation of the quantum perturbation theory for the density matrix of the system [41]. As will be seen later, the diagram method provides an effective tool for the derivation of the master equations. Moreover, in some cases the diagrams can be calculated directly, this being in essence equivalent to the solution of the master equation.

At the beginning of this chapter we formulate a simple version of the diagram method of the non-stationary perturbation theory for the density matrix. Using the diagrams we describe the fundamentals of the spontaneous decay theory for a single atom and for two initially excited atoms. Then we present the quantum theory of super-radiance of an extended three-dimensional polyatomic system within the Markov and the rotating wave approximations and neglecting retardation. Note that in some works on super-radiance the Markov and rotating wave approximations are not used (see e.g. the series of papers [Se86a, Se86b, SR87, SR89, Se91b, Se93]).

Our main objective is to discuss the spatial properties of super-radiance, including the angular correlation of photons, which have been mentioned in earlier works by Dicke [D54] and Ernst and Stehle [ES68], and were investigated later also by Mostowski and Sobolewska [MS83b], Richter [R83], Prasad and Glauber [PG85b], Duncan and Stehle [DS87], and by Białynicki-Birula and Białynicka-Birula [8].

Below we shall mainly follow our papers [Sv74, SST72, ST74, ST75].

3.1 Formulation of the problem

We shall consider super-radiance, i.e. the spontaneous decay of a polyatomic system consisting of identical two-level atoms, as a non-stationary quantum electrodynamical problem of the evolution of an excited system arising from the elements' mutual interaction and their interaction with the transverse electromagnetic field. The initial condition is the vacuum state of the electromagnetic field and fully excited states of the atomic system.

The system of atoms plus field will be described by the total Hamiltonian

$$\widehat{H} = \widehat{H}_{at} + \widehat{H}_{f} + \widehat{H}_{af} \tag{3.1.1}$$

where $\widehat{H}_{at}$ is the energy operator for free atoms (see (1.2.10)), $\widehat{H}_{f}$ is the Hamiltonian for a free electromagnetic field, and $\widehat{H}_{af}$ is the interaction Hamiltonian of atoms and field in the electric dipole approximation [50] (see also, the next chapter)

$$\widehat{H}_{at} = \hbar\omega_0 \sum_{i=1}^{N} \widehat{R}_3^{(i)} \tag{3.1.2}$$

$$\widehat{H}_{f} = \sum_{k\lambda} \hbar\omega_k (\widehat{a}_{k\lambda}^{\dagger}\widehat{a}_{k\lambda} + \tfrac{1}{2}) \tag{3.1.3}$$

$$\widehat{H}_{af} = -\sum_{i=1}^{N} d^{(i)}(\widehat{R}_+^{(i)} + \widehat{R}_-^{(i)})\widehat{\mathcal{E}}(r_i). \tag{3.1.4}$$

Here ω_0 is the resonant frequency of a single atom, $R_3^{(i)}$, $R_{\pm}^{(i)}$ are the quasi-spin operators for the ith atom introduced in section 1.2 , $\widehat{a}_{k\lambda}$, $\widehat{a}_{k\lambda}^{\dagger}$ the annihilation and creation operators of a photon state with wavevector k and polarization state $\lambda = 1, 2$, $\omega_k = kc$, $d^{(i)}$ is the transition matrix element of the electric dipole moment of the ith atom. In (3.1.4) $\widehat{\mathcal{E}}(r_i)$ is the operator for the transverse electric field at the location of the ith atom

$$\widehat{\mathcal{E}}(r) = i \sum_{k\lambda} \epsilon_{k\lambda} \sqrt{\frac{2\pi\hbar\omega_k}{V_Q}} (\widehat{a}_{k\lambda}e^{ikr} - \widehat{a}_{k\lambda}^{\dagger}e^{-ikr}) \tag{3.1.5}$$

where $\epsilon_{k\lambda}$ is the polarization unit vector orthogonal to the vector k, i.e. $(k\epsilon_{k\lambda}) = 0$, V_Q is the volume of quantization, and the summation is performed over permitted values of the vector k, and over two polarizations, $\lambda = 1, 2$, for each k. The operators $\widehat{a}_{k\lambda}$ and $\widehat{a}_{k\lambda}^{\dagger}$ obey the commutation relations

$$[\widehat{a}_{k\lambda}, \widehat{a}_{k\lambda}^{\dagger}] = \delta_{kk'}\delta_{\lambda\lambda'} \qquad [\widehat{a}_{k\lambda}, \widehat{a}_{k'\lambda'}] = [\widehat{a}_{k\lambda}^{\dagger}, \widehat{a}_{k'\lambda'}^{\dagger}] = 0 \quad (3.1.6)$$

from which follow the relations

$$\widehat{a}_{k\lambda}|n_{k\lambda}\rangle = \sqrt{n_{k\lambda}}\,|n_{k\lambda} - 1\rangle \qquad \widehat{a}_{k\lambda}^{\dagger}|n_{k\lambda}\rangle = \sqrt{n_{k\lambda} + 1}\,|n_{k\lambda} + 1\rangle \quad (3.1.7)$$

where $|n_{k\lambda}\rangle$ are eigenstates of a field oscillator with $k\lambda$.

We shall fully obtain the information about the evolution we are interested in if we calculate the density matrix of the system 'atoms plus field' as a function of time, and then by a trace operation obtain separately the reduced density matrices for the atomic system and the field. The most effective method for doing this is the diagram method of non-stationary perturbation theory [41], which we first consider in general form in the next section.

3.2   Diagram method of non-stationary perturbation theory

Consider a stationary non-perturbed quantum system with Hamiltonian $\widehat{H}^{(0)}$, eigenstates φ_n, and eigenvalues E_n

$$\widehat{H}^{(0)}\varphi_n = E_n\varphi_n. \tag{3.2.1}$$

If a perturbation with the operator $\widehat{V}(t)$ (in the general case, time-dependent) is applied to the system, the quantum evolution is governed by the Schrödinger equation

$$\dot{\psi} = -\frac{i}{\hbar}(\widehat{H}^{(0)} + \widehat{V})\psi. \tag{3.2.2}$$

Expanding the wavefunction ψ in terms of the basis consisting of the eigenfunctions φ_n, we obtain the equation for the quantum amplitudes $c_n(t)$ in the Schrödinger representation

$$\psi = \sum_n c_n\varphi_n \qquad \dot{c}_n = -\frac{i}{\hbar}\left(E_n c_n + \sum_m V_{nm}c_m\right). \tag{3.2.3}$$

Here $\widehat{V}_{nm} = \langle n|\widehat{V}(t)|m\rangle$ is the matrix element of the perturbation taken in the Schrödinger picture also.

Let us go over to the interaction representation, where the amplitudes and the matrix elements (labelled with a tilde) are defined as

$$c_n(t) = e^{-i\omega_n t}\tilde{c}_n(t) \qquad \widetilde{V}_{nm}(t) = e^{+i\omega_n t}V_{nm}(t)e^{-i\omega_m t} \tag{3.2.4}$$

where $\omega_n = E_n/\hbar$. For the amplitudes in the interaction picture equation (3.2.3) reduces to

$$\dot{\tilde{c}}_n = -\frac{i}{\hbar} \sum_m \tilde{V}_{nm} \tilde{c}_m. \qquad (3.2.5)$$

In the absence of perturbation, i.e. $\widehat{V} = 0$, the amplitudes in the interaction picture are time-independent, $\tilde{c}_n(t) = \tilde{c}_n(t_0)$. In perturbation theory, equation (3.2.5) is solved by using a series expansion of the quantum amplitudes in powers of the perturbation $\widehat{V}$

$$\tilde{c}_n(t) = \tilde{c}_n(t_0) + \tilde{c}_n^{(1)}(t) + \tilde{c}_n^{(2)}(t) + \cdots \qquad (3.2.6)$$

here $\tilde{c}_n^{(n)} \sim V^n$. From equation (3.2.5) we obtain

$$\dot{\tilde{c}}_n^{(l)} = -\frac{i}{\hbar} \sum_m \tilde{V}_{nm} \tilde{c}_m^{(l-1)} \qquad (3.2.7)$$

and after integration

$$\tilde{c}_n^{(l)}(t) = -\frac{i}{\hbar} \int_{t_0}^t dt_l \, \tilde{V}_{nm}(t_l)\tilde{c}_m^{(l-1)}(t_l) \qquad\qquad \tilde{c}_n^{(0)}(t) = \tilde{c}_n(t_0). \qquad (3.2.8)$$

Taking into account all powers of the perturbation, we come to the general solution of the problem

$$\tilde{c}_n(t) = \sum_{n_0} \tilde{S}_{nn_0}(t_0, t)\tilde{c}_{n_0}(t_0). \qquad (3.2.9)$$

The evolution operator $\tilde{S}$ in the interaction representation is

$$\tilde{S}_{nn_0}(t_0, t) = \sum_{l=0}^{\infty} \tilde{S}_{nn_0}^{(l)}(t_0, t)$$

$$= \delta_{nn_0} + \sum_{l=1}^{\infty} \left[-\frac{i}{\hbar}\right]^l \int_{t_0}^t dt_l \int_{t_0}^{t_l} dt_{l-1} \cdots \int_{t_0}^{t_2} dt_1$$

$$\times \left\{\widehat{\tilde{V}}(t_l) \cdots \widehat{\tilde{V}}(t_1)\right\}_{nn_0}. \qquad (3.2.10)$$

Using equation (3.2.4) we find the quantal amplitudes and the evolution matrix in the Schrödinger picture

$$c_n(t) = \sum_{n_0} S_{nn_0}(t_0, t)c_{n_0}(t_0)$$

$$S_{nn_0}(t_0, t) = e^{-i\omega_n t}\tilde{S}_{nn_0}(t_0, t)e^{i\omega_{n_0} t_0}. \qquad (3.2.11)$$

The term in the series (3.2.10) proportional to the *l*th power of the perturbation and modified in accordance with the transformation (3.2.11) can be set in correspondence with the diagram below. The horizontal time axis is directed to the right from the initial time t_0 towards the final time of evolution t (see arrow). The vertices $t_1, \ldots, t_l$ correspond to the instants of the elementary events (transitions)

$$S_{nn_0}^{(l)}(t_0, t) = \underset{\substack{t_0 \quad t_1 \quad t_2 \qquad\qquad t_l \quad t}}{\underbrace{\quad n_0 \quad n_1 \quad n_2 \qquad\qquad n \quad}} \qquad (3.2.12)$$

Let us list the rules for correspondence between diagrams and terms in the series (3.2.10)

Every vertex t_p, $p = 1, \ldots, l$, represents the factor under the integral in (3.2.10) of the form

$$-\frac{i}{\hbar} \tilde{V}_{n_p.n_{p-1}}(t_p) = -\frac{i}{\hbar} V_{n_p.n_{p-1}}(t_p) \, e^{i(\omega_{n_p} - \omega_{n_{p-1}})t_p}. \qquad (3.2.13)$$

The factors $e^{i\omega_{n_0}t_0}$ and $e^{-i\omega_n t}$ arising as a result of the transformation (3.2.11) to the Schrödinger picture correspond to the beginning and the end of the time axis. Note that the pairs of exponentials combine to construct the free propagators of the basis eigenfunctions

$$g_p(t_p, t_{p+1}) = e^{-i\omega_p(t_{p+1} - t_p)}. \qquad (3.2.14)$$

Thus we can say that the intervals $(t_0, t_1), \ldots, (t_l, t)$ of the time axis represent the propagators (3.2.14).

The summation over the intermediate states $n_1, \ldots, n_l$ and the integration over the ordered instants $t_1 < \cdots < t_l$ then have to be carried out.

Consider the density matrix $\hat{\rho}(t)$ of the system

$$\hat{\rho}(t) = \overline{|\psi(t)\rangle\langle\psi(t)|} \qquad \rho_{nm} = \overline{c_n(t)c_m^*(t)}. \qquad (3.2.15)$$

The statistical average over the initial mixed state is denoted by a horizontal overbar. The quantum evolution of the density matrix is governed by the evolution operator

$$\hat{\rho}(t) = \widehat{S}(t_0, t)\hat{\rho}(t_0)\widehat{S}^\dagger(t_0, t)$$

$$\rho_{nm}(t) = \sum_{n_0 m_0} S_{nn_0}(t_0, t)\rho_{n_0 m_0}(t_0)S_{mm_0}^*(t_0, t). \qquad (3.2.16)$$

Both evolution matrices in equation (3.2.16) (i.e. S and S^*) are given by the series (3.2.10). Let us set in correspondence to the density matrix element $\rho_{nm}(t)$ the diagram

$$\rho_{nm}(t) \sim \left\{ \begin{array}{l} \underset{\substack{t_0 \quad t_1 \qquad\qquad t_l}}{\underbrace{\quad n_0 \qquad\qquad\qquad n \quad}} \\[2em] \underset{\substack{t_0 \quad t_1' \qquad\qquad t_k' \\ m_0 \qquad\qquad\qquad m}}{\underbrace{\quad \qquad\qquad\qquad\qquad \quad}} \end{array} \right. \qquad (3.2.17)$$

The factor $S_{nn_0}^{(1)}(t_0, t)$ is recovered by using the upper time axis, in accordance with the rules formulated above. The lower time axis is introduced in order to represent the complex conjugate evolution matrix element $S_{mm_0}^{(k)*}(t_0, t)$ also occurring in the definition (3.2.16) as a factor.

An additional rule is introduced, implying that inverting the direction of the arrow on the lower time axis means the complex conjugation of the evolution matrix element.

The large bracket in the left-hand side of the figure represents the initial condition (in the sense that one has to take into account the factor $\rho_{n_0 m_0}(t_0)$ and to sum with respect to the labels n_0, m_0).

We shall assume the interaction Hamiltonian of a system of identical atoms with transverse quantized electromagnetic field to be of the form

$$\widehat{V} = \sum_{k\lambda} \left\{ f_{k\lambda} \widehat{R}_k^\dagger \widehat{a}_{k\lambda} + f_{k\lambda}^* \widehat{R}_k \widehat{a}_{k\lambda}^\dagger \right\} \tag{3.2.18}$$

where

$$f_{k\lambda} = -i \left[\frac{2\pi \hbar \omega_k}{V_Q} \right]^{1/2} d\epsilon_{k\lambda} \tag{3.2.19}$$

$$\widehat{R}_k = \sum_i \widehat{R}_-^{(i)} e^{-ikr_i}, \qquad \widehat{R}_k^\dagger = \sum_i \widehat{R}_+^{(i)} e^{ikr_i}. \tag{3.2.20}$$

Here we have eliminated from the interaction (3.1.4) the non-resonant terms of the form $\widehat{R}\widehat{a}$, $\widehat{R}^\dagger\widehat{a}^\dagger$ (the rotating wave approximation, RWA). The excited and ground states of the ith atoms with coordinates r_i and energy levels E_e and E_g are connected by the lowering and raising operators $\widehat{R}_+^{(i)}$ and $\widehat{R}_-^{(i)}$ by

$$\widehat{R}_+^{(i)}|g_i\rangle = |e_i\rangle \qquad \widehat{R}_-^{(i)}|e_i\rangle = |g_i\rangle. \tag{3.2.21}$$

Let us introduce an additional rule for interpreting the diagrams for the specific interaction (3.2.18). For this purpose, consider, independently, the contributions to the interaction Hamiltonian, proportional to the operators $\widehat{a}_{k\lambda}$, $\widehat{a}_{k\lambda}^\dagger$. The vertices in the diagram (3.2.12) coming from these contributions now look like

$$\tag{3.2.22}$$

The factors (3.2.13) for these vertices reduce to

$$-\frac{i}{\hbar} f_{k\lambda} e^{-i(\omega_k + \omega_g - \omega_e)t_p} \langle n_p | \widehat{R}_k^\dagger \widehat{a}_{k\lambda} | n_{p-1}\rangle$$

$$\tag{3.2.23}$$

$$-\frac{i}{\hbar} f_{k\lambda}^* e^{+i(\omega_k + \omega_g - \omega_e)t_p} \langle n_p | \widehat{R}_k \widehat{a}_{k\lambda}^\dagger | n_{p-1}\rangle.$$

In what follows, our purpose is to consider collective spontaneous emission in free space. The initial state of the field is the vacuum

$$\rho_f(t_0) = |\{0\}\rangle\langle\{0\}|. \qquad (3.2.24)$$

By arbitrary expansion of the quantization box, $V_Q \rightarrow \infty$, the coupling constant goes to zero, $|f_{k\lambda}| \sim V_Q^{-1/2} \rightarrow 0$. This dependence is compensated by the sum over quantized values of the wavevectors of field oscillators in the continuous limit

$$\sum_k \rightarrow \frac{V_Q}{(2\pi)^3} \int dk. \qquad (3.2.25)$$

Non-vanishing contributions to the values of physical variables (to the average values of the corresponding operators) are given only by the diagrams for the density matrix in which a pair of the vertices for the electric–dipole interaction is accompanied by an independent sum, as in (3.2.25). For this reason we consider only the diagrams in which the number of photons in any one of the field oscillators is equal to 0 or 1, $n_{k\lambda} = 0, 1$. The diagrams with $n_{k\lambda} = 2, 3, \ldots$ disappear with $V_Q \rightarrow \infty$. This assumption will become clearer from the examples of diagrams to be given below.

The state of the field with l quanta, in which $n_{k_1\lambda_1} = \cdots = n_{k_l\lambda_l} = 1$, will be denoted by $|k_1\lambda_1, \ldots, k_l\lambda_l\rangle$. The factors $\sqrt{n_{k\lambda}}$ and $\sqrt{n_{k\lambda}+1}$ in the matrix elements of the annihilation and creation operators (see (3.1.7)) are always equal to unity in our diagrams.

3.3   The spontaneous decay of a single two-level atom

In this section we consider an elementary, and at the same time fundamental, problem of the interaction of a single atom with a quantized transverse electromagnetic field: the spontaneous decay of an excited state of a single atom. We introduce the basic approximations that lead to the exponential decay law by using the diagram method, which is important for formulating the polyatomic problem in the remainder of the chapter.

Let us retain only the single-atom contribution to the Hamiltonian (3.2.18), assuming that the atom is located at $r_1 = 0$, then $\widehat{R}_k = \widehat{R}$ (the atomic label is omitted). In the RWA only the states $|e\rangle|\{0\}\rangle$ and $|g\rangle|k\lambda\rangle$ of the atom–field system can arise as intermediate states. Consider the quantum amplitude $S_{e,0;e,0}(t_0, t)$ the square modulus of which determines the probability of finding the atom still in the excited state up to the time t. The perturbation theory series (3.2.10) reduces to

$$(3.3.1)$$

The pairs of photon lines describing the emission and subsequent absorption of the photon $k\lambda$ (and correspondingly the photon exponentials from (3.2.23)) are combined into the photon propagators similar to the propagators (3.2.14).

All the terms of the perturbation theory series for quantum amplitudes have a convolution structure, namely

$$g(t - t_0) = \int_{t_0}^{t} dt_l \int_{t_0}^{t_l} dt_{l-1} \cdots \int_{t_0}^{t_2} dt_1 \prod_{p=0}^{l} g_p(t_{p+1} - t_p). \qquad (3.3.2)$$

It turns out to be convenient to apply the Laplace transform

$$g(p) = \int_{0}^{\infty} g(\tau) e^{-p\tau} d\tau \qquad g(\tau) = \frac{1}{2\pi i} \int_{-i\infty+a}^{i\infty+a} g(p) e^{p\tau} dp. \qquad (3.3.3)$$

The contour for the inverse transform lies in the right half-plane of the complex variable p (with respect to all possible singularities). It follows from (3.3.2) and (3.3.3) that

$$g(p) = \prod_{s=0}^{l} g_s(p). \qquad (3.3.4)$$

It is evident that the Laplace transformed series (3.3.1) has the structure of a geometrical progression. Hence the sum of the series satisfies an equation analogous to the Dyson equation, the diagram form of which is

$$S_{e,0\,;\,e,0} \equiv \quad \xrightarrow{\quad e \quad e \quad} \quad = \quad \xrightarrow{\quad e \quad e \quad} \quad + \quad \xrightarrow{\ e\ } \overset{g}{\frown} \xrightarrow{\ e\ } . \qquad (3.3.5)$$

Here the bold line represents the sum of the series. Let us introduce the notation $\Sigma_{e,e}$ for the self-energy part of the diagrams under consideration, which can be drawn as

$$\Sigma_{e,e} = \quad \xrightarrow{\ e\ } \overset{g}{\frown} \xrightarrow{\ e\ } . \qquad (3.3.6)$$

Equation (3.3.5) reads analytically as

$$S_{e,0\,;\,e,0} = \{1 + S_{e,0\,;\,e,0}(p) \Sigma_{e,e}(p)\}(p + i\omega_e)^{-1}. \qquad (3.3.7)$$

The explicit expression for $\Sigma_{e,e}$ includes the matrix elements for two vertices, and the propagator for the intermediate state $g, k\lambda$ (which is a product of the matter and field propagators), and is given by

$$\Sigma_{e,e} = \left[-\frac{i}{\hbar}\right]^2 \sum_{k\lambda} |f_{k\lambda}|^2 \frac{1}{p + i(\omega_g + \omega_k)}. \qquad (3.3.8)$$

In the absence of interaction $\Sigma_{e.e} \to 0$ and the amplitude, (3.3.7) has a pole at $p = -i\omega_e$. It is natural to assume that for a weak perturbation the neighbourhood of this pole gives the main contribution to the sum. Consequently we approximate the self-energy factor (3.3.8) by taking its value at $p = -i\omega_e + \varepsilon$, where $\varepsilon \to 0$. In the summation over the frequency of intermediate photons in (3.3.8) we use the identity

$$\frac{1}{i(\omega_k - \omega_0) + \varepsilon} \xrightarrow{\varepsilon \to 0} \pi\delta(\omega_k - \omega_0) + \text{Pr}\frac{1}{i(\omega_k - \omega_0)} \qquad (3.3.9)$$

where $\omega_0 = \omega_e - \omega_g$ is the atomic transition frequency. As usual, the symbol Pr means the principal value of the integral. Consider the real and imaginary parts of the self-energy factor

$$\Sigma_{e.e} = -\tfrac{1}{2}\gamma - \frac{i\alpha}{\hbar} \qquad (3.3.10)$$

where α is the quantum correction to the upper atomic energy level resulting from the interaction with the quantized electromagnetic field (the Lamb shift). We note that the shift α contains a high-frequency divergence and it has first to be renormalized (see, e.g. [AE75]). Therefore we have to go beyond the RWA in order to obtain a correct value for α.

Let us now consider the radiation damping constant γ for the upper atomic level. The sum over the polarization states is reduced to

$$\sum_{\lambda=1}^{2}(d\epsilon_{k\lambda})^2 = k^{-2}[d \times k]^2 \qquad (3.3.11)$$

where the orthogonality of the vectors ϵ_{k1}, ϵ_{k2} and k has been taken into account. With the help of (3.3.8)–(3.3.11) we obtain the following result for the constant γ

$$\gamma = -2\,\text{Re}\,\Sigma_{e.e} = \frac{1}{2\pi\hbar}\int\frac{dk}{k}[d \times k]^2\delta(k - k_0) = \frac{4}{3}\frac{d^2 k_0^3}{\hbar}. \qquad (3.3.12)$$

Here $k_0 = \omega_0/c$.

The approximation of a constant self-energy factor $\Sigma_{e.e}(p)$ which assumes its independence of p, is essential. It implies that in the time domain

$$\Sigma_{e.e}(t_2 - t_1) = -\left(\frac{i}{\hbar}\alpha + \tfrac{1}{2}\gamma\right)\delta(t_2 - t_1 - 0). \qquad (3.3.13)$$

Hence in the second-order contribution to the series (3.3.1) one integration (for instance, with respect to t_2) can be performed using (3.3.13). Actually, the second-order contribution reduces to a first-order term, in which the quantity $\alpha - i\hbar\gamma/2$ plays the rôle of a perturbation matrix element. A similar transformation occurs in all terms of the series.

The summation in (3.3.8), where the intermediate states $|g, k\lambda\rangle$ belong to the continuum, is taken over the frequencies ω_k, which, roughly speaking, belong to the optical band. This makes the self-energy factor $\Sigma_{e,e}(p)$ almost independent of the variable p when $|p + i\omega_e| \ll \omega_0$. As far as the perturbed pole of the amplitude $S_{e,0;e,0}(p)$ is located in the neighbourhood of the unperturbed pole, that is for $|p + i\omega_e| \sim \gamma$, the accuracy of the approximation is determined by the small parameter $\gamma/\omega_0 \ll 1$.

By using the rules formulated above, the diagram equation (3.3.5) reduces, in the time domain, to the integral equation

$$S_{e,0;e,0}(t_0, t) = e^{-i\omega_e(t-t_0)} + \int_{t_0}^{t} dt_2 \int_{t_0}^{t_2} dt_1 e^{-i\omega_e(t-t_2)} \Sigma_{e,e}(t_2 - t_1) S_{e,0;e,0}(t_0, t_1)$$

$$(3.3.14)$$

which may be transformed into the differential equation

$$\dot{S}_{e,0;e,0} = -\left\{i(\omega_e + \alpha/\hbar) + \tfrac{1}{2}\gamma\right\} S_{e,0;e,0}. \qquad (3.3.15)$$

The quantal amplitude of the evolution of the excited atomic state in the presence of spontaneous decay is found to be of the form

$$S_{e,0;e,0}(t_0, t) = e^{-[i(\omega_e+\alpha/\hbar)+\gamma/2](t-t_0)}. \qquad (3.3.16)$$

In what follows we imply that the Lamb shift is included in the excited energy level $E_e = \hbar\omega_e$, so that the shift is omitted from the subsequent equations.

It can be seen that equation (3.3.16) expresses the law of *exponential damping* of excitation. For spontaneous emission at optical frequencies this is a good approximation as it is determined by the small parameter $\gamma/\omega_0 \ll 1$. The deviations from the exponential law at very small and very large time intervals were studied elsewhere [39, F66, DK70a, DK70b, PS71, SH88, SH89].

Let us turn to the diagram representation of the quantal amplitude for the state $|g, k\lambda\rangle$ of the atom–field system, which is a result of spontaneous emission from the excited atom. The perturbation series is

$$(3.3.17)$$

After Laplace transformation, each term of this sequence becomes a product of the corresponding term of the series (3.3.1) and a factor which arises from the last right-hand vertex, together with the propagator to the right of that vertex. Hence the summation of the series (3.3.1) can be carried out, and we obtain the following result

$$S_{g,k\lambda;e,0}(p) = -\frac{i}{\hbar} f_{k\lambda}^*(p + i(\omega_g + \omega_k))^{-1} S_{e,0;e,0}(p) \qquad (3.3.18)$$

and, in the time domain

$$S_{g,k\lambda;e,0}(t_0, t) = -\frac{i}{\hbar} f_{k\lambda}^* \int_{t_0}^{t} dt_1 \, e^{-i(\omega_g+\omega_k)(t-t_1)} S_{e,0;e,0}(t_0, t_1). \qquad (3.3.19)$$

The probability of finding the atom in the excited or ground state is given by the corresponding element of the atomic density matrix. We go over to the density matrix of the atomic subsystem ρ_a by eliminating the field variables in the standard way

$$\rho_a = \mathrm{Tr}_f \, \rho \qquad (3.3.20)$$

i.e. by taking the trace with respect to the labels of the photon subsystem. The probability that the initially excited atom is still in the upper state is given by

$$\rho_{e,e}(t) = |S_{e,0;e,0}(t_0, t)|^2 = e^{-\gamma(t-t_0)}. \qquad (3.3.21)$$

The diagram for this density matrix element looks like

$$\rho_{e,e} \; (t) = \left\{ \begin{array}{c} \text{diagram} \end{array} \right. \qquad (3.3.22)$$

The probability of finding the initially excited atom in the ground state is

$$\rho_{g,g}(t) = \sum_{k\lambda} \rho_{g,k\lambda;g,k\lambda}(t) = \sum_{k\lambda} |S_{g,k\lambda;e,0}(t_0, t)|^2. \qquad (3.3.23)$$

This density matrix element of the atom is represented by the diagram

$$\rho_{g,g} \; (t) = \left\{ \begin{array}{c} \text{diagram} \end{array} \right. \qquad (3.3.24)$$

The pair of photon lines describing the interaction of the atom with one and the same field oscillator $k\lambda$ on the upper and lower time axes have been combined, in order to construct the photon line propagating from the vertex on the upper time axis to the lower vertex. Summation over the photon labels $k\lambda$ is implied.

In the preceding part of this section we have mentioned the relation between the continuous spectrum of the field subsystem and the exponential law of spontaneous decay. Now we shall discuss the consequences of summing over the photon frequency ω_k in the diagram (3.3.24) for the density matrix element of a single atom.

For this purpose we represent the double integral over the independent time arguments t_1 and t_1' as a sum of two time-ordered integrals, where $t_1' > t_1$ and $t_1 > t_1'$ respectively. Since the integrand is constructed out of the exponential functions of time (free and decaying propagators), the time-ordered integrals have a convolution structure (see (3.3.2)). After taking the Laplace transform we obtain for the diagram (3.3.24)

$$\rho_{g.g}(p) = (p + \gamma)^{-1} \frac{1}{\hbar^2} \sum_{k\lambda} |f_{k\lambda}|^2$$

$$\times \left\{ \frac{1}{p + i(\omega_k - \omega_0) + \gamma/2} + \frac{1}{p - i(\omega_k - \omega_0) + \gamma/2} \right\} p^{-1}. \quad (3.3.25)$$

If the interaction is weak ($\gamma \to 0$) there is a double pole at $p = 0$. Similarly to the case of the self-energy diagram, we neglect above the dependence of the sum over the photon energy ω_k upon the Laplace variable p. Taking the sum at the pole value $p = \varepsilon$, $\varepsilon \to 0$, by using (3.3.7) and (3.3.9) we find that

$$\frac{1}{\hbar^2} \sum_{k\lambda} |f_{k\lambda}|^2 \left\{ \frac{1}{\varepsilon + i(\omega_k - \omega_0)} + \frac{1}{\varepsilon - i(\omega_k - \omega_0)} \right\} = \gamma. \quad (3.3.26)$$

In the time representation the fragment of the diagram (3.3.24) corresponding to the expression (3.3.26) takes the form

$$= \gamma \delta(t_1 - t_1') . \quad (3.3.27)$$

The photon line becomes vertical and contributes the δ-like factor (3.3.27) to the integrand. One integration can be performed immediately, and the diagram (3.3.24) gives

$$\rho_{g.g}(t) = \gamma \int_{t_0}^{t} dt_1 \, e^{-\gamma(t_1 - t_0)} = 1 - e^{-\gamma(t - t_0)}. \quad (3.3.28)$$

The atomic density matrix is normalized according to

$$\rho_{e.e}(t) + \rho_{g.g}(t) = 1. \quad (3.3.29)$$

The basic approximations that we have applied for calculating the density matrix are, in essence, equivalent to using the master equation approach.

The probability of the emission of a photon $k\lambda$ is described by the diagonal element of the density matrix for the total system

$$\rho_{g,k\lambda;g,k\lambda}\,(t) = \begin{cases} \end{cases} \qquad (3.3.30)$$

For a given direction and state of polarization the spontaneous emission spectrum relates to a mean number of photons in the field mode $k\lambda$ as follows

$$\langle n_{k\lambda}\rangle = \lim_{t\to\infty} \rho_{g,k\lambda;g,k\lambda}(t) = \frac{1}{\hbar^2}|f_{k\lambda}|^2\left[\left(\tfrac{1}{2}\gamma\right)^2 + (\omega_0 - \omega_k)^2\right]^{-1}. \quad (3.3.31)$$

By using the definition of the radiation damping constant γ we obtain the normalization condition

$$\sum_{k\lambda} \langle n_{k\lambda}\rangle = 1. \qquad (3.3.32)$$

3.4 The interaction of two-level atoms via the transverse electromagnetic field

Let us consider a system of n identical atoms of which the position vectors are $r_1, \ldots, r_n$. We are interested in the quantal amplitude of the transition between the states of the atom–field system that contains m excited atoms and no photons both in the initial and final states. The perturbation series has a form analogous to (3.3.1) and it satisfies an equation similar to equation (3.3.5), namely

$$S_{b,a} \equiv \qquad = \qquad \delta_{a,b} + \qquad . \qquad (3.4.1)$$

Here, for brevity, we denote the basis states of the system with m excited atoms by a single letter. It is clear that the number of excited atoms in an intermediate state d is equal to $m - 1$, hence $\omega_a - \omega_d = \omega_0$, etc. The self-energy part of the diagrams above is

$$\Sigma_{b,c} = \qquad . \qquad (3.4.2)$$

Similarly to the elementary problem of a single atom, we calculate the self-energy diagram neglecting the dependence upon the Laplace variable p, i.e. upon the pole $p = -i\omega_b + \varepsilon$, $\varepsilon \to 0$. This can be justified, as above, by noticing that the summation in (3.4.2) is carried out within a band of the continuous photon frequency ω_k. We obtain

$$\Sigma_{b,c} = \left(-\frac{i}{\hbar}\right)^2 \sum_{d,k\lambda} |f_{k\lambda}|^2 \frac{\langle b|\widehat{R}_k^\dagger|d\rangle\langle d|\widehat{R}_k|c\rangle}{i(\omega_k - \omega_0) + \varepsilon}. \qquad (3.4.3)$$

Here

$$\sum_d \widehat{R}_k^\dagger|d\rangle\langle d|\widehat{R}_k = \widehat{R}_k^\dagger \widehat{R}_k = \sum_{i,j} \widehat{R}_+^{(i)} \widehat{R}_-^{(j)} e^{ikr_{ij}} \qquad (3.4.4)$$

where $r_{ij} = r_i - r_j$. If the atoms are separated from each other by a distance greatly exceeding the wavelength, the exponent in (3.4.3) and (3.4.4) can be considered as frequency-independent only within the interval $|\omega_k - \omega_0| < c/r_{ij}$. Consequently the corresponding self-energy diagram does not depend upon the Laplace variable p in the frequency range of the order of c/r_{ij}.

As a result, the approach that we consider in this chapter is applicable only to 'slow' super-radiant systems. Indeed, the super-radiance time scale T_R is assumed to be large compared with the time that light takes to propagate across the atomic system, $T_R \gg L/c$, where L is the characteristic linear dimension of the active volume.

Let us take the self-energy factor in the form

$$\Sigma_{b,c} = -\sum_{i,j} \langle b|\widehat{R}_+^{(i)} \widehat{R}_-^{(j)}|c\rangle \left(\frac{iu_{ij}}{\hbar} + \tfrac{1}{2}\gamma_{ij}\right) \qquad (3.4.5)$$

where

$$\tfrac{1}{2}\gamma_{ij} + \frac{iu_{ij}}{\hbar} = (4\pi^2\hbar)^{-1} \int \frac{dk}{k} \frac{[d \times k]^2}{i(k - k_0) + \varepsilon} e^{ikr_{ij}}. \qquad (3.4.6)$$

The calculation of the integral is elementary and we omit the details. Note that it turns out to be useful first to integrate over the directions of the wavevector k, and then to take the integral with respect to the photon energy, between the formally extended limits $-\infty < k < \infty$, by using Cauchy's theorem.

The matrices $u_{ij}(r)$ and $\gamma_{ij}(r)$ (here $r = r_{ij}$) are real and symmetric. The matrix element $u_{ij}(r)$ describes the retarded Coulomb interaction between the atoms and is found to be of the form

$$u_{ij}(r) = -d^2 k_0^3 \left\{ \sin^2\theta \frac{\cos k_0 r}{k_0 r} \right.$$
$$\left. + (2\cos^2\theta - \sin^2\theta) \left(\frac{\sin k_0 r}{(k_0 r)^2} + \frac{\cos k_0 r}{(k_0 r)^3}\right) \right\} \qquad (3.4.7)$$

here $k_0 = \omega_0/c$, and the angle between the vectors d and k is denoted by θ.

At small distances, $r < \lambda$, the matrix element u_{ij} is proportional to r^{-3}

$$u_{ij}(r) \xrightarrow{r \to 0} \frac{d^2}{r^3}(1 - 3\cos^2\theta) = \sum_{\alpha,\beta=x,y,z} d_\alpha d_\beta \frac{r^2\delta_{\alpha,\beta} - 3r_\alpha r_\beta}{r^5} \qquad (3.4.8)$$

and describes the usual static Coulomb interaction of two dipoles. Thus we have explicitly shown that the assumed electric–dipole interaction with the transverse electromagnetic field actually accounts for the Coulomb interaction.

The matrix element $\gamma_{ij}(r)$ describes the radiation damping of one atom caused by the light field emitted by the other one. We obtain

$$\gamma_{ij}(r) = \frac{2d^2k_0^3}{\hbar}\left\{ \sin^2\theta \frac{\sin k_0 r}{k_0 r} + (2\cos^2\theta - \sin^2\theta)\left(\frac{\sin k_0 r}{(k_0 r)^2} + \frac{\cos k_0 r}{(k_0 r)^3}\right)\right\}.$$
$$(3.4.9)$$

At small distances, $r \ll \lambda$, the off-diagonal element of the radiation damping matrix is equal to the radiation damping constant γ of a single atom

$$\gamma_{ij}(r) \xrightarrow{r \to 0} \frac{4}{3}\frac{d^2k_0^3}{\hbar} = \gamma. \qquad (3.4.10)$$

In the forthcoming sections of this chapter we shall also use the integral form for the real part of the self-energy diagram. This representation immediately follows, from equations (3.3.9) and (3.4.6), in the form

$$\gamma_{b,c} = -2\,\mathrm{Re}\,\Sigma_{b,c} = (2\pi\hbar k_0)^{-1}\int dk\,\delta(k_0 - k)[d \times k]^2 \langle b|R_k^\dagger R_k|c\rangle. \qquad (3.4.11)$$

The value of the self-energy diagram in the time representation is given by

$$\Sigma_{b,c}(t_2 - t_1) = -\sum_{i,j}\langle b|\widehat{R}_+^{(i)}\widehat{R}_-^{(j)}|c\rangle\left(\frac{iu_{ij}}{\hbar} + \tfrac{1}{2}\gamma_{ij}\right)\delta(t_2 - t_1 - 0). \qquad (3.4.12)$$

The transformation of the diagram equation (3.4.1) for the quantal amplitude into the time domain is easily carried out, as in section 3.3. We obtain a Schrödinger-like equation for the atomic system

$$\dot{S}_{b,a} = -i\omega_b S_{b,a} - \sum_{c,i,j}\langle b|\widehat{R}_+^{(i)}\widehat{R}_-^{(j)}|c\rangle\left(\frac{iu_{ij}}{\hbar} + \tfrac{1}{2}\gamma_{ij}\right)S_{c,a}. \qquad (3.4.13)$$

The effective interaction Hamiltonian can be adopted in the form

$$\sum_{i,j}\widehat{R}_+^{(i)}\widehat{R}_-^{(j)}\left(u_{ij} - \tfrac{1}{2}i\hbar\gamma_{ij}\right). \qquad (3.4.14)$$

However, owing to the radiation damping, the non-Hermitian matrix (3.4.14) cannot be considered to be a 'true' interaction Hamiltonian. The self-consistent quantum description of the atomic subsystem should be based on the density matrix representation.

Assume the atoms to be in an arbitrary quantum state at the initial moment t_0. The initial state of a field is the vacuum. Consider the density matrix element for the atom–field system which is diagonal with respect to the photon quantum numbers and corresponds to the emission of photons $k_1\lambda_1, \ldots, k_m\lambda_m$ (abbreviated to $\{k_i\lambda_i\}$). The lowest order diagram is

$$\rho_{a.\{k_i\lambda_i\}:b.\{k'_i\lambda'_i\}}(t) \sim \sum_{P\{k_i\lambda_i\}} \sum_{P\{k'_i\lambda'_i\}} \left\{ \begin{array}{c} \end{array} \right. \qquad (3.4.15)$$

Here a and b are the arbitrary states of the atomic subsystem. The summation is carried out independently over the permutations of photon labels $k_i\lambda_i$ on the upper and lower time axes. The sets $\{k'_i\lambda'_i\}$ and $\{k_i\lambda_i\}$ coincide.

In order to obtain the atomic density matrix we have to take the trace with respect to the field quantum numbers

$$\mathrm{Tr}_{\mathrm{f}}(\cdots) = \frac{1}{m!} \sum_{k_1\lambda_1\cdots k_m\lambda_m} (\cdots). \qquad (3.4.16)$$

As every state of the field is counted $m!$ times in the summation over m photon labels the corresponding compensation factor $(m!)^{-1}$ has been introduced in equation (3.4.16).

Similarly to the method of section 3.3, we combine the pairs of photon lines with coincident photon labels in order to construct the photon propagators. The summation (3.4.16) makes these photon propagators vertical, because in the time domain δ-like factors arise under the time integral. The calculation gives the following result for the vertical photon line

$$= \sum_{i,j} \langle b|\widehat{R}^{(i)}_{+}|d\rangle \langle g|\widehat{R}^{(j)}_{-}|c\rangle \gamma_{ij}(r)\delta(t_l - t'_l). \qquad (3.4.17)$$

All diagrams with intersecting photon propagators, arising from the basic diagram (3.4.15) as the result of permutations, are neglected as we are interested in the evolution on the time scale of the order of γ^{-1}. Indeed, the δ-like function of time in (3.4.17) has a spread of the order of ω_0^{-1} (or l/c for an extended system) for any simple approximation chosen for this function so that the diagrams with the intersections are small, cf $\gamma/\omega_0 \ll 1$ (or $l/cT_R \ll 1$ for the extended system). The $m!$ permuted diagrams without intersections are equal to each other and, with the normalization constant in (3.4.16), the diagram for the atomic density matrix with m vertical photon lines is be taken into account only once.

In order to account for the radiation damping of quantal amplitudes it suffices to include in our diagrams the self-energy parts to all perturbation orders. Hence, instead of free atomic propagators we have to use the dressed ones, satisfying equation (3.4.13). These are represented by the bold horizontal lines. (Recall that a bold line on the lower time axis takes the complex conjugate value.)

As a result, an arbitrary element of the density matrix for an atomic subsystem is represented by the series

$$\rho_{ab}(t) \equiv \left\{ \cdots \right\} = \sum_{m=0}^{\infty} \left\{ \cdots \right\}. \tag{3.4.18}$$

The box denotes the sum of the series for the density matrix element $\rho_{ab}(t)$ arising from the special initial condition $\rho(t_0) = |a_0\rangle\langle b_0|$. We shall use the notation $\rho_{ab}^{a_0 b_0}(t_0, t)$ when it is necessary to specify an initial condition of this kind. In accordance with the rules of our diagram method, the large bracket in the figure denotes the convolution (3.2.16) with the initial condition.

The series (3.4.18) satisfies the diagram equation

$$\rho_{ab}(t) = \left\{ \cdots \right\} = \left\{ \cdots \right\} + \left\{ \cdots \right\}. \tag{3.4.19}$$

This equation shows that if we add to the diagrams of the series a vertical photon line and, respectively, two atomic propagators on the time interval (t_1, t) in the upper and the lower time axis, we arrive at the same series. The missing term (without the vertical photon line) is accounted for explicitly. In the time domain it follows from equation (3.4.19) that

$$\rho_{ab}(t) = \sum_{\substack{a_0, a_1, a_2 \\ b_0, b_1, b_2}} \left\{ S_{a, a_0}(t_0, t) \rho_{a_0, b_0}(t_0) S_{b, b_0}^*(t_0, t) \right.$$

$$+ \int_{t_0}^{t} dt_1 \, S_{a.a_2}(t_1, t) \left[\sum_{i,j} \langle a_2 | \widehat{R}_-^{(j)} | a_1 \rangle \gamma_{ij} \rho_{a_1 b_1}(t_1) \langle b_1 | \widehat{R}_+^{(i)} | b_2 \rangle \right]$$

$$\times \, S_{b.b_2}^*(t_1, t) \Big\} . \tag{3.4.20}$$

By differentiating with respect to time we obtain the *master equation* for the atomic density operator $\rho = \sum_{a,b} |a\rangle \rho_{ab} \langle b|$ that takes the form

$$\dot{\rho} = -\frac{i}{\hbar} [\widehat{H}_{at}, \rho] - \frac{i}{\hbar} \sum_{i,j}{}' u_{ij} [\widehat{R}_+^{(i)} \widehat{R}_-^{(j)}, \rho]$$

$$+ \tfrac{1}{2} \sum_{i,j} \gamma_{ij} \{ 2 \widehat{R}_-^{(j)} \rho \widehat{R}_+^{(i)} - \widehat{R}_+^{(i)} \widehat{R}_-^{(j)} \rho - \rho \widehat{R}_+^{(i)} \widehat{R}_-^{(j)} \}. \tag{3.4.21}$$

In this derivation, equation (3.4.13) and the initial condition for the propagators

$$S_{a.b}(t, t) = \delta_{ab} \tag{3.4.22}$$

have been taken into account. We note that in the retarded Coulomb interaction in (3.4.21) the atomic labels i, j do not coincide.

We now consider the frequency spectrum of emission, $\langle n_{k\lambda} \rangle$. Let us assume the field oscillator $k\lambda$ to be a separate field subsystem. The spectrum via the diagonal density matrix elements is defined as follows

$$\langle n_{k\lambda} \rangle = \lim_{t \to \infty} \sum_{a} \sum_{f'} \rho_{a.f'.k\lambda.a.f'.k\lambda}(t) \tag{3.4.23}$$

where the trace is taken over the states of the atoms and over the states f' of all field oscillators, except the given one. In the lowest order the interaction with the $k\lambda$ oscillator should be accounted for only once in the upper time axis and, respectively, just once in the lower axis. The interaction of atoms with a sea of field oscillators can be described by the Wigner–Weisskopf approximation. The justification for this is that the elimination of one field oscillator does not affect the summations with respect to the continuous frequency spectrum of fields provided that the emission takes place into free space and the quantization volume is arbitrarily large. We obtain just the same atomic propagators and vertical photon lines as before.

Denote the summation with respect to the final states of the atomic subsystem in (3.4.23) by the vertical solid line connecting the ends of the upper and lower time axes, and directed downwards. Then a diagram for the spectrum

arises in the form

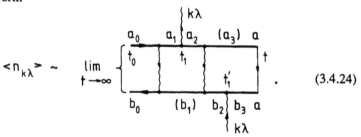

$$\langle n_{k\lambda} \rangle \sim \lim_{t \to \infty} \quad \text{(3.4.24)}$$

The complete diagram series for the spectrum includes the diagrams with an arbitrary number of the vertical photon lines on the time intervals (t_0, t_1), (t_1, t_1'), (t_1', t) (where the corresponding summation is implied). To every time-ordered diagram of the form (3.4.24), where $t_1 < t_1'$, there corresponds the complementary diagram in which $t_1' < t_1$. This diagrammatic counterpart is obtained by transferring all vertices and state indices from the upper time axis to the lower and vice versa. In accordance with the rules of the diagram method, such a transfer results in complex conjugation. For this reason one has to calculate for the spectrum only the real part of the time-ordered diagrams of the form (3.4.24).

The atomic propagators can always be factorized into factors corresponding to adjacent time intervals which depend on the time differences $(t - t_1')$, $(t_1' - t_1)$, $(t_1 - t)$. This is evident from the exponential form of the general solution of the equation (3.4.13)

$$S_{b,a}(t_0, t) = \langle b| \exp\left\{ \left(-\frac{i\hat{H}_{\text{at}}}{\hbar} - \sum_{i,j} \hat{R}_{+}^{(i)} \hat{R}_{-}^{(j)} \left(\frac{iu_{ij}}{\hbar} + \tfrac{1}{2}\gamma_{ij} \right) \right)(t - t_0) \right\} |a\rangle$$

(3.4.25)

which satisfies the identity

$$S_{b,a}(t_0, t) = \sum_{c} S_{b,c}(t_1, t) S_{c,a}(t_0, t_1).$$
(3.4.26)

The factorization of quantum propagators in the diagram (3.4.24) at the time instants t_1, t_1' results in additional summations with respect to the intermediate labels, which are shown in the diagram in brackets. Using the diagram representation (3.4.18) for the atomic density matrix elements, we find the emission spectrum in the form

$$\langle n_{k\lambda} \rangle = \lim_{t \to \infty} 2\text{Re} \quad \text{(3.4.27)}$$

The boxes describe the quantum evolution of the matrix elements of the atomic density on adjacent time intervals, as specified above.

The expression (3.4.27) for the spectrum can be further simplified by using the conservation of the trace of the density matrix in the quantum evolution process. The trace of the last box in (3.4.27) can be taken with the initial condition for this box. This yields

$$\sum_a \rho_{aa}^{a_3 b_3}(t_1', t) = \mathrm{Tr}_a(|a_3\rangle\langle b_3|) = \delta_{a_3 b_3} \qquad (3.4.28)$$

for the corresponding factor in the integrand. After summation over the intermediate indices a_3, b_3 we obtain the diagram

$$\langle n_{k\lambda}\rangle = \lim_{t\to\infty} \; 2\mathrm{Re} \left\{ \cdots \right\} \qquad (3.4.29)$$

and, analytically,

$$\langle n_{k\lambda}\rangle = 2\,\mathrm{Re}\,\frac{|f_{k\lambda}|^2}{\hbar^2} \sum_{a,a_1,a_2,b_1,b_2} \langle a_2|R_k|a_1\rangle\langle b_2|R_k^\dagger|a\rangle$$

$$\times \int_{t_0}^{\infty} dt_1' \int_{t_0}^{t_1'} dt_1 \, e^{-i\omega_k(t_1'-t_1)} \rho_{a_1 b_1}(t_1)\rho_{ab_2}^{a_2 b_1}(t_1, t_1'). \qquad (3.4.30)$$

The diagram representation for other physical quantities, describing dynamics and statistical properties of the atom–field system (for instance, the light–field correlation functions), can also be elaborated on the basis of our approach.

3.5 Super-radiance of two two-level atoms

In this section we consider collective spontaneous emission by two excited atoms located at an arbitrary distance from each other. This elementary problem can be solved exactly within the framework of the diagram method, and the results can be used to discuss the frequency and angular dependence of the emission, the decay law, as well as the integral light flux power as a function of time. We shall also examine the effect of energy splitting and Coulomb interaction upon the collective features of the emission. The analysis will be useful for further investigation of the collective effects in the emission from extended polyatomic systems.

Consider a pair of two-level atoms, $i = 1, 2$, separated by a distance $r = |r_{12}|$, $r_{12} = r_1 - r_2$, which interact with the transverse electromagnetic field. We assume that the transition frequencies ω_1, ω_2, are sufficiently close to each other, $|\omega_1 - \omega_2| \ll c/r$. It is easily seen that the main estimates which support the approximations (3.3.13) and (3.4.17) (δ-like self-energy parts and vertical photon lines) are not affected by the small frequency mismatch, so the theory developed above can be applied to the system.

Assume both atoms to be excited at the initial instant $t_0 = 0$. Using simplified notations in this section, we denote the eigenstates of the two-atom system by indicating only the labels of the excited atoms, namely

$$|e_1, e_2\rangle = |12\rangle \quad |e_1, g_2\rangle = |1\rangle \quad |g_1, e_2\rangle = |2\rangle \quad |g_1, g_2\rangle = |g\rangle. \quad (3.5.1)$$

For the energy of the ground state we assume $\omega_g = 0$, and we take the average transition frequency to be $\omega_0 = \frac{1}{2}(\omega_1 + \omega_2)$

The spectrum $\langle n_{k\lambda} \rangle$ of the collective spontaneous emission is represented by the diagrams

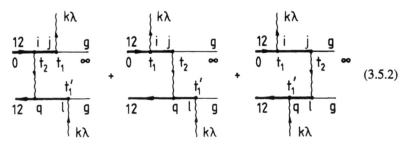

$$(3.5.2)$$

where all the possibilities of drawing one vertical photon line (in addition to the emission of the 'detected' photon $k\lambda$) are taken into account. The diagrams with no vertical lines are neglected, since such diagrams include only dressed (and hence decaying as $t \to \infty$) atomic propagators. Note that $i, j, q, l = 1, 2$ and $t_1 < t_1'$. The contribution of the diagrams with the opposite time ordering is accounted for by complex conjugation.

The initial and final states of the two-atom subsystem are uniquely specified as $|12\rangle$ and $|g\rangle$, and the sums over the atomic labels at $t = 0$ and $t = \infty$ are not needed. The population of the atomic states at a finite time t is given by the diagonal elements $\rho_{12,12}(t)$ and $\rho_{jj}(t)$ of the atomic density matrix, where $i, j = 1, 2$ and $\rho_{gg}(t)$. The diagrams corresponding to these quantities look like

$$(3.5.3)$$

Owing to the conservation of the sum of the atomic and field excitations, we can express the average number of emitted photons as

$$n(t) = \rho_{11}(t) + \rho_{22}(t) + 2\rho_{gg}(t). \tag{3.5.4}$$

The normalization condition is

$$\rho_{12,12}(t) + \rho_{11}(t) + \rho_{22}(t) + \rho_{gg}(t) = 1. \tag{3.5.5}$$

We apply Laplace transform to the calculation of atomic propagators. Then the general equations (3.4.1) and (3.4.13) reduce to

$$
\begin{aligned}
S_{12,12} &= S_{12,12}^{(0)}(1 + \Sigma_{12,12}S_{12,12}) \\
S_{i,i} &= S_{i,i}^{(0)}(1 + \Sigma_{i,i}\,S_{i,i} + \Sigma_{i,j}S_{j,i}) \\
S_{j,i} &= S_{j,j}^{(0)}(\Sigma_{j,j}S_{j,i} + \Sigma_{j,i}S_{i,i}) \\
S_{g,g} &= S_{g,g}^{(0)}
\end{aligned}
\tag{3.5.6}
$$

where $i \neq j$. Here, for example, the free propagator for the state j is

$$S_{j,j}^{(0)} = \frac{1}{p + i\omega_j}. \tag{3.5.7}$$

The self-energy factors are

$$
\begin{aligned}
&\Sigma_{12,12} = -\gamma \qquad\qquad \Sigma_{1,1} = -\tfrac{1}{2}\gamma \\
&\Sigma_{1,2} = -\tfrac{1}{2}\gamma_{12}(r) - \frac{iu_{12}(r)}{\hbar} \qquad \Sigma_{2,1} = \Sigma_{1,2}.
\end{aligned}
\tag{3.5.8}
$$

The quantities $\gamma_{12}(r)$ and $u_{12}(r)/\hbar$ (denoted below as $\tilde{\gamma}$ and V) are responsible for the collective radiation damping and retarded Coulomb interaction of the atomic dipoles (see section 3.4). The solution of the system above is

$$
S_{12,12} = \frac{1}{(p + 2i\omega_0 + \gamma)} \qquad S_{i,i} = \frac{p + i\omega_j + \gamma/2}{(p - p_1)(p - p_2)}
\tag{3.5.9}
$$

$$
S_{j,i} = -\frac{(\tilde{\gamma}/2 + iV)}{(p - p_1)(p - p_2)} \qquad S_{g,g} = \frac{1}{p}
$$

where

$$p_{1,2} = -i\omega_0 - \tfrac{1}{2}\gamma \pm \sqrt{(\tfrac{1}{2}\tilde{\gamma} + iV)^2 - \Delta^2} \tag{3.5.10}$$

$$2\Delta = \omega_1 - \omega_2.$$

When $|\Delta| \gg \bar{\gamma}, V$, the poles of the propagators correspond to the non-interacting atoms, and there are no collective effects. However, if the Coulomb interaction is strong, $|V| \gg |\Delta|$, then the radiative constants at the poles are equal to $\gamma \pm \bar{\gamma}$, independently of the ratio of the frequency mismatch Δ and the spontaneous emission rate γ. The reason is that in this case the damping matrix γ_{ij} and the energy matrix are simultaneously diagonalized.

We consider first the spectrum of the collective spontaneous emission of identical ($\omega_1 = \omega_2$) atoms, and discuss the general case later on. We use a symmetrized basis for two identical atoms

$$|s\rangle = \frac{1}{\sqrt{2}}(|1\rangle + |2\rangle) \qquad |a\rangle = \frac{1}{\sqrt{2}}(|1\rangle - |2\rangle) \qquad (3.5.11)$$

in which both the damping matrix and the energy matrix reduce to diagonal form. It is now necessary to sum over $i, j, k, l = s, a$ in the diagrams (3.5.2) and (3.5.3). The atomic propagators in this basis are diagonal

$$S_{s,s} = \frac{1}{(p - p_2)} \qquad S_{a,a} = \frac{1}{(p - p_1)} \qquad S_{a,s} = S_{s,a} = 0$$

$$(3.5.12)$$

$$p_{1,2} = -i\omega_0 - \tfrac{1}{2}\gamma \pm (\tfrac{1}{2}\bar{\gamma} + iV).$$

The vertical photon line containing simultaneously both indices a and s vanishes in the symmetrized basis. For this reason the intermediate indices s, a do not appear simultaneously in any of the diagrams. The spectrum is a sum of two groups of terms, one of which corresponds to decay via the symmetrical state, and the other to decay via the anti-symmetrical state.

We calculate, for example, the first diagram (3.5.2) for the intermediate state s

$$\langle n_{k\lambda} \rangle \sim 2 \operatorname{Re} (\gamma + \bar{\gamma})(1 + \cos k r)\frac{|f_{k\lambda}|^2}{\hbar^2}$$

$$\times \int_0^\infty dt_1' \int_0^{t_1'} dt_1 \int_0^{t_1} dt_2 \exp(-2\gamma t_2 + i\omega_k(t_1 - t_1'))$$

$$\times S_{ss}(t_1 - t_2)S_{ss}^*(t_1' - t_2). \qquad (3.5.13)$$

Regarding these integrals as the limiting case of a Laplace transform for $p \to 0$, and taking into account (3.5.12), we obtain

$$\langle n_{k\lambda} \rangle \sim 2 \operatorname{Re} (\gamma + \bar{\gamma})(1 + \cos k r)\frac{|f_{k\lambda}|^2}{\hbar^2}$$

$$\times \{2\gamma(\gamma + \bar{\gamma})[i(\omega_k - \omega_0 + V) + \tfrac{1}{2}(\gamma + \bar{\gamma})]\}^{-1}. \qquad (3.5.14)$$

The remaining diagrams are calculated in a similar way. The spectrum is then obtained in the form

$$\langle n_{k\lambda} \rangle = \frac{\pi}{\gamma} \frac{|f_{k\lambda}|^2}{\hbar^2} \{(1 + \cos kr)[(1 - \alpha)L_{\gamma(3+\delta)/2}(v + V)$$

$$+ (1 + \alpha)L_{\gamma(1+\delta)/2}(v - V) + \beta(X_{\gamma(3+\delta)/2}(v + V) - X_{\gamma(1+\delta)/2}(v - V))]$$

$$+ (1 - \cos kr)[(1 - \alpha')L_{\gamma(3-\delta)/2}(v - V) + (1 + \alpha')L_{\gamma(1-\delta)/2}(v + V)$$

$$- \beta'(X_{\gamma(3-\delta)/2}(v - V) - X_{\gamma(1-\delta)/2}(v + V))]\}. \tag{3.5.15}$$

We have here introduced the following notation

$$L_\gamma(\omega) = \frac{\gamma}{\pi(\gamma^2 + \omega^2)} \qquad X_\gamma(\omega) = \frac{\omega}{\pi(\gamma^2 + \omega^2)} \qquad \beta = \frac{2v(1 + \delta)}{1 + 4v^2}$$

$$\beta' = \frac{2v(1 - \delta)}{1 + 4v^2} \qquad \alpha = \frac{1 + \delta}{1 + 4v^2} \qquad \alpha' = \frac{1 - \delta}{1 + 4v^2}$$

$$\delta = \frac{\tilde{\gamma}}{\gamma} \qquad v = \frac{V}{\gamma} \qquad v = \omega_k - \omega_0. \tag{3.5.16}$$

To interpret the frequency spectrum of collective spontaneous emission let us consider the level scheme of the radiating system, shown in figure 3.1. The

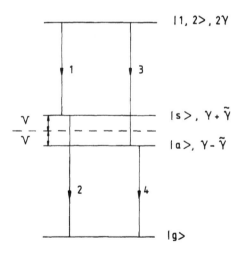

Figure 3.1. Collective spontaneous emission. Possible transitions in a system of two atoms. The shift $V = u_{12}(r)/\hbar$ is due to the interaction of atomic dipoles, equation (3.5.9).

terms with the factor $(1 + \cos kr)$ result from the decay via the symmetrical channel, with the transitions 1 and 2 of figure 3.1 corresponding to Lorentzian contributions with widths $\gamma(3+\delta)/2$ and $\gamma(1+\delta)/2$ respectively. The positions and widths of the Lorentzians are determined by the splitting and broadening of the energy levels involved. The quantum interference of the transitions, which results from the nearly equidistant spacing of the levels of the system, is described by the second diagram (3.5.2). It results in the redistribution of the weights of the narrow and the broad Lorentzian components, and in the appearance of the interference terms X. These terms bring the spectral lines closer together and disturb their symmetry. For $V \to 0$ the odd terms X vanish and quantum interference only redistributes the weights of the Lorentzian components in favour of the narrow one.

The terms in the spectrum with the factor $(1 - \cos \boldsymbol{kr})$ result from decay via the antisymmetric intermediate state a and are set in correspondence with the transitions 3 and 4 in figure 3.1.

The angular dependence of the spectrum is also easily explained. The field oscillators interacting most effectively with the state s are those for which the phase difference $\boldsymbol{kr}$ is equal to $0, 2\pi, \dots$. In the case of the antisymmetric state a these are the field oscillators for which $\boldsymbol{kr} = \pm\pi, \pm3\pi, \dots$. The spectrum observed in an arbitrary direction is a sum of two independent spectra. In the limit $kr \ll 1$ (the Dicke problem) the system decays only via the symmetrical channel, since the interaction becomes symmetrical as well. The characteristic form of the spectrum (3.5.15), averaged over the radiation directions, is shown in figure 3.2 for $|V| \gg \gamma$ and $\bar{\gamma}/\gamma \sim 1$.

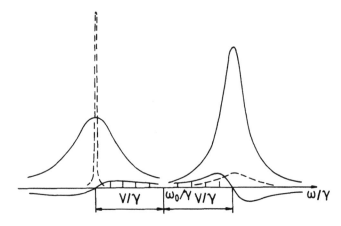

Figure 3.2. Components of collective emission spectrum averaged over the directions at $V/\gamma = 5$ and $\delta = 0.9$. The broken curves represent the contribution of the antisymmetric channel.

For the case of frequency mismatch ($\omega_1 \neq \omega_2$) the frequency spectrum also contains four Lorentzian curves affected by the interference terms X with positions and widths corresponding to the poles (3.5.10). Here we discuss the limiting cases that provide, together with the spectrum (3.5.15), a sufficiently full physical picture. When the atoms are separated by a small distance ($kr \ll 1$), so that $\bar{\gamma} = \gamma$ and $|V| \gg \gamma$, the difference in the atomic frequencies only slightly modifies the spectrum. If the condition $|V| \gg |\Delta|$ is satisfied, then the weights, positions and widths of the spectral components differ from the spectrum calculated above only in the second and highest orders of the small parameter $|\Delta/V| \ll 1$. The only new (with respect to (3.5.15)) spectral component owed to the frequency mismatch is given by

$$\frac{\pi}{\gamma} \frac{|f_{k\lambda}|^2}{\hbar^2} \frac{\Delta^2}{V^2} L_{\gamma(\Delta/V)^2/2} \left\{ v + V \left(1 + \frac{1}{2}\left(\frac{\Delta}{V}\right)^2\right)\right\}. \qquad (3.5.17)$$

Although the system decays at $\bar{\gamma} = \gamma$ only via the state s, the difference in the atomic frequencies leads to quantum oscillations between the states s and a. The admixture of the nearly stationary antisymmetric state a at the intermediate stage of decay slows down the decay and creates a narrow spectral component (3.5.17).

Let us consider the spectrum for the case of medium distances between the atoms $(kr > 1)$. Since the Coulomb interaction V decreases with distance more rapidly than $\bar{\gamma}$ (i.e. as r^{-3}, compared with r^{-1}) we set V equal to zero. For $|\Delta| > \gamma/2$ the spectrum averaged over directions takes the form

$$\overline{\langle n_{k\lambda}\rangle} = \frac{\pi}{\gamma} \frac{\overline{|f_{k\lambda}|^2}}{\hbar^2} \{\alpha_1[L_{\gamma/2}(v + \Omega_0) + L_{\gamma/2}(v - \Omega_0)]$$

$$- \alpha_2[L_{3\gamma/2}(v + \Omega_0) + L_{3\gamma/2}(v - \Omega_0)]$$

$$+ \beta_1[X_{\gamma/2}(v + \Omega_0) - X_{\gamma/2}(v - \Omega_0)]$$

$$+ \beta_2[X_{3\gamma/2}(v + \Omega_0) - X_{3\gamma/2}(v - \Omega_0)]\} \qquad (3.5.18)$$

where

$$\Omega_0 = \left(\Delta^2 - \left(\tfrac{1}{2}\bar{\gamma}\right)^2\right)^{1/2} \qquad \alpha_1 = 2 + \frac{\delta^2(1 - \delta^2)}{1 + 4\lambda^2} \qquad \alpha_2 = \frac{\delta^2(1 - \delta^2)}{1 + 4\lambda^2}$$

$$\beta_1 = \frac{\delta^2}{2\lambda}\left(2 + \frac{1 - \delta^2}{1 + 4\lambda^2}\right) \qquad \beta_2 = \frac{\delta^2}{2\lambda}\left(1 + \frac{\delta^2 + 4\lambda^2}{1 + 4\lambda^2}\right) \qquad (3.5.19)$$

$$\lambda = \frac{\Omega_0}{\gamma}.$$

In the spectrum (3.5.18) the Lorentz curves are shifted towards each other relative to the atomic frequencies. The broad negative Lorentzians ensure a faster decrease of lines on the wings. The shift and the distortion of the spectral lines are owed to the interference terms X. As the ratio γ/Δ goes to zero, the spectrum (3.5.18) reduces to the emission spectrum of independent atoms.

Now let us consider the emission dynamics for the case of identical atoms $(\omega_1 = \omega_2)$ with an allowance for the Coulomb interaction. Let $n^{(s)}(t)$ and $n^{(a)}(t)$

be the average numbers of photons emitted in the decay via the symmetrical and antisymmetrical states. These quantities are found from the relations

$$n^{(s)}(t) = \rho_{ss}(t) + 2\rho_{gg}^{(s)}(t) \qquad n^{(a)}(t) = \rho_{aa}(t) + 2\rho_{gg}^{(a)}(t) \qquad (3.5.20)$$

where $\rho_{gg}^{(s)}$ is the contribution to ground-state population caused by decay via the symmetrical state.

By using the second diagram (3.5.3) with the symmetrical intermediate state we obtain

$$\begin{aligned}\rho_{ss}(t) &= (\gamma + \bar{\gamma}) \int_0^t dt_1 \, e^{-2\gamma t_1} |S_{s.s}(t - t_1)|^2 \\ &= \frac{1 + \delta}{1 - \delta} e^{-2\gamma t} (e^{(\gamma - \bar{\gamma})t} - 1).\end{aligned} \qquad (3.5.21)$$

Similar calculations give

$$n^{(s)} = (1 + \delta)\{1 - (1 - \delta)^{-1} e^{-2\gamma t} (e^{(\gamma - \bar{\gamma})t} - \delta)\}$$

$$n^{(a)} = (1 - \delta)\{1 - (1 + \delta)^{-1} e^{-2\gamma t} (e^{(\gamma + \bar{\gamma})t} + \delta)\} \qquad (3.5.22)$$

It is seen that the Coulomb interaction plays no rôle in the radiation decay of identical atoms. If $kr \ll 1$, the antisymmetric intermediate state is not involved in the decay process, and the emitted light energy is

$$n(t) = 2\{1 - e^{-2\gamma t}(1 + \gamma t)\}. \qquad (3.5.23)$$

For large distances $kr \gg 1$, $\delta \to 0$ and we come to the decay law for independent atoms

$$n(t) = 2(1 - e^{-\gamma t}). \qquad (3.5.24)$$

Our analysis shows that collective effects in the emission of two atoms come into play when the distance between the atoms is smaller than, or comparable with, the radiation wavelength. At distances greatly exceeding the wavelength the off-diagonal damping matrix elements tend to zero, and the atoms decay independently. If the Coulomb interaction between the atoms is small, as in the case for relatively large distances, then the collective effect depends upon the degree of resonance. The difference between the frequencies of the atoms must not exceed the natural width of the upper level. If, however, the Coulomb interaction is larger than the inhomogeneity, then the energy eigenstates will coincide with the collective radiative states. This case is apparently the most realistic for the manifestation of collective radiative properties of a two-atom system.

Very recently DeVoe and Brewer [DVB96] have observed collective super-radiant and sub-radiant (see section 11.4) emission by two identical ions in a trap. We present the details of their experiment at the end of the book, in section 11.5 and compare it with the results of the present section.

3.6 Super-radiance by extended polyatomic systems

In this section we shall consider collective spontaneous emission by a completely excited, extended $(L > \lambda)$ polyatomic system. We discuss the possibility of diagonalizing the radiation damping matrix and obtain an approximate solution of the master equation for the polyatomic system. The time and angular dependences of the emission are investigated for a volume of arbitrary shape. The dynamics of radiative decay and the distribution of the emission spectrum are obtained for the isotropic case.

An important discussion of angular correlation of photons in super-radiance is also presented in this section.

3.6.1 Collective radiation constants for extended polyatomic systems

Let us consider a system of N identical atoms with transition frequency ω_0 and position vectors r_i, $i = 1, \ldots, N$. The collective spontaneous decay is governed by the radiation damping matrix (3.4.9), $\hat{\gamma} = \{\gamma_{ij}\}$. Let G be the matrix of the unitary transformation that diagonalizes the damping matrix

$$(G\hat{\gamma}G^{\dagger})_{ij} = \sum_{p,l} G_{ip}\gamma_{pl}G_{jl}^{*} = \gamma_i \delta_{ij}. \tag{3.6.1}$$

Consider the singly excited state of the atomic system $|i\rangle$, $i = 1, \ldots, N$. Here the label i identifies the excited atom. The transformed states

$$|\tilde{i}\rangle = \sum_j G_{ij}|j\rangle \tag{3.6.2}$$

decay exponentially (if we discard the Coulomb interaction). The equation (3.4.13) for atomic propagators reduces to

$$\dot{S}_{\tilde{i},\tilde{i}} = -(i\omega_0 + \tfrac{1}{2}\gamma_i)S_{\tilde{i},\tilde{i}}. \tag{3.6.3}$$

Since the trace of the damping matrix is invariant, we have

$$\sum_{i=1}^{N} \gamma_i = \mathrm{Tr}\,\hat{\gamma} = N\gamma. \tag{3.6.4}$$

The collective radiation constant γ_i depends, in general, upon the position of every atom. In the case of a macroscopic system with random distribution of the atoms in the volume, we are interested in the probability density $P(\gamma)$ (per unit frequency interval) of finding an eigenstate of the damping matrix equal to γ, rather than in the exact knowledge of the radiation damping constants.

The normalization condition has the form

$$\int_0^{N\gamma_{\mathrm{sp}}} d\gamma\, P(\gamma) = 1. \tag{3.6.5}$$

To avoid a possible confusion, in this subsection we denote the radiation constant of a single atom by γ_{sp}. From the invariance of the trace of the matrix $\hat{\gamma}$ it follows that

$$\int_0^{N\gamma_{sp}} d\gamma \; \gamma P(\gamma) = \frac{1}{N} \operatorname{Tr} \hat{\gamma} = \gamma_{sp}. \tag{3.6.6}$$

We shall now investigate the higher statistical momenta of the distribution $P(\gamma)$. The reason for applying the momentum method in our consideration is that in the present treatment of super-radiance we are interested in the large collective radiation constants (i.e. those greatly exceeding γ_{sp}). These give the main contribution to the momenta. For the momentum of an arbitrary order n we have

$$\int_0^{N\gamma_{sp}} d\gamma \; \gamma^n P(\gamma) = \frac{1}{N} \langle \operatorname{Tr} \hat{\gamma}^n \rangle_r . \tag{3.6.7}$$

Here $\langle \cdots \rangle_r$ denotes the statistical average over the positions of the atoms. For the second-order momentum equation (3.6.7) reduces to

$$\frac{1}{N} \langle \operatorname{Tr} \hat{\gamma}^2 \rangle_r = \gamma_{sp}^2 + (N - 1)\langle \gamma_{12}\gamma_{21} \rangle_r. \tag{3.6.8}$$

The first and second terms above are related to the diagonal and off-diagonal matrix elements. Since $N \gg 1$, we replace $N - 1$ by N and neglect the first term in (3.6.8). The higher momenta will be considered in the same approximation.

It turns out to be more convenient to exploit the integral representation (3.4.11) of the radiation damping matrix rather than using the explicit expression (3.4.9). We find

$$\gamma_{12}(r_{12}) = \frac{1}{2\pi\hbar} \int dk \; k\delta(k - k_0) \left\{ \sum_{\lambda=1,2} (d_1 \epsilon_{k\lambda})(d_2 \epsilon_{k\lambda}) \right\}$$

$$\times \exp(-ikr_{12}). \tag{3.6.9}$$

The directions of the dipole momenta of different atoms are here allowed to be independent and random. The corresponding averaging is denoted by $\langle \cdots \rangle_d$.

In the highest order in $N \gg 1$ the nth momentum of the distribution $P(\gamma)$ is evaluated as

$$\frac{1}{N} \langle \operatorname{Tr} \hat{\gamma}^n \rangle_{r,d} = N^{n-1} \langle \gamma_{12}\gamma_{23} \cdots \gamma_{n1} \rangle_{r,d}$$

$$= \frac{1}{N} \left(\frac{Nk_0}{2\pi\hbar V} \right)^n \int_V dr_1 \ldots dr_n \int dk_1 \ldots dk_n$$

$$\times \prod_{i=1}^n \delta(k_i - k_0) \exp \left\{ -i \sum_{j=1}^n k_j r_{j,j+1} \right\} \langle F^{(n)}(\{d\}, \{k\}) \rangle_d$$

$$\tag{3.6.10}$$

where $r_{n,n+1} \equiv r_{n,1}$, and V is the volume of the active medium. The factors depending upon the polarization vectors and the atomic dipole momenta (see sums in the curly brackets in (3.6.9)) have been absorbed into the quantities $F^{(n)}$. It is convenient to rewrite the exponential in (3.6.10) in the form

$$\exp\left\{-i \sum_{j=2}^{n} k_{j1} r_{j.j+1}\right\} \tag{3.6.11}$$

where $k_{j1} = k_j - k_1$. The averaging over the atomic coordinates in an extended volume ($V^{1/3} \gg \lambda$) generates the δ-like functions of k_{j1}, $j = 2, \ldots, n$. For a fixed k_1 this allows us to restrict the integration with respect to $k_2 \ldots k_n$ to a small area on the resonant sphere k_0. Taking first the integrals in momentum space in the range $k_0 \gg |k_{j1}|$ (where the values $|k_{j1}| \gg V^{-1/3}$ are allowed, since $k_0 V^{1/3} \gg 1$ for an extended system) we obtain

$$\int dk_j\, \delta(k_j - k_0) \exp\{-ik_{j1} r_{j.j+1}\} = (2\pi)^2 \delta(r_{j.j+1}). \tag{3.6.12}$$

Here $\delta(r_{j.j+1})$ is the two-dimensional δ-like function in the plane perpendicular to k_1. The external integration in (3.6.10) gives

$$\int dr_1 \ldots dr_n \prod_{j=2}^{n} \delta(r_{j.j+1}) = \int dr_1 l^{n-1}(k_1, r_1). \tag{3.6.13}$$

The length within the active volume, measured along the line parallel to k_1 and crossing the point r_1, is denoted as $l(k_1, r_1)$. For a spherical volume of radius R

$$\int dr_1 l^{n-1}(k_1, r_1) = \frac{\pi (2R)^{n+2}}{2(n+2)}. \tag{3.6.14}$$

The remaining averaging is taken over the independent and random orientations of the atomic dipoles. For the reasons discussed above we assume $k_1 = \ldots = k_n$, and subsequently obtain

$$\langle F^{(n)}(\{d\}, \{k\})\rangle_d = 2 \left(\tfrac{1}{3} d^2\right)^n. \tag{3.6.15}$$

Taking into account the value (3.3.12) of the radiation damping constant for a single atom, we find the nth momentum of the distribution $P(\gamma)$ in the form

$$\frac{1}{N} \langle \mathrm{Tr}\, \hat{\gamma}^n \rangle_{r.d} = \frac{3(K \gamma_{\text{sp}})^n}{(n+2)K} \tag{3.6.16}$$

where $K = N_0 \lambda^2 R/4\pi$, and $N_0 = N/V$ is the mean number of atoms per unit volume (the atomic concentration). The condition for super-radiance in an extended polyatomic system reads as $K \gg 1$, and is easily met for a system with average inter-atomic distance not exceeding the wavelength, i.e. $N_0 \lambda^3 > 1$.

The distribution $P(\gamma)$ with momenta (3.6.16) is readily found in the form

$$
P(\gamma) = \begin{cases} \delta(\gamma)\left(1 - \dfrac{3}{2K}\right) + \dfrac{3\gamma}{K^3\gamma_{sp}^2} & \text{for } \gamma \le K\gamma_{sp} \\[4mm] 0 & \text{for } \gamma > K\gamma_{sp}. \end{cases}
\tag{3.6.17}
$$

The relatively small values of the collective radiation constant γ, responsible for radiation trapping in a low-excited medium with phase memory, are represented by a δ-like contribution with statistical weight close to unity. For more detailed information about small values of γ one has to go beyond the estimates based on the momentum method.

Regarding the large values of the collective radiation constant which correspond to super-radiant states of the atomic subsystem, we come to the following general conclusions.

(i)   The most probable values are those of the order of $K\gamma_{sp} \gg \gamma_{sp}$.
(ii) In comparison with the Dicke problem (atoms in a small volume), the spatial extent of the system reduces the super-radiant constant by the diffraction factor $K/N \sim (\lambda/R)^2 \ll 1$.
(iii) The number of singly excited super-radiant states is of the order of $N/K \sim (R/\lambda)^2 \gg 1$, and depends upon the size of the system.
(iv) The probability of finding a collective radiation constant that greatly exceeds $K\gamma_{sp}$ is negligible.

Coulomb interaction and inhomogeneous broadening resulting in the effective atomic frequency bandwidth ω^* do not change the decay law of a super-radiant single excited state if $K\gamma_{sp} \gg \omega^*$. This condition can be rewritten as

$$
\frac{N_0\lambda^2 R\gamma_{sp}}{4\pi\omega^*} \gg 1.
\tag{3.6.18}
$$

In the linear theory of light propagation, this inequality indicates strong absorption of incident resonant light in a volume of radius R.

3.6.2   The super-radiant states

The problem of determining the singly excited super-radiant states consists in diagonalizing the radiation damping matrix. Consider the states

$$
|k\rangle = \frac{1}{\sqrt{N}} \sum_{i=1}^{N} \exp(ikr_i)|i\rangle = \frac{1}{\sqrt{N}} R_k^\dagger|g\rangle
\tag{3.6.19}
$$

where $k = k_0$, and $|g\rangle$ is the ground state of the atomic system. In a state of this type the atomic dipoles are phase correlated, providing constructive interference of light emitted in the direction given by k.

The off-diagonal element of the matrix $\tilde{\gamma}$ in this basis is given by

$$\gamma_{k_1,k_2} = \frac{1}{N} \sum_{i,j} \gamma_{ij} \exp\{-i(k_1 r_i - k_2 r_j)\}. \tag{3.6.20}$$

It will be assumed in this and the following paragraphs that the atomic dipoles d_i are parallel. Replacing the summation over the positions of atoms in (3.6.20) by integration within the volume V of the medium and taking into account, (3.6.9), we find that

$$\gamma_{k_1,k_2} = \frac{(2\pi)^5 N}{\hbar V^2 k_0} \int d\mathbf{k}\, \delta(k-k_0)[\mathbf{d} \times \mathbf{k}]^2 \Delta(\mathbf{k}-\mathbf{k}_1)\Delta^*(\mathbf{k}-\mathbf{k}_2) \tag{3.6.21}$$

where

$$\Delta(\mathbf{k}-\mathbf{k}_1) = \frac{1}{(2\pi)^3} \int_V d\mathbf{r}\, \exp\{i\mathbf{r}(\mathbf{k}-\mathbf{k}_1)\} \tag{3.6.22}$$

is a three-dimensional δ-like function in momentum space with a spread of the order of $V^{-1/3}$. The convolution of two δ-like functions in (3.6.21) tends to zero at $|k_1 - k_2| V^{1/3} \gg 1$. This means that for the single excited states under consideration the radiation damping matrix is approximately diagonal.

The collective radiation constant (the diagonal matrix element) is

$$\begin{aligned}
\gamma_{k_1,k_1} &\equiv N\gamma(k_1) \\
&= \frac{N}{2\pi\hbar V^2 k_0} \int_V d\mathbf{r}_1 d\mathbf{r}_2 \int d\mathbf{k}\, \delta(k-k_0)[\mathbf{d} \times \mathbf{k}]^2 \exp\{i\mathbf{r}_{12}(\mathbf{k}-\mathbf{k}_1)\} \\
&= \frac{(2\pi)^2 N}{\hbar V k_0} \int d\mathbf{k}\, \delta(k-k_0)[\mathbf{d} \times \mathbf{k}]^2 \tilde{\Delta}(\mathbf{k}-\mathbf{k}_1) \tag{3.6.23}
\end{aligned}$$

where

$$\tilde{\Delta}(\mathbf{k}-\mathbf{k}_1) = \frac{(2\pi)^3}{V} |\Delta(\mathbf{k}-\mathbf{k}_1)|^2. \tag{3.6.24}$$

The normalized (i.e. divided by the total number of atoms) collective radiation constant $\gamma(k)$ introduced above depends only upon the geometry of the system. The δ-like function $\tilde{\Delta}(\mathbf{k}-\mathbf{k}_1)$ cuts off a small plane area on the resonant sphere $k = k_0$ in the integral (3.6.23). If, for example, it is assumed that the volume containing the atoms is a rectangular parallelepiped with edges L_x, L_y and L_z, then

$$\gamma(k) = \frac{2\pi}{\hbar k_0} [\mathbf{d} \times \mathbf{k}]^2 S. \tag{3.6.25}$$

Here S is the cross-sectional area of the rectangular parallelepiped in momentum space with edges L_x^{-1}, L_y^{-1} and L_z^{-1}, centred at the position k_1, of a sphere $k = k_0$.

In the general case we first integrate (3.6.23) over the momentum k within the area $k_0 \gg |k - k_1|$, and then with respect to r and r_2 in analogy with the calculation in the preceding section. This gives

$$\gamma(k) = \frac{2\pi}{\hbar V k_0} [d \times k]^2 \langle l(k, r)\rangle_r \qquad (3.6.26)$$

where $\langle l(k, r)\rangle_r$ is the average linear dimension of the system in the direction of k, as defined in (3.6.13).

The maximum value of the collective radiation constant $\gamma(k)$ corresponds to the direction of the greatest mean extent of the volume. For a spherical volume of radius $R \gg \lambda$ the radiation constant $N\gamma(k)$ is of the order of magnitude of $K\gamma$, in agreement with the statistical estimate based on the momentum method.

Now let us assume that all the atoms are initially excited. The lowest order probability of emission of photons with wavevectors $k_1 \ldots k_l$ by the time t is given by the diagram

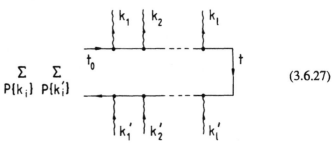

$$\begin{array}{c} \Sigma \quad \Sigma \\ P\{k_i\} \ P\{k_i'\} \end{array} \qquad\qquad (3.6.27)$$

where summation over all permutations of the vertices on the upper and lower time axes is implied, as well as summation over the polarization indices. The sets $\{k_i\}$ and $\{k_i'\}$ coincide. If Φ_N is the initial fully excited state of N atoms, the state arising after the first vertex is

$$\Phi_{k_1} = \frac{R_{k_1}\Phi_N}{\sqrt{Z_{k_1}}} \qquad (3.6.28)$$

where Z_k is the normalizer

$$Z_k = \langle R_k\Phi_N|R_k\Phi_N\rangle. \qquad (3.6.29)$$

Following the lth interaction point we obtain the state

$$\Phi_{k_1\ldots k_l} = \frac{R_{k_l}\ldots R_{k_1}\Phi_N}{\sqrt{Z_{k_1\ldots k_l}}}$$

$$\qquad (3.6.30)$$

$$Z_{k_1\ldots k_l} = \langle R_{k_l}\ldots R_{k_1}\Phi_N|R_{k_l}\ldots R_{k_1}\Phi_N\rangle.$$

The radiation damping constant $\gamma_{k_1\ldots k_l}$ of this state is related to the corresponding diagonal self-energy diagram. Using (3.4.11) we find

$$\gamma_{k_1\ldots k_l} = \frac{1}{2\pi\hbar k_0}\int dk\,\delta(k-k_0)[d \times k]^2 \frac{\langle R_{k_l}\ldots R_{k_1}\Phi_N|R_k^\dagger R_k|R_{k_l}\ldots R_{k_1}\Phi_N\rangle}{\langle R_{k_l}\ldots R_{k_1}\Phi_N|R_{k_l}\ldots R_{k_1}\Phi_N\rangle}$$

$$= \frac{1}{2\pi\hbar k_0} \int dk \, \delta(k - k_0)[d \times k]^2 \frac{Z_{k_1...k_l k}}{Z_{k_1...k_l}}. \tag{3.6.31}$$

The ratio of two normalizers in (3.6.31) can be evaluated explicitly. For this purpose we reduce the symmetric quantity $Z_{k_0 k_1...k_l}$ (where k_0 stands for k) to

$$Z_{k_0...k_l} = \sum_{\{a_0...a_l\}} \sum_{P\{a_0...a_l\}} \exp\left\{i \sum_{j=0}^{l} k_j(a_j - a_{p_j})\right\}. \tag{3.6.32}$$

Here we sum over all possible sets $\{a_0 \ldots a_l\}$ of $(l+1)$ atoms and over $(l+1)!$ permutations. For brevity we have set $a_j \equiv r_{a_j}$. It is readily seen that the summation over $(l+1)!$ permutations can be represented as the explicit sum over $(l+1)$ permutations of the index a_0 with indices $a_0 \ldots a_l$, with a subsequent summation over $l!$ permutations within the set $\{a_1 \ldots a_l\}$. After some simple transformations this gives

$$Z_{k_0...k_l} = \sum_{\{a_1...a_l\}} \sum_{P\{a_1...a_l\}} \left\{ \left[\sum_{a_0 \neq a_1...a_l}^{N} \sum_{s=0}^{l} \exp\{i(k_0 - k_s)(a_0 - a_{p_s})\} \right] \right.$$
$$\left. \times \exp\left\{i \sum_{j=1}^{l} k_j(a_j - a_{p_j})\right\} \right\}. \tag{3.6.33}$$

Since $N_0 \lambda^3 > 1$, the sum with respect to the position a_0 can be replaced by the volume integration. As a result, a δ-like function of the difference $k_0 - k_s$ arises, and the dependence of the corresponding exponential upon the coordinate a_{p_s} becomes weak. This allows us to average the first exponential factor in (3.6.33) over these coordinates. At the final stage of radiative decay the number of close or coincident wavevectors k_s is large. This also justifies averaging over the coordinate a_{p_s} in the continuous limit.

As a result we come to the following approximate recurrence relation

$$Z_{k_0...k_l} = (N - l) \left\{1 + \frac{(2\pi)^3}{V} \sum_{s=1}^{l} \tilde{\Delta}(k_0 - k_s)\right\} Z_{k_1...k_l}. \tag{3.6.34}$$

The radiation damping constant (3.6.31) reduces to

$$\gamma_{k_1...k_l} = \int dk \delta(k - k_0)[d \times k]^2 \frac{(N-l)}{2\pi\hbar k_0} \left\{1 + \frac{(2\pi)^3}{V} \sum_{s=1}^{l} \tilde{\Delta}(k - k_s)\right\}$$

$$= (N - l)\left(\gamma + \sum_{s=1}^{l} \gamma(k_s)\right). \tag{3.6.35}$$

We can also evaluate the off-diagonal element of the radiation damping matrix (3.4.11) in the basis of states $\Phi_{k_1...k_l}$. A simple but cumbersome estimate, similar to the estimate (3.6.21), shows that, at least for small numbers of emitted photons, these states approximately diagonalize the damping matrix, as in the case of the single excited states $|k\rangle$.

3.6.3   Dynamics and angular dependence of super-radiance

The basic approximation that we assume in the present quantum electrodynamic treatment of super-radiance consists of neglecting the off-diagonal elements of the radiation damping matrix computed in the basis of super-radiant states $\Phi_{k_1...k_l}$. These matrix elements have been estimated and found, in subsection 3.6.2, to be small for the initial stage of super-radiance. However, it is well known that it is the initial (linear) evolution of atoms and field that, to a large extent, governs the dynamical and statistical properties of collective spontaneous emission.

Our main purpose here is to give a qualitative description of the angular distribution of super-radiance and to investigate such interesting statistical features of collection spontaneous emission as the angular correlation of photons and the emission of multi-photon beams.

As will be evident further on, the present approach qualitatively accounts for the strong nonlinearity by the evolution of the super-radiant system. The shape of the predicted super-radiance pulse has the well known 'sech' form. This is also the case for Dicke super-radiance from a small volume as well as for super-radiance from a pencil-shaped volume (Fresnel number ~ 1). On the other hand, some dynamical details, such as, for instance, the complex form of the super-radiance pulse from an extended system ('ringing'), which was observed experimentally and successfully described by the theory based on the Maxwell–Bloch equations, are not accounted for by the present theory.

Let us consider the probability of emission of photons with wavevectors $k_1 \ldots k_l$, given by the diagram (3.6.27), with the radiation damping taken into account in the quantum propagators. Since the radiation damping matrix is assumed to be diagonal, the propagators become exponentials with decay rate $\gamma_{k_1...k_l}$, and have to be depicted in a diagram by using bold horizontal lines.

Let us integrate the diagram with respect to the frequencies (but not the directions) of emitted photons. As is clear from the discussion in section 3.4, pairs of photon lines combine into vertical photon lines. The diagram for the probability $W_{k_1...k_l}(t)$ of emission of l photons in the directions $k_1 \ldots k_l$ (per unit solid angle, where, now, $k_i = k_0$) looks like

$$W_{k_1\cdots k_l}(t) = \sum_{P\{k_i\}} \qquad (3.6.36)$$

In order to avoid explicitly calculating the value of the vertical photon line k_i note that after integration over the direction k_i this value should be equal to the collective radiation constant, $\gamma_{k_1...k_{i-1}}$, of the preceding state $\Phi_{k_1...k_{i-1}}$. To make sure that this is true, it suffices to examine the analogy between the self-energy diagram (3.4.2) and the vertical photon line (3.4.17) for $c = b = \Phi_{k_1...k_{i-1}}$, $d = g = \Phi_{k_1...k_i}$, and for a fixed direction (but not frequency) of the photon k_i.

As a result we find that the value of the vertical photon line k_i is given by the integrand in the integral representation (3.6.35) of the collective radiation constant $\gamma_{k_1 \ldots k_{i-1}}$, and has the form

$$\int_{t_i'}^{t_i} k_i = (N - i + 1) \left\{ \kappa(k_i) + \sum_{s=1}^{i-1} \gamma(k_s) \bar{\delta}(\Omega_i - \Omega_s) \right\} \delta(t_i - t_i'). \qquad (3.6.37)$$

The quantity $\kappa(k)$ describes the angular distribution (the probability per second per unit solid angle) of spontaneous emission from an isolated atom

$$\kappa(k) = \frac{k_0 [d \times k]^2}{2\pi\hbar} \qquad \int_{4\pi} d\Omega \kappa(k) = \gamma. \qquad (3.6.38)$$

The angular distribution of collective emission stimulated by the earlier emitted photon k_s is given by

$$\gamma(k_s) \bar{\delta}(\Omega_i - \Omega_s) = \frac{(2\pi)^2 k_0}{\hbar V} [d \times k_s]^2 \tilde{\Delta}(k_i - k_s)$$

$$\int_{4\pi} d\Omega_i \, \bar{\delta}(\Omega_i - \Omega_s) = 1. \qquad (3.6.39)$$

The rate of collective spontaneous emission is proportional to the direction-dependent radiation constant $\gamma(k)$ defined in (3.6.23)–(3.6.26). In the case of an extended system, $V^{1/3} \gg \lambda$, sharp angular correlation arises between the earlier emitted and secondary photons. The spread of the δ-like function of solid angle, $\bar{\delta}(\Omega_i - \Omega_s)$, is related to diffraction.

We are now in a position to derive the master equation for the angular multi-photon probability $W_{k_1 \ldots k_l}(t)$. By differentiating the diagram (3.6.36) with respect to time we obtain

$$\dot{W}_{k_1 \ldots k_l} = -\gamma_{k_1 \ldots k_l} W_{k_1 \ldots k_l}$$

$$+ (N - l + 1) \sum_{i=1}^{l} \left\{ \left(\kappa(k_i) + \sum_{s \neq i}^{l} \gamma(k_s) \bar{\delta}(\Omega_i - \Omega_s) \right) W_{k_1 \ldots k_{i-1} k_{i+1} \ldots k_l} \right\}.$$

$$(3.6.40)$$

The first term describes the collective radiation decay state of the state $\Phi_{k_1 \ldots k_l}$. This contribution is owed to the differentiation of exponential propagators. The second term is responsible for the spontaneous emission as well as for the collective emission initiated by the earlier emitted quanta. Formally, this contribution comes from the substitution of the integrand in equation (3.6.36) at the upper limit $t_l = t$, and is proportional to the value (3.6.37) of the vertical photon line. We can refer the reader interested in details to section 3.4, since the master equation (3.6.40) is derived from diagrams similar to those for the general master equation (3.4.21).

In agreement with (3.6.40)

$$\sum_{l=0}^{N} \dot{W}_l(t) = 0 \qquad (3.6.41)$$

and we can define the normalization condition. Here $W_l(t)$ is the probability that l photons have been emitted in arbitrary directions

$$W_l(t) = \frac{1}{l!} \int_{4\pi} d\Omega_1 \ldots d\Omega_l \, W_{k_1\ldots k_l}(t). \qquad (3.6.42)$$

Let us consider the angular distribution of radiation averaged over the angular photon statistics. Let $n(k, t)$ be the mean number of photons emitted up to the time t in the direction k (per unit solid angle, and $k = k_0$). By using the multi-photon probabilities, define

$$n(k, t) = \frac{1}{\varepsilon} \sum_{l=1}^{N} \sum_{p=1}^{l} \frac{p}{p!(l-p)!} \int_{\varepsilon} \prod_{i=1}^{p} d\Omega_i \int_{4\pi-\varepsilon} \prod_{j=p+1}^{l} d\Omega_j \, W_{k_1\ldots k_l}(t). \quad (3.6.43)$$

where ε is a small solid angle ($\varepsilon \ll 1$) in the direction of k. The sum has to be extended over all possibilities of emitting p photons out of the total number l into the given solid angle ε.

By differentiating the quantity $n(k, t)$ with respect to time and using (3.6.40), we can obtain the rate equation for the mean angular distribution of super-radiance. In the general case this equation is not closed: its right-hand side does not reduce to any function of $n(k, t)$. Nevertheless, in a quasi-classical approximation (i.e. assuming that the number of emitted photons l in (3.6.43) is equal to the mean number of photons $l = n(t)$) we find that

$$\dot{n}(k, t) = (N - n(t))\{\kappa(k) + n(k, t)\gamma(k)\} \qquad (3.6.44)$$

where

$$n(t) = \int_{4\pi} d\Omega \, n(k, t). \qquad (3.6.45)$$

Representing this equation in the form

$$\frac{d}{dt} \ln\{\kappa(k) + n(k, t)\gamma(k)\} = \gamma(k)(N - n(t)) \qquad (3.6.46)$$

and integrating with respect to time, we obtain the angular distribution of photons emitted during the super-radiance pulse

$$n(k, t = \infty) = \frac{\kappa(k)}{\gamma(k)} \{\exp(\gamma(k)N\tau) - 1\}. \qquad (3.6.47)$$

The effective lifetime τ of the super-radiant atomic system is

$$\tau = \frac{1}{N} \int_{t_0}^{\infty} dt \, (N - n(t)). \tag{3.6.48}$$

The angular distribution of super-radiance depends exponentially upon the collective radiation constant $\gamma(k)$, which is proportional to the mean length of the system in the given direction. If the active volume is pencil shaped, then almost all photons are radiated in the direction of maximum length. The dynamics of super-radiance is governed by the collective radiation constant, $\gamma(k)$, for these directions.

Let us discuss super-radiance dynamics in the limiting cases of isotropic and strongly non-isotropic active systems. In the first case we assume the constant $\gamma(k)$ to be independent of direction, $\gamma(k) = \gamma_m$. In the case of a strongly non-isotropic system with a transverse size L_{tr}, we assume that $\gamma(k) = \gamma_m$ for a solid angle Ω_0, where $(\lambda/L_{tr})^2 \ll \Omega_0 \leq 4\pi$, and $\gamma(k) = 0$ for the other directions, neglecting thereby collective emission outside the solid angle Ω_0. It is evident from (3.6.47) that the number of photons spontaneously emitted in the directions for which $\gamma(k)N\tau \to 0$ is relatively small. For this reason we neglect spontaneous emission outside the solid angle Ω_0 as well.

The angular distribution should satisfy the normalization condition $n(t = \infty) = N$. The effective super-radiance lifetime τ can be evaluated by using this condition and the equations (3.6.45) and (3.6.47).

For an isotropic system this readily gives

$$\tau^i = \frac{1}{N\gamma_m} \ln \left(\frac{N\gamma_m}{\gamma} + 1 \right) \tag{3.6.49}$$

and for a strongly non-isotropic system, respectively

$$\tau^n = \frac{1}{N\gamma_m} \ln \left(\frac{N\gamma_m}{\kappa\Omega_0} + 1 \right). \tag{3.6.50}$$

The effective lifetime (3.6.50) is slightly larger, compared with the lifetime (3.6.49). The physical reason is that in the strongly non-isotropic system the collective emission is initiated by only a fraction of spontaneously emitted photons. These initializing photons, emitted into the solid angle Ω_0, appear (on average) later than the photons in arbitrary directions.

In both cases (isotropic and strongly non-isotropic) under consideration the master equation (3.6.40) reduces to a simple master equation for the multi-photon probabilities (3.6.42). In the isotropic case the integration in (3.6.40) with respect to the emission angles gives

$$\dot{W}_l(t) = - (N - 1)(\gamma + l\gamma_m)W_l(t)$$
$$+ (N - l + 1)(\gamma + (l - 1)\gamma_m)W_{l-1}. \tag{3.6.51}$$

If we here formally set $\gamma_m \rightarrow \gamma$, which corresponds to the Dicke problem $(V^{1/3} \ll \lambda)$, we arrive at the decay dynamics for a system of small linear dimensions. The integration in (3.6.40) within only the solid angle Ω_0 (the strongly non-isotropic system) leads to a master equation similar to (3.6.51), where $\gamma \rightarrow \kappa\Omega_0$.

In the case of a greatly elongated pencil-shaped volume, having a characteristic transverse size L_{tr}, the solid angle of collective emission can achieve the diffraction-limited value $\Omega_0 \sim (\lambda/L_{tr})^2$, when $\kappa\Omega_0 \sim \gamma_m$. This case of emission into a single direction, which is formally analogous to Dicke's problem, has been investigated in detail in [BSH71a, BSH71b] (see Chapter 1).

Neglecting the spread of the distribution $W_l(t)$, we transform the difference equation (3.6.51) into a differential equation

$$\frac{\partial}{\partial t} W(x, t) = \gamma \frac{\partial}{\partial x} \left\{ \left[\frac{N\gamma_m}{2\gamma} (1 - x^2) + (1 + x) \right] W(x, t) \right\} \quad (3.6.52)$$

where $W(x, t) = W_l(t)$, $x = 1 - 2l/N$ and $-1 \le x \le 1$. The quantity x is analogous to the projection of classical 'energy spin' in Dicke's problem, which describes the population of the excited states of the atomic system or the relative number of emitted photons.

The solution of (3.6.52) for the initial condition $W(x, 0) = \delta(x-1)$ (excited atoms at $t = 0$) is

$$W(x, t) = \delta(x - x(t)) \quad (3.6.53)$$

$$x(t) = \left\{ \frac{\gamma}{N\gamma_m} - \left(1 + \frac{\gamma}{N\gamma_m} \right) \tanh \left[\tfrac{1}{2}(N\gamma_m + \gamma)(t - t_D) \right] \right\}. \quad (3.6.54)$$

The delay time of the super-radiance pulse, t_D, is evaluated as

$$t_D = \frac{1}{N\gamma_m} \ln \frac{N\gamma_m}{\gamma} \quad (3.6.55)$$

which is almost equal to the effective lifetime of τ of (3.6.49). The emission power $I(t)$ is found in the form

$$I(t) = \tfrac{1}{2}\dot{x}N = \frac{(N\gamma_m + \gamma)^2}{4\gamma_m} \operatorname{sech}^2 \left\{ \tfrac{1}{2}(N\gamma_m + \gamma)(t - t_D) \right\}. \quad (3.6.56)$$

As is evident from the discussion above, the super-radiance from a strongly non-isotropic system is described by equations similar to (3.6.51)–(3.6.56), where we set $\gamma \rightarrow \kappa\Omega_0$.

3.7 Angular correlation of photons in super-radiance

Dicke was the first to indicate that the spatial coherence of atomic states in super-radiance from an extended system leads to the angular correlation of photons.

But the explicit form of the correlation dependence, which is determined by the shape of an active volume and has to be obtained as a result of summing over the positions of the atoms, had not been considered.

Ernst and Stehle, in their paper [ES68] devoted to a generalization of the Wigner–Weisskopf method to include polyatomic systems, proposed that the entire radiation from the totally inverted system is produced in the form of a single beam with diffraction angular dimensions. The averaging over the directions of the beam gives a qualitatively correct distribution of the light flux from a non-isotropic system. Nonetheless, it follows from the master equation (3.6.40) that for extended systems with an angular divergence of light flux which is large compared with the diffraction angle ($\Omega_0 \gg (\lambda/L_{tr})^2$) the probability of emission of all photons in one beam is small.

Let us consider a strongly non-isotropic system. Assume $\gamma(k) = \gamma_m$, $\kappa(k) = \kappa$ for the directions of emission within the solid angle Ω_0 and neglect emission in other directions.

Consider the probability density of emission of photons in the directions $k_1, \ldots, k_N$ ($k_i = k_0$) up to the final instant $t \to \infty$. By explicit integration over time instants $t_1, \ldots, t_N$ in the diagram (3.6.36), where the initial state corresponds to the totally excited atomic system, we find

$$W_{k_1 \ldots k_N} = \sum_{P\{k_i\}} \left\{ \prod_{s=1}^{N} \frac{\kappa + \gamma_m \sum_{i=1}^{s-1} \tilde{\delta}(\Omega_s - \Omega_i)}{\kappa \Omega_0 + (s-1)\gamma_m} \right\}. \qquad (3.7.1)$$

This expression is constructed out of the product of values of vertical photon lines contained in the diagram, divided by the radiation damping constants of the intermediate super-radiant states $\Phi_{k_1 \ldots k_l}$. Estimating, in what follows, the probability of the observation of a multi-photon beam, we shall neglect in (3.7.1) the diffraction variance of δ-like functions of solid angle. The diffraction divergence of a multi-photon beam will be discussed later.

Note that the product in brackets in (3.7.1) becomes symmetrical, and that the permuted terms are equal to each other. The normalization condition is

$$\frac{1}{N!} \int_{\Omega_0} d\Omega_1 \ldots d\Omega_N \, W_{k_1 \ldots k_N} = 1. \qquad (3.7.2)$$

The probability that l photons are emitted into a solid angle $\varepsilon \ll \Omega_0$ is represented in the form

$$W_{l,\varepsilon} = \frac{1}{l!(N-l)!} \int_{\varepsilon} d\Omega_1 \ldots d\Omega_l \int_{\Omega_0 - \varepsilon} d\Omega_{l+1} \ldots d\Omega_N \, W_{k_1 \ldots k_N}. \qquad (3.7.3)$$

The contribution, linear in ε, to the probability $W_{l,\varepsilon}$ is apparently owed to the totally correlated emission of l photons. We use the term l-photon beam for the process of correlated emission of l photons into a small solid angle ε with probability $W_{l,\varepsilon}^{(b)}$ proportional to ε. The linearity in ε is interpreted as a result of statistical averaging over the direction of the beam within the range of the solid angle ε. Looking into the structure of (3.7.1) and (3.7.3), and discarding the higher powers of ε, we find

$$
\begin{aligned}
W_{l.\varepsilon}^{(b)} &= \frac{N!}{l!(N-l)!} \frac{\kappa\gamma_m^{l-1}}{\prod_{s=N-l}^{N-1}(\kappa\Omega_0 + s\gamma_m)} \\
&\quad \times \int_\varepsilon d\Omega_1 \ldots d\Omega_l \prod_{j=2}^{l}\left(\sum_{i=1}^{j-1}\delta(\Omega_j - \Omega_i)\right) \\
&= \frac{N!}{l(N-l)!} \frac{\varepsilon\kappa\gamma_m^{l-1}}{\prod_{s=N-l}^{N-1}(\kappa\Omega_0 + s\gamma_m)}.
\end{aligned}
\tag{3.7.4}
$$

The probability of the emission of an l-photon beam derived above gives the correct value of the average number of photons emitted into the solid angle ε

$$
\sum_{l=1}^{N} l W_{l.\varepsilon}^{(b)} = \frac{\varepsilon}{\Omega_0} N.
\tag{3.7.5}
$$

The integral probability of the emission of the l-photon beam during the super-radiance pulse (into the solid angle Ω_0) is

$$
W_{l.\Omega_0}^{(b)} = \frac{1}{\varepsilon}\Omega_0 W_{l.\varepsilon}^{(b)}.
\tag{3.7.6}
$$

For the average energy contribution of l-photon beams it follows from the relation

$$
(l+1)W_{l+1.\Omega_0}^{(b)} = l W_{l.\Omega_0}^{(b)}\left(1 + \frac{\kappa\Omega_0 - \gamma_m}{(N-1)\gamma_m}\right)^{-1}
\tag{3.7.7}
$$

that at $l \ll N$

$$
l W_{l.\Omega_0}^{(b)} \approx W_{1.\Omega_0}^{(b)} e^{-l/l_0} \qquad W_{1.\Omega_0}^{(b)} = \frac{\kappa\Omega_0}{\gamma_m} \qquad l_0 = \frac{N\gamma_m}{\kappa\Omega_0}.
\tag{3.7.8}
$$

It is seen that the energy contribution of multi-photon beams, with the number of photons of the order of l_0, is significant. The total number of bright beams with more than l photons, where $l > l_0$, is estimated as

$$
\sum_{s>l} W_{s.\Omega_0}^{(b)} \approx \int_l^\infty ds \frac{\kappa\Omega_0}{s\gamma_m} e^{-s/l_0} \leq \frac{\kappa\Omega_0}{\gamma_m}\frac{l_0}{l} e^{-l/l_0}.
\tag{3.7.9}
$$

Since $\kappa\Omega_0/\gamma_m \gg 1$, the observation of intense beams with a number of photons greatly exceeding l_0 is possible in a super-radiance pulse.

Now let us estimate the angular divergence of a multi-photon beam. Since the divergence results from the uncertainty in momenta of the emitted photons, we have to take into account the finite variance of the δ-like functions of solid angle in (3.7.1).

Consider the normalized angular density of the radiated energy $n_l^{(b)}(\Omega)$ in an l-photon beam, initiated by spontaneous emission in the direction of the axis '3' of the coordinate system, where

$$\int d\Omega \, n_l^{(b)}(\Omega) = l. \tag{3.7.10}$$

This angular density satisfies (see [ST74]) the recurrence relation

$$n_{l+1}^{(b)}(\Omega) = n_l^{(b)}(\Omega) + \frac{1}{l} \int d\Omega' \, n_l^{(b)}(\Omega') \tilde{\delta}(\Omega - \Omega'). \tag{3.7.11}$$

This relation is easily understood from a physical point of view. The angular distribution for the $(l + 1)$th photon in the beam, coherently 'stimulated' by all earlier emitted photons (in the sense of the phase memory of the atomic system), is similar to the angular distribution of light flux at the preceding stage of the radiation decay, but suffers an additional diffraction spread (the convolution in (3.7.11)).

A simple analysis shows that the main intensity flux in an l-photon beam propagates within the solid angle $\delta\varepsilon \sim (\lambda L_{\mathrm{tr}}^{-1} \ln l)^2$. This angle determines the characteristic divergence of the beam. With an increasing number of photons in the beam, the diffraction divergence of the beam increases very slowly (as $\ln l$). As a result, in a divergent ($\Omega_0 \gg (\lambda/L_{\mathrm{tr}})^2$) super-radiance pulse emitted from an extended polyatomic system the observation of light spots, related to the brightest multi-photon beams, is possible in principle. The measurement should be carried out at a large distance from the source, or in the focal plane of the lens (in order to resolve the multi-photon beams and to avoid interference of their field patterns).

The grainy transverse structure in the far zone has been observed in pulsed emission from gas lasers working at low pressures. Under optimal conditions, the coherence time of the emission is of the order of the pulse duration and shorter than the reciprocal Doppler broadening [A71]. The measured spatial coherence occurs only within the limits of each grain, whilst the positions of the grains are random. The observed angular dimensions of the grains are of the order of the estimates above.

In order to examine experimentally the theory of the angular structure of super-radiance pulses presented here, the basic criteria of collective spontaneous emission (total initial population inversion, no significant transverse relaxation) should be met in a possible test experiment. The super-radiance pulse should be emitted into a solid angle Ω_0 greatly exceeding the diffraction solid angle $(\lambda/L_{\mathrm{tr}})^2$. Detailed investigation of grainy structure (for example, of the statistics of the grain brightness) might provide useful information.

Chapter 4

Quantum fluctuations and self-organization in super-radiance

This chapter is devoted to the relation between quantum and semiclassical approaches in super-radiance theory. This connection is most easily established by using the Heisenberg picture for the quantum electrodynamical description of the coupled atom–field problem. This approach is complementary to the treatment of the previous chapter, where the dynamics of the system was followed by the master equation for the density matrix in the Schrödinger picture. The main result we present here was obtained by Glauber, Haake and co-workers [GH78, HKSHGH79, HKSHG79] (see also [PSV79]), who showed that the quantum expectation value of the super-radiance intensity, as well as the correlation functions for the radiation field and the polarization, can be evaluated as averages over ensembles of classical trajectories obtained by integrating the c-number Maxwell–Bloch equations, using stochastic initial conditions for the polarization. Following [HKSHG79], we shall give a foundation for this statement and present some statistical characteristics of the super-radiance pulse.

4.1 Quantum Maxwell–Bloch equations, Heisenberg picture

Let us consider N identical two-level atoms distributed uniformly in a certain volume. Each atom sited at a position x_j is described by the quasi-spin operators $\widehat{R}_1^{(j)}$, $\widehat{R}_2^{(j)}$, $\widehat{R}_3^{(j)}$, where j is the number of the atom (see section 1.2). We can define the corresponding operator densities

$$\widehat{\mathcal{R}}_\pm(x) = \sum_j^N (\widehat{R}_1^{(j)} \pm i\widehat{R}_2^{(j)})\delta(x - x_j)$$

$$\widehat{\mathcal{R}}_3(x) = \sum_j^N \widehat{R}_3^{(j)}\delta(x - x_j) \qquad (4.1.1)$$

obeying the commutation relations

$$[\widehat{\mathcal{R}}_3(x), \widehat{\mathcal{R}}_\pm(x')] = \pm\widehat{\mathcal{R}}_\pm(x)\delta(x - x')$$
$$[\widehat{\mathcal{R}}_+(x), \widehat{\mathcal{R}}_-(x')] = 2\widehat{\mathcal{R}}_3(x)\delta(x - x') \qquad (4.1.2)$$

which follow easily from the commutation relations (1.2.9) for $\widehat{R}_\alpha^{(j)}$.

The atomic energy operator is then

$$\widehat{H}_{\text{at}} = \hbar\omega_0 \sum_{j=1}^{N} \widehat{R}_3^{(j)} = \hbar\omega_0 \int \widehat{\mathcal{R}}_3(x)\, d^3x. \qquad (4.1.3)$$

Assuming the matrix element of the atomic dipole moment, d, to be real, we can express the polarization operator as

$$\widehat{P}(x) = d(\widehat{\mathcal{R}}_+(x) + \widehat{\mathcal{R}}_-(x)). \qquad (4.1.4)$$

To describe the radiation field we shall use the operator of the vector potential, $\widehat{A}$, in the Coulomb gauge [29, 50]. With the notations of (3.1.5)

$$\widehat{A} = \sum_{k,\lambda} \epsilon_{k\lambda} \sqrt{\frac{2\pi\hbar c}{kV_Q}}\, (\widehat{a}_{k\lambda}^\dagger e^{-ikx} + \widehat{a}_{k\lambda} e^{ikx}). \qquad (4.1.5)$$

As is well known, the Hamiltonian for a system of bound charges and a radiation field may be written in the so-called *minimal coupling form*, i.e. the 'p–A' form

$$\widehat{\mathcal{H}} = \sum_j \frac{1}{2m_j} \left(\widehat{p}_j - \frac{e_j}{c}\widehat{A}(x_j)\right)^2 + \widehat{\mathcal{U}} + \sum_{k,\lambda} \hbar k c(\widehat{a}_{k\lambda}^\dagger \widehat{a}_{k\lambda}) \qquad (4.1.6)$$

where $\widehat{\mathcal{U}}$ is the potential energy operator of the Coulomb interaction between the charges e_j.

Power and Zienau [57] (see also [12] and [M76]) showed that the unitary operator

$$\widehat{S} = \exp\left(-\frac{i}{\hbar c} \int \widehat{P}(x)\widehat{A}(x)\, d^3x\right) \qquad (4.1.7)$$

transforms the Hamiltonian (4.1.6), in the dipole approximation, into the 'd–$\mathcal{E}$' form

$$\widehat{H} = \widehat{S}\widehat{\mathcal{H}}\widehat{S}^{-1}$$
$$= \widehat{H}_0 - \int \widehat{P}\widehat{\mathcal{E}}\, d^3x + 2\pi \int |\widehat{P}(x)|^2\, d^3x. \qquad (4.1.8)$$

Here $\widehat{H}_0$ is the unperturbed Hamiltonian of the atoms and the radiation field

$$\widehat{H}_0 = \hbar\omega_0 \int \widehat{\mathcal{R}}_3(x)\, \mathrm{d}^3 x + \sum_{k,\lambda} \hbar k c (\hat{a}_{k\lambda}^\dagger \hat{a}_{k\lambda} + \tfrac{1}{2}) \qquad (4.1.9)$$

and $\widehat{\mathcal{E}}$ is the operator of the transversal part of the free electric field

$$\widehat{\mathcal{E}} = \frac{\mathrm{i}}{c\hbar}[\widehat{A}, \widehat{H}_0] = \sum_{k\lambda}(\widehat{\mathcal{E}}_{k\lambda}^- + \widehat{\mathcal{E}}_{k\lambda}^+) \qquad (4.1.10)$$

where the square brackets $[\,,]$ denote the commutator of the operators and $\widehat{\mathcal{E}}_{k\lambda}^-$ and $\widehat{\mathcal{E}}_{k\lambda}^+$ are operators for the negative and positive frequency parts of the electric field of the k, λ mode

$$\widehat{\mathcal{E}}_{k\lambda}^- = -\mathrm{i}\epsilon_{k\lambda}\sqrt{\frac{2\pi\hbar k c}{V_Q}}\,\hat{a}_{k\lambda}^\dagger e^{-\mathrm{i}kx} \qquad \widehat{\mathcal{E}}_{k\lambda}^+ = (\widehat{\mathcal{E}}_{k\lambda}^-)^\dagger.$$

We see that the Hamiltonian (4.1.8), with the last self-polarization term omitted, coincides with the Hamiltonian introduced in section 3.2.

 First we shall obtain the inhomogeneous wave equation for the electric field $\widehat{\mathcal{E}}$. To do that we shall apply the commutator of the Hamiltonian (4.1.8) with the negative frequency part of a single $k\lambda$ mode, $\widehat{\mathcal{E}}_{k\lambda}^-$. Using the commutation relations (4.1.2) for the operators $\widehat{R}(x)$ and (3.1.6) for the operators $a_{k\lambda}^\dagger$, $a_{k\lambda}$, we obtain

$$\frac{\partial \widehat{\mathcal{E}}_{k\lambda}^-}{\partial t} = -\frac{\mathrm{i}}{\hbar}[\widehat{\mathcal{E}}_{k.\lambda}^-, \widehat{H}]$$

$$= \mathrm{i}kc\,\widehat{\mathcal{E}}_{k.\lambda}^- - \epsilon_{k\lambda}(d\epsilon_{k\lambda})\frac{2\pi k c}{V_Q}$$

$$\times \int (\widehat{R}_+(x') + \widehat{R}_-(x'))e^{\mathrm{i}k(x'-x)}\,\mathrm{d}^3 x' \qquad (4.1.11)$$

$$\frac{\partial^2 \widehat{\mathcal{E}}_{k.\lambda}^-}{\partial t^2} = -\frac{\mathrm{i}}{\hbar}[\partial\widehat{\mathcal{E}}_{k.\lambda}^-/\partial t, \widehat{H}]$$

$$= -k^2 c^2 \widehat{\mathcal{E}}_{k.\lambda}^- + \frac{2\pi}{V_Q}k^2 c^2 \epsilon_{k.\lambda}(d\epsilon_{k.\lambda})$$

$$\times \int (\widehat{R}_-(x') + \widehat{R}_+(x'))e^{\mathrm{i}k(x'-x)}\,\mathrm{d}^3 x'$$

$$+ \frac{2\pi}{V_Q}\omega_0 k c \int \epsilon_{k.\lambda}(d\epsilon_{k\lambda})(\widehat{R}_-(x') - \widehat{R}_+(x'))e^{\mathrm{i}k(x'-x)}\,\mathrm{d}^3 x'.$$

$$(4.1.12)$$

By Hermitian conjugation of (4.1.12) we obtain the second derivative of the positive frequency part of this field mode. Then summing over k, λ, the third term in (4.1.12) being cancelled, we obtain

$$\frac{\partial^2 \widehat{\mathcal{E}}}{\partial t^2} - c^2 \nabla^2 \widehat{\mathcal{E}} = -4\pi c^2 \nabla^2 \widehat{\mathcal{P}}_\perp \tag{4.1.13}$$

where $\widehat{\mathcal{P}}_\perp$ is the operator for the transversal part of the electric polarization density, which is determined as

$$\widehat{\mathcal{P}}_\perp = \frac{1}{V_\varrho} \sum_{k.\lambda} \epsilon_{k.\lambda} (\epsilon_{k.\lambda} d) \int (\widehat{\mathcal{R}}_+(x') + \widehat{\mathcal{R}}_-(x')) e^{ik(x'-x)} \, d^3x'. \tag{4.1.14}$$

Let us now introduce the operator of the transversal electric field in the medium, which corresponds to the total Hamiltonian (4.1.8)

$$\widehat{\mathcal{E}}' = \frac{i}{\hbar c} [\widehat{A}, \widehat{H}]. \tag{4.1.15}$$

Having calculated the commutator, we obtain

$$\widehat{\mathcal{E}}' = \widehat{\mathcal{E}} - 4\pi \widehat{\mathcal{P}}_\perp \tag{4.1.16}$$

so that $\widehat{\mathcal{E}}$, already used in the previous chapter (cf (3.1.5)), is the operator for the transversal part of the displacement vector. It can be shown [57] that

$$\widehat{\mathcal{E}}' = \widehat{S} \widehat{\mathcal{E}} \widehat{S}^{-1}. \tag{4.1.17}$$

With the help of (4.1.16), equation (4.1.13) can be rewritten in the form

$$\nabla^2 \widehat{\mathcal{E}}' - \frac{1}{c^2} \frac{\partial^2}{\partial t^2} \widehat{\mathcal{E}} = \frac{4\pi}{c^2} \frac{\partial^2}{\partial t^2} \widehat{\mathcal{P}}_\perp \tag{4.1.18}$$

which is the same as in classical theory (cf equation (1.4.6)).

In order to obtain equations of motion for the operators $\widehat{\mathcal{R}}_\pm$ and $\widehat{\mathcal{R}}_3$, we need to form the commutators of these operators with the Hamiltonian (4.1.8)

$$\begin{aligned}
\frac{\partial \widehat{\mathcal{R}}_\pm}{\partial t} &= -\frac{i}{\hbar} [\widehat{\mathcal{R}}_\pm, \widehat{H}] \\
&= \pm i\omega_0 \widehat{\mathcal{R}}_\pm \pm 2\frac{i}{\hbar} d\widehat{\mathcal{E}} \widehat{\mathcal{R}}_3 \\
&\quad \mp \frac{4\pi i d^2}{\hbar} \{(\widehat{\mathcal{R}}_+ + \widehat{\mathcal{R}}_-)\widehat{\mathcal{R}}_3 + \widehat{\mathcal{R}}_3(\widehat{\mathcal{R}}_+ + \widehat{\mathcal{R}}_-)\} \\
&= \pm i\omega_0 \widehat{\mathcal{R}}_\pm \pm \frac{i}{\hbar} d(\widehat{\mathcal{E}} \widehat{\mathcal{R}}_3 + \widehat{\mathcal{R}}_3 \widehat{\mathcal{E}})
\end{aligned} \tag{4.1.19}$$

$$\frac{\partial \widehat{\mathcal{R}}_3}{\partial t} = -\frac{i}{\hbar} [\widehat{\mathcal{R}}_3, \widehat{H}]$$

$$= \frac{i}{\hbar} d\widehat{\mathcal{E}}(\widehat{\mathcal{R}}_+ - \widehat{\mathcal{R}}_-)$$

$$- \frac{i}{\hbar} 2\pi d^2 \{(\widehat{\mathcal{R}}_+ + \widehat{\mathcal{R}}_-)(\widehat{\mathcal{R}}_+ - \widehat{\mathcal{R}}_-) + (\widehat{\mathcal{R}}_+ - \widehat{\mathcal{R}}_-)(\widehat{\mathcal{R}}_+ + \widehat{\mathcal{R}}_-)\}$$

$$= \frac{i}{2\hbar} d\{\widehat{\mathcal{E}}(\widehat{\mathcal{R}}_+ - \widehat{\mathcal{R}}_-) + (\widehat{\mathcal{R}}_+ - \widehat{\mathcal{R}}_-)\widehat{\mathcal{E}}\}. \qquad (4.1.20)$$

These equations have the same form as the semiclassical equations (1.4.11) if the operators $\mathcal{E}$ and $\mathcal{R}$ are interpreted as c numbers. For two-level atoms, equations (4.1.19) and (4.1.20) can be reduced to a more simple form [M76]. Indeed, by virtue of the identities

$$(\widehat{\mathcal{R}}_+ + \widehat{\mathcal{R}}_-)(\widehat{\mathcal{R}}_+ - \widehat{\mathcal{R}}_-) + (\widehat{\mathcal{R}}_+ - \widehat{\mathcal{R}}_-)(\widehat{\mathcal{R}}_+ + \widehat{\mathcal{R}}_-) = 0$$

$$(\widehat{\mathcal{R}}_+ + \widehat{\mathcal{R}}_-)\widehat{\mathcal{R}}_3 + \widehat{\mathcal{R}}_3(\widehat{\mathcal{R}}_+ + \widehat{\mathcal{R}}_-) = 0 \qquad (4.1.21)$$

we obtain

$$\frac{\partial \widehat{\mathcal{R}}_\pm}{\partial t} = \pm i\omega_0 \widehat{\mathcal{R}}_\pm - \frac{2d}{\hbar}\widehat{\mathcal{E}}\widehat{\mathcal{R}}_3$$

$$\frac{\partial \widehat{\mathcal{R}}_3}{\partial t} = \frac{i}{\hbar} d\widehat{\mathcal{E}}(\widehat{\mathcal{R}}_+ - \widehat{\mathcal{R}}_-). \qquad (4.1.22)$$

These equations were derived in [HKSHG79], using the reduced form (4.1.8) of the Hamiltonian without the polarization energy term.

4.2   The initial conditions for the quantum Maxwell–Bloch equations

Further on in this chapter we shall consider the one-dimensional problem by treating propagation of radiation as being along only one axis, and neglecting the transversal variation of the fields. The two- and three-dimensional problems will be considered in Chapter 6.

We shall assume that the atoms are confined in a pencil-shaped volume of length L, $L \gg \lambda$, and that the concentration of atoms, N_0, is large enough for there to be many atoms in an element of length λ of the volume, $N_0 \mathcal{D}^2 \lambda \gg 1$, where $\mathcal{D}$ is the diameter of the sample.

It is useful to describe such a system in terms of *smoothly varying* fields, by averaging the microscopic fields over a transversal section with volume $\Delta V = \mathcal{D}^2 \Delta x$ and thickness $\Delta x \ll \lambda$

$$\hat{\rho}_\alpha(x) = \frac{1}{\Delta V N_0} \sum_{x_i \in \Delta V} \widehat{R}_\alpha^{(i)} \qquad (4.2.1)$$

where the index α is '$\pm$' or 3. Here we choose the x axis to be that of the sample, and we shall assume that the directions of the polarization and the electric field are the same and perpendicular to it (omitting the $\perp$ subscript and the vector notation). The equation of motion for the smoothed field can be constructed by carrying out the averaging operation $(\Delta V)^{-1} \int_{\Delta V} d^3x \ldots$ in equations (4.1.13) and (4.1.22) and replacing the averages of the products of operators by products of the averages of the individual operators.

We now need to complete the Heisenberg equations by adding the initial quantum state of the system 'atoms plus field'. For the super-radiance problem, the initial quantum state for the field is the vacuum state $|\{0\}\rangle$ in the 'p–A' representation, and consequently $\widehat{S}|\{0\}\rangle$ is in the 'd–$\mathcal{E}$' representation. The initial state of the atoms is the fully excited state, $|\{e\}\rangle$, which also has to be determined in the 'd–$\mathcal{E}$' representation (for a discussion, see [44]), hence the initial state for super-radiance in the 'd–$\mathcal{E}$' representation is

$$|i\rangle = |\{e\}\rangle\widehat{S}|\{0\}\rangle. \tag{4.2.2}$$

Then with the help of (4.1.7) the second-order correlation function for the electric field at the initial instant $t = 0$ is

$$\langle i|\widehat{\mathcal{E}}'^+(x,0)\,\widehat{\mathcal{E}}'^-(x',0)|i\rangle = \langle\{0\}|S^\dagger\widehat{S}\,\widehat{\mathcal{E}}'^+(x,0)\widehat{S}^\dagger\widehat{S}\widehat{\mathcal{E}}'^-(x',0)\widehat{S}^\dagger\widehat{S}|\{0\}\rangle$$

$$= \langle\{0\}|\widehat{\mathcal{E}}^+(x,0)\widehat{\mathcal{E}}^-(x,0)|\{0\}\rangle = 0. \tag{4.2.3}$$

In order to calculate the correlation function for the initial polarization we use the smoothed polarization operators (4.2.1)

$$\langle\{e\}|\hat{\rho}_+(x',0), \hat{\rho}_-(x',0)|\{e\}\rangle = \frac{1}{(\Delta V N_0)^2}\left\langle \sum_{i\in\Delta x} \widehat{R}_+^{(i)} \sum_{j\in\Delta x'} \widehat{R}_-^{(j)} \right\rangle. \tag{4.2.4}$$

Then using the well known expectation value for individual spin-$\frac{1}{2}$ operators

$$\langle e|\widehat{R}_+^{(i)}\,\widehat{R}_-^{(j)}|e\rangle = \delta_{ij} \tag{4.2.5}$$

we obtain

$$\langle\{e\}|\hat{\rho}_+(x,0)\hat{\rho}_-(x',0)|\{e\}\rangle = \frac{\delta_{xx'}}{N_0\Delta V} = \frac{L\delta_{xx'}}{N\Delta x} \tag{4.2.6}$$

where, we recall, L is the length of the sample, and N is the total number of atoms in a system. In a macroscopic description, $\delta_{xx'}/\Delta x$ can be replaced by $\delta(x - x')$. This result for the pair correlation function has been generalized in [HKSHG79] for the multi-point correlation function, as follows

$$\langle\{e\}|\hat{\rho}_+(x_1,0)\ldots\hat{\rho}_+(x_j,0)\hat{\rho}_-(x'_1,0)\ldots\hat{\rho}_-(x'_j,0)|\{e\}\rangle$$

$$= \sum_p \langle\hat{\rho}_+(x_1,0)\hat{\rho}_-(x'_{p_1},0)\rangle \cdots \langle\hat{\rho}_+(x_j,0)\hat{\rho}_-(x'_{p_j},0)\rangle$$

$$= \left(\frac{L}{N}\right)^j \sum_p \delta(x_1 - x'_{p_1})\ldots\delta(x_j - x'_{p_j}) \tag{4.2.7}$$

where the sums run over all $j!$ permutations of primed coordinates. This correlation function for the quantum field is formally identical with a correlation function of a classical Gaussian stochastic field with complex amplitude $\rho(x, 0)$

$$\langle \rho^*(x_1, 0) \cdots \rho^*(x_j, 0)\rho(x'_1, 0) \cdots \rho(x'_j, 0)\rangle$$

$$= \left(\frac{L}{N}\right)^j \sum_p \delta(x_1 - x'_{p_1}) \ldots \delta(x_j - x'_{p_j}). \qquad (4.2.8)$$

Thus we can calculate initial expectation values of operator products, normally ordered in the sense indicated in (4.2.7), as moments of the distribution

$$W(\rho_\pm(x, 0)) \sim \prod_i \exp\left\{-\frac{N\Delta x}{L} |\rho_\pm(x_i, 0)|^2\right\} \qquad (4.2.9)$$

for corresponding classical fields.

The randomness of initial polarization seen in (4.2.9) is the consequence of quantum mechanical uncertainty. Since operators for population inversion $\hat{\rho}_3$ and polarization $\hat{\rho}_\pm$ do not commute with each other, $\hat{\rho}_\pm$ must have non-vanishing dispersion in the atomic initial state if the latter is an eigenstate of $\hat{\rho}_3$.

4.3   Averaging over semiclassical solutions

The statistical behaviour of the system under consideration can be represented in terms of classical stochastic fields not only initially, but also at all later times. Following [HKSHG79] we shall show this for the early (linear) stage of the radiation process, where depletion of the atomic excitation can still be neglected. We shall take the field operators $\widehat{\mathcal{E}}$ and $\hat{\rho}_\pm$ in the form of a plane wave propagating in the positive direction

$$\widehat{\mathcal{E}}^\pm(x, t) = \frac{\hbar}{2d T_R} e^{\pm i(kx - \omega t)} \widehat{\varepsilon}^\pm(x, t)$$

$$\hat{\rho}_\pm(x, t) = \tfrac{1}{2} e^{\mp i(kx - \omega t)} \widehat{R}_\pm(x, t) \qquad (4.3.1)$$

where the amplitudes $\hat{\varepsilon}^\pm$, $\widehat{R}_\pm$ are dimensionless, and

$$T_R = \left(\frac{3}{8\pi} \gamma\lambda^2 N_0 L\right)^{-1}.$$

Inserting (4.3.1) into (4.1.13) and (4.1.22) and using the SVEA and RWA we obtain the system of reduced Maxwell–Bloch equations

$$\frac{\partial}{\partial t}\widehat{R}_\pm = \frac{\widehat{Z}\hat{\varepsilon}^\mp}{T_R}$$

$$\frac{\partial \widehat{Z}}{\partial t} = -\frac{1}{2T_R}(\widehat{R}_+\widehat{\varepsilon}^+ + \widehat{R}_-\widehat{\varepsilon}^-) \tag{4.3.2}$$

$$\left(\frac{\partial}{\partial x} + \frac{1}{c}\frac{\partial}{\partial t}\right)\widehat{\varepsilon}^{\pm} = \frac{1}{L}\widehat{R}_{\mp}$$

where

$$\widehat{Z} = 2\widehat{R}_3.$$

For the linear stage of the process we may take $\widehat{Z} = 1$, to obtain a system of linear equations formally similar to those of the semiclassical theory

$$\frac{\partial}{\partial t}\widehat{R}_{\pm} = \frac{1}{T_R}\widehat{\varepsilon}^{\mp}$$

$$\left(\frac{\partial}{\partial x} + \frac{1}{c}\frac{\partial}{\partial t}\right)\widehat{\varepsilon}^{\pm} = \frac{1}{L}\widehat{R}_{\mp}. \tag{4.3.3}$$

The solution of (4.3.3) can be obtained in the form of a linear functional of the initial field operators. In particular, for the electric field $\varepsilon(x, t)$ the result is

$$\widehat{\varepsilon}(x, t) = \frac{1}{L}\int_0^x dx' \, \widehat{R}_{\pm}(x - x', 0)\theta(t - x'/c)I_0\{2\sqrt{x'(t - x'/c)/LT_R}\}$$

$$+ \frac{1}{L}\int_0^x dx'\widehat{\varepsilon}^{\mp}(x - x', 0)\theta(t - x'/c)\left(\frac{T_R x'}{L(t - x'/c)}\right)^{1/2}$$

$$\times I_1\{2\sqrt{x'(t - x'/c)/LT_R}\} + \theta(x - ct)\widehat{\varepsilon}^{\mp}(x - ct, 0) \tag{4.3.4}$$

where θ is the step function and I_n is the modified Bessel function of order n. The same result holds for the c-number fields (for details of the derivation of (4.3.4), see Chapter 5).

We define the intensity of the radiation as the expectation value of the number of photons per active atom passing through the cross-section of the sample per unit time. At the end of the active volume, $x = L$, it is

$$\langle I(t)\rangle = \frac{1}{4T_R}\langle\{0\}\{e\}|\widehat{\varepsilon}^-(L, t)\widehat{\varepsilon}^+(L, t)|\{0\}\{e\}\rangle. \tag{4.3.5}$$

Inserting (4.3.4) into (4.3.5) and using (4.2.3) and (4.2.5) for the initial correlation functions, we obtain

$$\langle I(t)\rangle = \frac{1}{NT_R L}\int_0^L dx'\theta(t - x'/c)\{I_0(2\sqrt{x'(t - x'/c)/LT_R})\}^2. \tag{4.3.6}$$

The radiated intensity as given by (4.3.6) can be obtained by interpreting the linear Maxwell–Bloch equations as classical ones for the c-number fields, and by replacing the quantum mechanical initial-state average with an average over initial configurations of the fields according to the weight functional

$$W(\varepsilon, R, Z) \sim \prod_x \delta^2[\varepsilon(x)]\delta[Z(x) - 1]\exp\{-\tfrac{1}{4}N\Delta x|R(x)|^2\}. \quad (4.3.7)$$

This result was obtained in [GH78] and then generalized in [HKSHG79] to the normally ordered higher correlation functions as well as to nonlinear stages of the process. The normal order for electric field operators means that the positive frequency operators are put to the right of the negative ones, as in (4.3.5). Thus, the main result [HKSHG79] can be formulated as the following (*Haake–Glauber*) theorem:

> All normally ordered correlation functions of the electric field and of the polarization can be calculated as if the radiation process were a classical stochastic process with deterministic dynamics, but with random initial values of the polarization obeying the distribution (4.3.7).

Strictly speaking, the semiclassical solution can be regarded only as an ingredient of a convenient scheme for calculating quantum mechanical expectation values. Indeed, at the earliest stages, when the electric field and the polarization have not yet risen above the level of quantum fluctuations, the radiative process is not a classical one and would be greatly disturbed by measurements. However, as soon as the early stage amplification has brought the field intensity up to a macroscopic level, each of the calculated trajectories corresponds to a possible experimental pulse. Thus a super-radiance pulse can be considered as an amplification of the initial quantum fluctuations, and therefore displays some stochastic features, such as fluctuations of the delay time and the intensity.

4.4 Statistical characteristics of super-radiance

Here we shall first present some statistical characteristics of the linear stage of super-radiance.

The mean intensity obtained in the linear approximation is given by (4.3.6). Further, since the field is a linear functional of a Gaussian random complex quantity (the initial polarization), it obeys the distribution

$$dW(\varepsilon(x, t)) = \frac{1}{\pi \overline{|\varepsilon(x, t)|^2}}\exp\left(-\frac{|\varepsilon(x, t)|^2}{\overline{|\varepsilon(x, t)|^2}}\right)d^2\varepsilon. \quad (4.4.1)$$

Whence one has for the intensity $I(x, t) \sim |\varepsilon(x, t)|^2$ the exponential distribution

$$dW(I(x, t)) = \frac{1}{\overline{I}}\exp\left(-\frac{I(x, t)}{\overline{I}(x, t)}\right)dI \quad (4.4.2)$$

and therefore, in the linear stage, the variance of the intensity is equal to the square of the mean intensity

$$\overline{(I - \bar{I})^2} = (\bar{I})^2. \tag{4.4.3}$$

A correlation of dipole moments of atoms located at different points of a sample already appears in the linear stage of super-radiance. To see this we shall present the polarization operators $\widehat{R}_\pm(x, t)$ as a linear functional of the initial polarization. The solution of (4.3.3) is

$$\widehat{R}_\pm(x, t) = \widehat{R}_\pm(x, 0)$$
$$+ \int_0^x dx' \, \widehat{R}_\pm(x - x', 0)\theta(t - x'/c) \left(\frac{L(t - x'/c)}{T_R x'} \right)^{1/2}$$
$$\times I_1\{2\sqrt{x'(t - x'/c)/LT_R}\} \tag{4.4.4}$$

where we have omitted the terms containing the initial electric field operators. Using (4.4.4) we can readily calculate the correlation function $f(x_1, x_2, t)$ within the linear approximation

$$f(x_1, x_2, t) = \langle\{e\}|[\widehat{R}_+(x_1, t) - \widehat{R}_+(x_1, 0)][\widehat{R}_-(x_2, t) - \widehat{R}_-(x_2, 0)]|\{e\}\rangle$$
$$= \frac{4}{NL} \int_0^{x_2} dx' \, \theta(t - (x_1 - x_2 + x')/c)$$
$$\times \frac{L}{T_R} \left(\frac{t - (x_1 - x_2 + x')/c}{x_1 - x_2 + x'} \right)^{1/2} \left(\frac{t - x'/c}{x'} \right)^{1/2}$$
$$\times I_1\{2[(x_1 - x_2 + x')(t - (x_1 - x_2 + x')/c)/LT_R]^{1/2}\}$$
$$\times I_1\{2[x'(t - x'/c)/LT_R]^{1/2}\}. \tag{4.4.5}$$

Owing to the step function in the integrand, the value of $f(x_1, x_2, t)$ is different from zero only if $t > (x_1 - x_2)/c$. This result is in accordance with the causality principle.

Figure 4.1 shows the normalized correlation function for polarization

$$\tilde{f}(x_1, x_2, t) = \frac{f(x_1, x_2, t)}{[f(x_1, x_1, t)f(x_2, x_2, t)]^{1/2}} \tag{4.4.6}$$

at two points of the sample, $x_1 = L$ and $x_2 = L/2$, as a function of time. It is seen that during the passage of light through the sample almost complete correlation is achieved, which accounts for the super-radiance effect: the creation of an in-phase state of atomic dipoles.

We can similarly find the second-order correlation function for the field at the same spatial point but at different times

$$\tilde{g}(x_1, t_1, t_2) = \frac{\langle\{0\}|\widehat{\varepsilon}^-(x, t_1)\widehat{\varepsilon}^+(x, t_2)|\{0\}\rangle}{\langle\{0\}|\widehat{\varepsilon}^-(x, t_1)\widehat{\varepsilon}^+(x, t_1)|\{0\}\rangle^{1/2}\langle\{0\}|\widehat{\varepsilon}^-(x, t_2)\widehat{\varepsilon}^+(x, t_2)|\{0\}\rangle^{1/2}}$$
$$\tag{4.4.7}$$

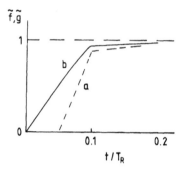

Figure 4.1. Normalized second-order correlation functions for the polarization and the field at the linear stage of super-radiance, $L/cT_R = 0.1$: (a) $\bar{f}(L, \frac{1}{2}L, t)$; (b) $\bar{g}(L, t, t + T_R)$.

where

$$\langle\{0\}|\widehat{\varepsilon}^-(x, t_1)\widehat{\varepsilon}(x, t_2)|\{0\}\rangle$$
$$= \frac{1}{L}\int_0^x dx'\,\theta(t - x'/c)I_0\{2[x'(t_1 - x'/c)/LT_R]^{1/2}\}$$
$$\times I_0\{2[x'(t_2 - x'/c)/LT_R]^{1/2}\}. \tag{4.4.8}$$

A plot of this correlation function for the spatial point $x = L$ and the time difference $t_2 - t_1 = T_R$ as a function of time t_1 is shown in figure 4.1. It is seen that second-order coherence rapidly increases during the passage of light through the sample.

The method described in section 4.3 for calculating the normally ordered correlation function may be applied to determine the statistical characteristics of the nonlinear stage of super-radiance. This requires a numerical solution of the complete system of the c-number Maxwell–Bloch equations corresponding to (4.3.2) for an ensemble with random initial polarization. Each realization of the initial conditions is associated with a classical super-radiant pulse, and the expectation value of the normally ordered product of operators is obtained by averaging the corresponding product of classical magnitudes over the pulse ensemble obtained. The mean super-radiance intensity and its variance have been obtained in [HKSHG79] by this method for relatively short samples ($L/cT_R = 0.3$) (see figure 4.2). In contrast to a single stochastic realization of a super-radiant pulse, the minima of the mean intensity are not zero, since the locations of minima in different stochastic realizations do not coincide. The root mean square (RMS) deviation also exhibits a structure having maxima near times at which the mean intensity changes most rapidly, and with minima near times at which the intensity is more stationary. In the early stage, as we mentioned above, the RMS deviation is equal to the mean intensity.

In order to perform the numerical calculation, the sample length was divided

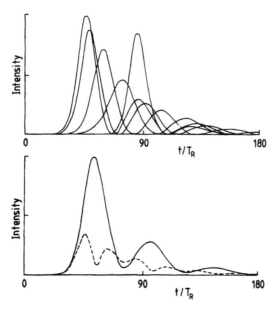

Figure 4.2. Intensity plots, for $L/cT_R = 0.3$, $N = 4 \times 10^9$, obtained by using a random-number realization of the initial polarization. Upper panel: typical realizations; lower panel: mean intensity (continuous curve) and RMS deviation of the intensity (broken curve) [HKSHG79].

into several sub-intervals of length Δx, and the initial polarization was chosen independently in each sub-interval according to the probability distribution (4.2.9).

The numerical calculation of these statistical characteristics shows that the results depend only slightly upon the size of interval Δx provided that time intervals are larger than L/c. Recall that in each interval Δx the initial polarization $R(x)$ is chosen independently according to the probability distribution (4.3.7) with variance $L/N\Delta x$. In particular, if we choose $\Delta x = L$, i.e. equal to the total length of the super-radiant sample, with the initial polarization spatially homogeneous of the order $N^{-1/2}$ this gives a qualitatively good representation of a super-radiant pulse, which is what we have used in Chapter 1.

Below we present the statistical characteristics of super-radiance obtained in [PT88] for a sample with $L/cT_R = 0.1$, $N = 10^6$. By numerically solving the system of c-number Maxwell–Bloch equations, an ensemble consisting of 100 pulses was obtained, over which averaging was performed. Using this pulse ensemble, the probability distribution for the intensity at different times was determined (see figure 4.3). At a time still corresponding to the linear stage, the intensity distribution is exponential, in agreement with (4.4.2). Another characteristic of the distribution is exhibited in the region of the first peak of

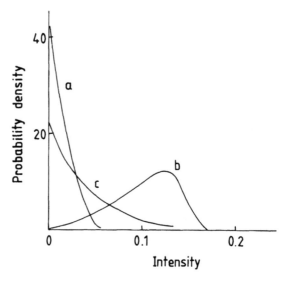

Figure 4.3. Probability distributions of the intensity, for $L/cT_R = 0.1$, $N = 10^6$, at different times (a) $t = 20T_R$; (b) $t = 40T_R$; (c) $t = 70T_R$.

the intensity, when the most probable value of the intensity is not zero. At the times near the minima of the intensity the distribution again has an exponential form.

Figure 4.4 shows the second-order correlation function (4.4.7) for the field in the nonlinear stage. It is seen that for the time difference $t_2 - t_1 = T_R$ there is a sizable region of coherence, where $g \simeq 1$. The minima of the correlation functions correspond to intensity minima. This is related to the behaviour of the phase of the field of the individual realization of the pulses in the ensemble: in the vicinity of an intensity minimum the phase of the field changes rapidly, whereas in the region of a maximum it remains unchanged. Therefore averaging over phases near a minimum causes a decrease of the magnitude of the correlation function.

Since the ensemble of classical trajectories generated by stochastic initial polarization constitutes an ensemble of experimentally observable pulses, we can draw not only the mean values, but also histograms for distributions of the directly measured macroscopic characteristic of super-radiance pulses; for example, for the distribution of the first intensity maximum I_{max}, and for the delay time at which the maximum occurs. Such histograms obtained in [HKSHG79, HHKSG81] are shown in figures 4.5 and 4.6.

The statistical analysis presented above is somewhat limited in that it has not included such important factors as de-phasing of the polarization, inhomogeneous broadening, diffraction and some others, which are usually present in real experiments. In addition to this, some experimental parameters,

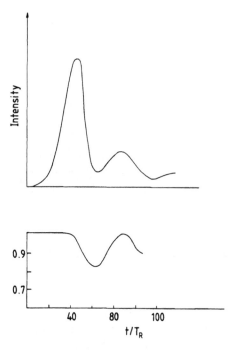

Figure 4.4. Mean intensity and second-order correlation function for the field at the nonlinear stage of super-radiance for $L/cT_R = 0.1$, $N = 10^6$ [HKSHG79].

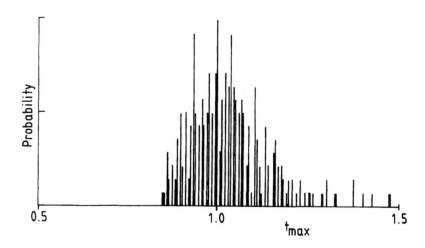

Figure 4.5. Distribution of delay times of the first intensity maxima, for $L/cT_R = 0.3$ $N = 4 \times 10^9$. The time is normalized to the delay time of the mean intensity [HKSHG79].

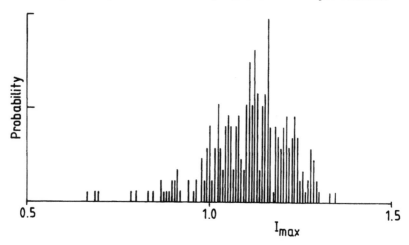

Figure 4.6. Distribution of heights of the first intensity maxima, for $L/cT_R = 0.3$, $N = 4 \times 10^9$. The height is normalized to the height of the mean intensity [HKSHG79].

such as the concentration of initially excited atoms, dynamics and spatial inhomogeneity of pumping, are not under sufficient control to permit an accurate comparison of theory with experiment. Nevertheless, the theoretical approach developed here incorporates the main statistical features of super-radiance which are found to be in good qualitative agreement with experiment.

4.5 Self-organization in super-radiance

As we have seen, super-radiance may be treated as an amplification of the field generated by quantum fluctuations of the polarization. Initially tiny, it grows to a macroscopic size and exhibits the statistics of super-radiance pulses. The parameters characterizing the shape of the pulse vary strongly enough within the ensemble for their RMS deviations are of the same order of magnitude as their mean values (as can be seen from figures 4.5 and 4.6).

Another aspect of super-radiance evolution is a build up of coherence both of the radiation field and the atomic polarization. In the classical simulation of super-radiance, used above, this process can be treated as self-organization in a system of atomic dipoles. Such computer simulations have been performed for relatively short ($L = 2cT_R$) as well as for much longer ($L = 200cT_R$) samples [KT90, K91].

In figure 4.7 phase distributions for the polarization over the sample length are presented at some consecutive instants of time for the relatively short system ($L = 2cT_R$). At an early stage coherent domains appear and they grow and join up as time passes. The time for field phasing to take place is of the order of the transient time, L/c. We note that the evolution must preserve the chaotic character of the polarization at the microscopic scale as suggested by the first term in (4.4.4). This is also confirmed by numerical calculation.

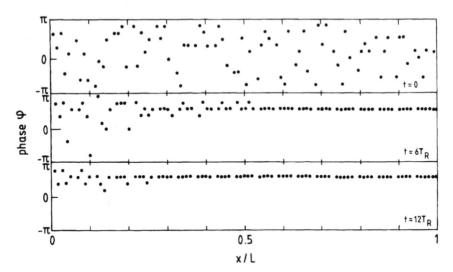

Figure 4.7. Phase distribution for the polarization along the length of a short sample at consecutive instants of time, $t = 0$, $t = 6T_R$, $t = 12T_R$, for a sample of length $L = 2cT_R$.

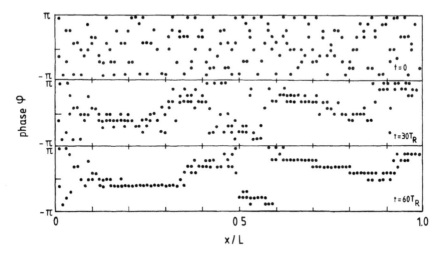

Figure 4.8. The same as figure 4.7 for a long sample, $L = 200cT_R$, at instants $t = 0$, $t = 30T_R$, $t = 60T_R$.

If the sample length exceeds a certain cooperation length L_c, the growth of coherent domains is limited. Figure 4.8 shows results of a computer simulation for a long system ($L = 200cT_R$), where we can easily see three coherent domains.

In section 1.5 we gave a simple estimate by equation (1.5.9) for the cooperative length, and used the relation $L_c = cT_R(L_c)$. A more accurate value of L_c can be found from the equation

$$L_c = cT_D(L_c) \qquad (4.5.1)$$

where T_D is the delay time for the first peak in the output power [MGF81a]

$$T_D = \tfrac{1}{4}T_R(\ln 2\pi \sqrt{N})^2. \qquad (4.5.2)$$

Using (1.4.29), the cooperative length determined in this way becomes

$$L_c = \left(\frac{2\pi c}{3\gamma N_0 \lambda^2}\right)^{1/2} \ln(2\pi \sqrt{N}) \qquad (4.5.3)$$

which is in good agreement with the average size of the domains obtained in the computer simulation.

Chapter 5

The semiclassical theory

5.1 The equations of the model

The formulation of the semiclassical theory of super-radiance has already been given in Chapter 1. The statistical interpretation of the semi-classical theory and its relation to the quantum electrodynamical description introduced in Chapter 3 have been presented in Chapter 4.

One of the advantages of the semiclassical method is that it allows one to consider the propagation effects of the field in extended systems. Another advantage is that it is not difficult to incorporate, in a phenomenological way, the effects of population and polarization relaxation into the equations that determine the dynamics of the atomic density matrix [AE75, SSL74]. In the present chapter we shall make use of these advantages, and will investigate the propagation problem and the effects of relaxation on SR in more detail.

The main reason for polarization relaxation (sometimes called phase relaxation) is the random phase variation of atomic wavefunctions which is caused by external perturbations. This has already been briefly discussed in section 1.1. The effect can be characterized by the relaxation time $T_2 \sim 1/\Gamma$ (i.e. the lifetime of the macroscopic polarization). The finite value of T_2 causes a broadening of the emission line which is known as *homogeneous broadening*.

Apart from that, the resonant frequency ω_0 of the individual atoms in a system usually shows a spread. In gases, this is due to the Doppler effect; in solids, mostly due to a different environment of the atoms. This spread causes a line broadening in the emission spectrum of the system known as *inhomogeneous broadening* [AE75, SSL74]. The significance of inhomogeneous broadening for observations of SR in solids can be seen from the experiments reviewed in Chapter 2. In crystals at low temperatures homogeneous broadening can be small and close to natural radiative broadening. However, inhomogeneous broadening resulting from the different environment of the centres cannot be overcome. The latter can significantly influence the experimental realization of super-radiant emission.

The distribution of the atoms according to their resonant frequencies is generally described by a function $g(\nu) = g(\omega_0 - \omega_c)$, where ω_c is a central frequency, which is assumed to coincide with the carrier frequency of the field. The macroscopic polarization density, which is the source of the field in the medium, will then be a weighted average of the dipoles oscillating with different frequencies; so, instead of (1.4.7), we have [AE75, SSL74]

$$\mathcal{P} = N_0 d \int [\rho_{eg}(r, t, \nu) + \rho_{ge}(r, t, \nu)]g(\nu)\, d\nu. \tag{5.1.1}$$

The most frequently used inhomogeneous distribution, the Gaussian, can be written in the form

$$g_G(\nu) = \frac{T_2^{*2}}{\sqrt{\pi}} \exp(-\nu^2 T_2^{*2}). \tag{5.1.2}$$

For the sake of simplicity it will sometimes be more convenient to approximate it by a Lorentzian

$$g_L(\nu) = \frac{T_2^*}{\pi} [1 + \nu^2 T_2^{*2}]^{-1}. \tag{5.1.3}$$

The width of the inhomogeneous line is characterized in both cases by the reciprocal of the time constant T_2^*.

In what follows, we shall limit ourselves to considering a model of one spatial dimension, and investigate the super-radiant properties of a pencil-shaped sample. First it is appropriate to use the super-radiant time T_R as the time scale in the equations. Let us recall, cf (1.5.4), that

$$T_R = \frac{\hbar c}{2\pi d^2 N_0 \omega_0 L}. \tag{5.1.4}$$

This is the time parameter that determines the evolution of the super-radiant process when retardation effects are not important, i.e. if the transit time of a light signal through the system is less than the time of the super-radiant emission.

With the aid of T_R we introduce a dimensionless field amplitude

$$\varepsilon = -i\frac{d}{\hbar} E T_R. \tag{5.1.5}$$

In terms of ε, the intensity of the field, which we defined in Chapter 1 as the number of photons passing through the cross-section of the sample, A, per unit time and per atom is given by

$$I = \frac{c}{8\pi}|E|^2\frac{A}{N\hbar\omega_0} = \frac{1}{4T_R}|\varepsilon|^2. \tag{5.1.6}$$

Following the procedure that led us in Chapter 1 from (1.4.7)–(1.4.10) to (1.4.11), but using the dimensionless field amplitude ε of (5.1.5) and the polarization density (5.1.1), we get the Maxwell–Bloch equations for unidirectional propagation

$$\frac{\partial \varepsilon(x,t)}{\partial x} + \frac{1}{c}\frac{\partial \varepsilon(x,t)}{\partial t} = \frac{1}{L}\int g(v)R(x,t,v)\,dv \qquad (5.1.7)$$

$$\frac{\partial R(x,t,v)}{\partial t} = \left(-iv - \frac{1}{T_2}\right)R(x,t,v) + \frac{1}{T_R}\varepsilon(x,t)Z(x,t,v) \qquad (5.1.8)$$

$$\frac{\partial Z(x,t,v)}{\partial t} = -\frac{1}{2T_R}\left(\varepsilon^*(x,t)R(x,t,v) + \varepsilon(x,t)R^*(x,t,v)\right) - \frac{1+Z}{T_1} \qquad (5.1.9)$$

where we have already included the terms containing the lifetimes T_1 and T_2 that describe the processes of population decay and dephasing respectively [AE75, SSL74]. Equations (5.1.7)–(5.1.9) are the semiclassical counterparts of the operator equations (4.3.2) of Chapter 4 complemented by the relaxation terms. As in most cases T_2 is shorter than T_1, the relaxation of the inversion will have a minor effect compared with the dephasing caused by polarization relaxation. Therefore in the following we set $T_1 = \infty$ and retain only T_2. The inhomogeneous broadening, as stipulated by (5.1.1), will be taken into account as well.

In the present chapter we first show how to obtain the rate equation description of the pulse amplification problem from the Maxwell–Bloch system if phase relaxation is sufficiently strong. Then in the subsequent sections we return to the more general situation. Before turning to discuss the semiclassical theory of super-radiance, we briefly summarize in section 5.3 some important results of the theory of coherent pulse propagation, namely the area theorem and self-induced transparency. The system of the Maxwell–Bloch equations can be used to describe several other resonant optical phenomena, photon echoes, optical bistability etc; reviews of these effects can be found in, e.g. [AE75] and [PS95]

We concentrate on super-radiance which corresponds to the initial condition, when all the atoms are completely inverted, and there is no external field at $t = 0$. According to the Maxwell–Bloch equations (5.1.7)–(5.1.9), for $T_1 \to \infty$ the system remains in the inverted state forever, because the quantum fluctuations that initiate the radiation process are not present in the semiclassical approach. As has been discussed in the previous chapter, this can be remedied by assuming a small stochastic initial polarization. This is one of the methods that will be applied in this chapter. We shall see later that in the presence of homogeneous broadening ($T_2 \neq \infty$) a stochastic source term for the polarization will also have to be added to the right-hand side of equation (5.1.8). The other possibility for initiating the dynamics—which is exploited in actual experiments as well (see Chapter 2)—is to inject a small external triggering pulse, which creates the necessary initial polarization. This effect, known as induced, or triggered, super-radiance will be discussed in detail in the present chapter.

At the end of the chapter we shall introduce a three-level atomic model and discuss two other coherent atomic phenomena; namely, coherent Raman scattering and lasing without inversion.

5.2 Transition to amplified spontaneous emission

Let us show that if the dephasing process is sufficiently strong, so that the relaxation time of the polarization T_2 is shorter than the duration of the radiation process, characterized by a parameter T_p, then the Maxwell–Bloch equations yield the ordinary balance equations for the intensity and the inversion. The corresponding regime of emission is called amplified spontaneous emission (ASE) or sometimes super-luminescence (SL).

We shall also ignore here the inhomogeneous broadening, and set $\varepsilon^* = \varepsilon$. Then equations (5.1.7)–(5.1.9) can be written in the following form

$$\frac{\partial \varepsilon(x,t)}{\partial x} + \frac{1}{c}\frac{\partial \varepsilon(x,t)}{\partial t} = \frac{1}{L}R(x,t) \tag{5.2.1}$$

$$\frac{\partial R(x,t)}{\partial t} = -\frac{1}{T_2}R(x,t) + \frac{1}{T_R}\varepsilon(x,t)Z(x,t) \tag{5.2.2}$$

$$\frac{\partial Z(x,t)}{\partial t} = -\frac{1}{T_R}\varepsilon(x,t)R(x,t). \tag{5.2.3}$$

If $T_2 < T_p$, we can neglect the time derivative $\partial R/\partial t$ in comparison with the relaxation term R/T_2 in equation (5.2.2). This allows us to express the amplitude of the off-diagonal density matrix element, R, in terms of the field amplitude ε and the inversion Z

$$R = \frac{T_2}{T_R}\varepsilon Z. \tag{5.2.4}$$

Thus, under conditions of effective dephasing the polarization at the spatial point x and at the instant of time t can be expressed in terms of the field amplitude at the same x and t. Substituting equation (5.2.4) into equations (5.2.1)–(5.2.3) we arrive at the following equations for ε and Z

$$\frac{\partial \varepsilon(x,t)}{\partial x} + \frac{1}{c}\frac{\partial \varepsilon(x,t)}{\partial t} = \tfrac{1}{2}\alpha \varepsilon Z \tag{5.2.5}$$

$$\frac{\partial Z}{\partial t} = -\frac{\alpha L}{2T_R}\varepsilon^2 Z \tag{5.2.6}$$

where the notation

$$\tfrac{1}{2}\alpha = \frac{T_2}{T_R L} = \frac{2\pi N_0 \omega_0 d^2}{\hbar c}T_2 \tag{5.2.7}$$

has been introduced. As is seen from (5.2.5), $\alpha/2$ is the small signal gain (absorption) coefficient for the field amplitude.

Equations (5.2.5) and (5.2.6) can be easily transformed into a pair of equations containing the field intensity $I = \varepsilon^2/4T_R$

$$\frac{\partial I}{\partial x} + \frac{1}{c}\frac{\partial I}{\partial t} = \alpha I Z \qquad (5.2.8)$$

$$\frac{\partial Z}{\partial t} = -2\alpha L I Z. \qquad (5.2.9)$$

To solve equations (5.2.8) and (5.2.9) we have to specify the initial and the boundary conditions. In general they are of the form

$$I(x, t = 0) = I_0(x) \qquad\qquad I(x = 0, t) = I_{in}(t)$$
$$Z(x, t = 0) = Z_0(x). \qquad\qquad\qquad\qquad\qquad (5.2.10)$$

If initially there is an incoming signal $I_{in}(t)$ at one of the ends of the system, then equations (5.2.8) and (5.2.9) describe either the incoherent amplification (if $Z_0 > 0$) or the absorption (if $Z_0 < 0$) of this input field [6, 20, 67].

Our aim here is to describe amplified spontaneous emission when there is no incoming field: $I_{in}(t) = 0$, the system is initially inverted: $Z_0(x) = 1$, and $I_0(x)$ is chosen to correspond to the rate of ordinary spontaneous emission. We denote the number of spontaneous photons emitted by one atom per unit time into the solid angle A/L^2 by γ_1, i.e. $I_0(x) = \gamma_1$.

In the absence of an incoming field the intensity can be expressed from (5.2.8) in terms of Z as

$$I = I_0(x - ct)\exp\left(\alpha c \int_0^t Z(x - ct + ct', t')\,dt'\right). \qquad (5.2.11)$$

Since $I_0(x)$ differs from zero only for $0 < x < L$, the factor $I_0(x - ct)$ ensures that for $ct > x$ the intensity $I = 0$, because there is no incoming signal at $x = 0$. This means that the field sweeps through the sample with the velocity of light, and the duration of the pulse at $x = L$ is

$$T_p = \frac{L}{c}. \qquad (5.2.12)$$

Assuming spatially homogeneous initial conditions, we can determine the shape of the ASE pulse at $x = L$ simply by omitting the spatial derivative from equation (5.2.8). The finite velocity of propagation ensures that until $t = L/c$ the spatial derivative remains zero at $x = L$, provided it was zero along the sample at $t = 0$. As we have seen in Chapter 1 for the Maxwell–Bloch equations, the solution with homogeneous initial conditions coincides with the spatially homogeneous solution until $t = L/c$ at $x = L$. This can also be proved by integrating the full set of equations (5.2.8) and (5.2.9), which can be performed analytically for arbitrary initial data as well [MMT82]. Here we restrict ourselves to homogeneous initial data, and thus we obtain the ordinary differential equations

$$\frac{dI}{dt} = \alpha c I Z \tag{5.2.13}$$

$$\frac{dZ}{dt} = -2\alpha L I Z \tag{5.2.14}$$

which can easily be integrated. The system has a first integral

$$Z + \frac{2L}{c} I = 1 + \frac{2L}{c} \gamma_1. \tag{5.2.15}$$

As follows from equations (5.2.13) and (5.2.14), $Z = 0$ is a stationary point of this model, and, since $Z_0 = 1$, $Z(t)$ cannot decrease below zero. Then according to (5.2.15) the maximal value of the radiated intensity is reached under saturation of the population inversion: $Z = 0$. If the populations of the atomic levels become equal, then the medium neither absorbs nor amplifies the radiation. Therefore after attaining saturation a constant flux equal to $\gamma_1 + c/2L \approx c/2L$ will be registered at the output of the system until the end of the process, $t = L/c$.

From (5.2.13) and (5.2.15) we obtain a single equation for I

$$\frac{dI}{dt} = aI(b - I) \tag{5.2.16}$$

where

$$a = 2\alpha L \qquad b = \frac{c}{2L} + \gamma_1 \approx \frac{c}{2L}. \tag{5.2.17}$$

The solution of (5.2.16) corresponding to the initial condition $I = \gamma_1$ is

$$I = \frac{\gamma_1 b}{\gamma_1 + (b - \gamma_1)\exp(-abt)} = \frac{\gamma_1}{(2L\gamma_1/c) + \exp(-\alpha ct)}. \tag{5.2.18}$$

The shape of this pulse is shown in figure 5.1.

From equation (5.2.18) it follows that the maximal value of the intensity, $c/(2L)$, is achieved essentially during a time interval t_0 determined by

$$\frac{2L\gamma_1}{c} = \exp(-\alpha ct_0) \tag{5.2.19}$$

giving

$$t_0 = \frac{1}{\alpha c} \ln \frac{c}{2L\gamma_1}. \tag{5.2.20}$$

Since the duration of the pulse is L/c, the intensity achieves saturation only if

$$t_0 < \frac{L}{c} \qquad \text{i.e.} \quad \frac{1}{\alpha L} \ln \frac{c}{2L\gamma_1} < 1. \tag{5.2.21}$$

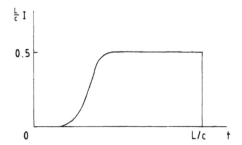

Figure 5.1. The shape of the amplified spontaneous emission pulse.

In this case t_0 is the duration of the actual amplification process.

Now the condition of fast relaxation mentioned at the beginning of the section can be given a more exact form. The incoherent process of ASE appears as soon as the phase relaxation time is shorter than the time delay obtained above

$$T_2 < \frac{1}{\alpha c} \ln \frac{c}{2L\gamma_1}. \tag{5.2.22}$$

This treatment gives a satisfactory description of the process only if the amplification is sufficiently strong, and the saturation régime is achieved. In the case of relatively weak amplification the correct description of ASE requires the introduction of a source of the spontaneous emission. On a longer time scale the effect of population relaxation, $T_1 < \infty$, also has to be taken into account. The details of the transition from super-radiation to amplified spontaneous emission with gradually decreasing values of T_2 can be followed by numerical calculations, which will be discussed in section 5.7.

5.3   Area theorem and self-induced transparency

The Maxwell–Bloch equations (5.1.7)–(5.1.9), that we are going to use for describing super-radiance, are actually applicable to a wider class of coherent optical phenomena: photon echo, coherent absorbers and amplifiers etc. Therefore in this section we first discuss a general result concerning the coherent propagation of light pulses, in particular that of a super-radiant pulse. It is the area theorem of McCall and Hahn [MH67, MH69] (see also [AE75]). In super-radiance the initial conditions correspond to a state where the atoms of the system are initially in their excited states. Therefore an extended super-radiant system can be called a coherent amplifier because during propagation the field amplitude is growing. Before turning to that question in the following sections, we shall discuss here the related problem of coherent pulse propagation in an attenuator, where the atoms are initially in their ground states. This latter problem is usually referred to as self-induced transparency (SIT) following the original investigators, McCall and Hahn.

Let us consider equations (5.1.7)–(5.1.9) and suppose that both population and phase relaxation are negligible, i.e. the process is so fast that the duration of the light pulse is much shorter than T_2 and T_1. Then the relaxation terms can be omitted and we have

$$\frac{\partial \varepsilon(x,t)}{\partial x} + \frac{1}{c}\frac{\partial \varepsilon(x,t)}{\partial t} = \frac{1}{L}\int g(v)R(x,t,v)\,dv \qquad (5.3.1)$$

$$\frac{\partial R(x,t,v)}{\partial t} = -iv R(x,t,v) + \frac{1}{T_R}\varepsilon(x,t)Z(x,t,v) \qquad (5.3.2)$$

$$\frac{\partial Z(x,t,v)}{\partial t} = -\frac{1}{2T_R}\left(\varepsilon^*(x,t)R(x,t,v) + \varepsilon(x,t)R^*(x,t,v)\right). \qquad (5.3.3)$$

For atoms in exact resonance, i.e. for those with $v = 0$, we can introduce the Bloch angle θ as in equation (1.4.18). Choosing the length of the Bloch vector $B = 1$ and supposing that ε is real, we have the following formal solution to equations (5.3.2) and (5.3.3)

$$Z = \pm\cos\theta \qquad R = \pm\sin\theta \qquad \theta(x,t) = \frac{1}{T_R}\int_{-\infty}^{t}\varepsilon(x,t')\,dt'. \qquad (5.3.4)$$

The upper sign is chosen for the amplifier, where initially the atoms are in their upper state: $Z(x,t=-\infty) = 1$. The lower one is valid for the attenuator where $Z(x,t=-\infty) = -1$.

Let us introduce now a key concept in coherent resonant optics, the notion of the pulse area, which is defined as

$$\mathcal{A}(x) = \theta(x,t=\infty) = \frac{1}{T_R}\int_{-\infty}^{\infty}\varepsilon(x,t')\,dt'. \qquad (5.3.5)$$

This quantity obeys a simple differential equation, as will now be shown. Integrating (5.3.1) with respect to t, we get for the space derivative of $\mathcal{A}(x)$

$$\frac{d\mathcal{A}}{dx} = \frac{1}{LT_R}\int_{-\infty}^{\infty}dt\int dv\, g(v)R(x,t,v) \qquad (5.3.6)$$

where we have exploited the fact that the field ε is necessarily of finite duration. As it is supposed that initially there is no polarization in the medium, (5.3.2) can be written in the integral form

$$R(x,t,v) = \frac{1}{T_R}\int_{-\infty}^{t}dt'\varepsilon(x,t')Z(x,t',v)e^{-iv(t-t')}. \qquad (5.3.7)$$

Substituting (5.3.7) into (5.3.6), interchanging the order of integration with respect to t and t' and introducing the variable $s = t - t'$ we get

$$\frac{d\mathcal{A}}{dx} = \frac{1}{LT_R}\int_{-\infty}^{\infty}dt'\varepsilon(x,t')\frac{1}{T_R}\int dv g(v)Z(x,t',v)\int_{0}^{\infty}ds\,e^{-ivs}. \qquad (5.3.8)$$

The last integral is equal to $\pi\delta(\nu) - i\text{Pr}(1/\nu)$ (where δ is Dirac's delta function and Pr denotes the principal value), and this enables us to readily perform the second integral in (5.3.8). Supposing that $g(\nu)$ is symmetric with respect to $\nu = 0$, we get

$$\frac{d\mathcal{A}}{dx} = \frac{\pi g(0)}{LT_R} \int_{-\infty}^{\infty} dt' \frac{1}{T_R} \varepsilon(x, t') Z(x, t', \nu = 0). \tag{5.3.9}$$

It can be seen that the x dependence of $\mathcal{A}$ is determined by the atoms which are in resonance with the field, i.e. for which $\nu = 0$. Now using (5.3.4) and (5.3.5) we arrive to the following equation, known as the area theorem

$$\frac{d\mathcal{A}}{dx} = \pm \frac{\alpha}{2} \sin \mathcal{A} \tag{5.3.10}$$

where now the coefficient α is defined as

$$\alpha = \frac{2\pi g(0)}{LT_R}. \tag{5.3.11}$$

Note that this definition of α coincides with that of (5.2.7) if $g(\nu)$ is a Lorentzian (see (5.1.3)), with T_2 instead of T_2^*.

Equation (5.3.10) admits the following solution

$$\tan(\mathcal{A}/2) = \tan(\mathcal{A}_0/2) \exp[\pm(\alpha/2)(x - x_0)]. \tag{5.3.12}$$

In the case of an attenuator, the asymptotic value $\mathcal{A}(\infty)$ is necessarily $2n\pi$, $n = 0, 1, 2, \ldots$ where n is determined by $\mathcal{A}_0$. In the case of an amplifier the asymptotic value is $(2n + 1)\pi$, depending again on $\mathcal{A}_0$. In super-radiance where initially there is no field present, the asymptotic value of the pulse area is π, as we have seen in Chapter 1 with respect to the auto-modelling solution.

For the case of the attenuator it is possible to find analytical solutions of the system (5.3.1)–(5.3.3). The simplest one is the famous '2π hyperbolic secant pulse' found by McCall and Hahn

$$\varepsilon = \frac{2T_R}{T} \text{sech} \left[\frac{1}{T} \left(t - \frac{x}{v} \right) \right]. \tag{5.3.13}$$

The corresponding analytic expressions for R and Z can be found in [MH69, AE75]. The pulse duration T and pulse velocity v are in the relation: $c/v = 1 + (\alpha/2\pi g(0))\langle 1/(\nu^2 + T^{-2})\rangle$. The angular brackets denote here the average over the inhomogeneous line shape. The particular solution (5.3.13) has a stable shape, and numerical calculations showed that all solutions for which initially $\pi < \mathcal{A}_0 < 3\pi$ were transformed into this type of pulse after sufficiently long propagation. In their pioneering work, McCall and Hahn demonstrated experimentally all these properties of SIT in ruby. Several other works connected with SIT are cited in [AE75].

The mathematical explanation [AKNS74] of the stability of the solution (5.3.13) lies in the fact that the system (5.3.1)–(5.3.3) belongs to the class of so-called integrable systems, allowing soliton solutions. Actually the theory of self-induced transparency contributed much to the development of the mathematical theory of the inverse scattering theory of solitons [43]. In the absence of inhomogeneous broadening, (5.3.1)–(5.3.3) can be transformed into the sine–Gordon equation, as we have already seen in section 1.4. The discussion there revealed that no simple analytic solution is available for the case of the initially inverted system, the amplifier, even in the absence of relaxation.

In subsequent sections we are going to analyse in more detail the amplifier, which is essentially an extended super-radiant system.

5.4   Linear régime of coherent amplification

As we have noted in section 5.1, super-radiance of a system of inverted atoms can be triggered by a small external pulse.   This process, which can be alternatively viewed as the coherent amplification of the triggering pulse, has been observed experimentally in several works [VS79, CJSGH80, VGKMMBT86]; see Chapter 2. The semiclassical theory is ideally suited to describe this process, because here the emission is initiated by a classical external field instead of the quantum fluctuations.

In this section we will consider the initial linear part of triggered super-radiance, while we can neglect the changes of the inversion, i.e. until the system remains near to its upper inverted state.

5.4.1   Solution of the linearized problem

We treat the problem here in the linear approximation where it is assumed that Z remains close to its initial value during the full process. Correspondingly we set $Z = 1$ in equations (5.1.8) and (5.1.9), and find two linear equations

$$\frac{\partial \varepsilon(x, t)}{\partial x} + \frac{1}{c}\frac{\partial \varepsilon(x, t)}{\partial t} = \frac{1}{L}\int g(v)R(x, t, v)\,dv \qquad (5.4.1)$$

$$\frac{\partial R(x, t, v)}{\partial t} = \left(-iv - \frac{1}{T_2}\right)R(x, t, v) + \frac{\varepsilon(x, t)}{T_R}. \qquad (5.4.2)$$

We take the amplitude of the incoming pulse at $x = 0$ to be a given function of time

$$\varepsilon(x = 0, t) = \varepsilon_0(t) \qquad \varepsilon(x, t = 0) = 0 \qquad x \neq 0. \qquad (5.4.3)$$

We also assume here that the polarization created by the external excitation is significantly larger than its spontaneous fluctuations, and therefore we set

$$R(x, t = 0, v) = 0. \qquad (5.4.4)$$

To solve the system (5.4.1) and (5.4.2) we introduce the following Laplace transforms

$$\varepsilon(q,s) = \int_0^\infty \int_0^\infty \varepsilon(x,t) \, e^{-qx} \, e^{-st} \, dx \, dt$$

$$R(q,s,v) = \int_0^\infty \int_0^\infty R(x,t,v) \, e^{-qx} \, e^{-st} \, dx \, dt \qquad (5.4.5)$$

$$\varepsilon_0(s) = \int_0^\infty \varepsilon_0(t) \, e^{-st} \, dt.$$

Then from equations (5.4.1) and (5.4.2) with the initial conditions (5.4.3) and (5.4.4) we find that

$$\varepsilon(q,s) = \frac{\varepsilon_0(s)}{q + s/c - \varphi(s)/L} \qquad (5.4.6)$$

where

$$\varphi(s) = \frac{1}{T_R} \int \frac{g(v)}{s + T_2^{-1} + iv} \, dv. \qquad (5.4.7)$$

We shall consider here the case where the inhomogeneous line shape is Lorentzian, see equation (5.1.3). Then the calculation can be performed analytically, and we obtain

$$\varphi_L(s) = \frac{1}{T_R} \frac{1}{s + \tau_2^{-1}} \qquad (5.4.8)$$

where

$$\frac{1}{\tau_2} = \frac{1}{T_2} + \frac{1}{T_2^*}. \qquad (5.4.9)$$

Taking the inverse Laplace transform of (5.4.6) in the variable q, we find

$$\varepsilon(x,s) = \varepsilon_0(s) \exp\left[x\left(\frac{\varphi_L(s)}{L} - \frac{s}{c}\right)\right]$$

$$= \varepsilon_0(s) \exp\left[x\left(\frac{1}{LT_R} \frac{1}{s + 1/\tau_2} - \frac{s}{c}\right)\right] \qquad (5.4.10)$$

The inverse Laplace transform of $\varepsilon(x,s)$ with respect to s can be obtained by considering the function

$$G(t) = e^{-t/\tau_2} \frac{d}{dt}(\varepsilon_0(t) \, e^{t/\tau_2}) = \frac{d}{dt}\varepsilon_0(t) + \frac{1}{\tau_2}\varepsilon_0(t). \qquad (5.4.11)$$

Its Laplace transform is

$$G(s) = \int_0^\infty G(t) \, e^{-st} \, dt = \left(s + \frac{1}{\tau_2}\right)\varepsilon_0(s). \qquad (5.4.12)$$

Therefore

$$\varepsilon(x, s) = \frac{G(s)}{s + 1/\tau_2} \exp\left[x\left(\frac{1}{LT_R}\frac{1}{s + 1/\tau_2} - \frac{s}{c}\right)\right]. \tag{5.4.13}$$

Using the Laplace transform [2]

$$\frac{1}{s} e^{x/s} = \int_0^\infty I_0\left(2\sqrt{tx}\right) e^{-st}\, dt \tag{5.4.14}$$

where I_0 is the modified Bessel function of zero order, we obtain

$$\varepsilon(x, t) = \int_0^t G(t - t') I_0\left(2\sqrt{x(t' - x/c)/LT_R}\right)$$
$$\times e^{-(t'-x/c)/\tau_2}\theta(t' - x/c)\, dt'. \tag{5.4.15}$$

Substituting (5.4.11) here, we can integrate (5.4.15) by parts. Assuming $\varepsilon_0(0) = 0$, and taking into account that $dI_0(x)/dx = I_1(x)$, where I_1 is the modified Bessel function of the first order, we have

$$\varepsilon(x, t) = \varepsilon_0(t - x/c)$$
$$+ \int_{x/c}^t \varepsilon_0(t - t') e^{-(t'-x/c)/\tau_2} \sqrt{\frac{x}{LT_R(t' - x/c)}}$$
$$\times I_1\left(2\sqrt{x(t' - x/c)/LT_R}\right)\, dt'. \tag{5.4.16}$$

This is the general formula describing coherent linear amplification in a fully inverted medium.

5.4.2 Lethargic gain

Let us now analyse the specific effects exhibited by the solution (5.4.16) in the absence of relaxation, $\tau_2^{-1} = 0$.

We first consider the case when the incoming pulse is a step pulse switched on at $t = 0$: $\varepsilon_0(t) = \varepsilon_0\theta(t)$. We then find from (5.4.16)

$$\varepsilon(x, t) = \varepsilon_0 I_0\left(2\sqrt{x(t - x/c)/LT_R}\right)\theta(t). \tag{5.4.17}$$

For a fixed value of the retarded time $t_r = t - x/c$ and for $xt_r \gg LT_R$, by using the asymptotic value of the Bessel function [2] I_0, we find

$$\varepsilon(x, t) = \frac{1}{\sqrt{2\pi\kappa\sqrt{x}}} \exp(\kappa\sqrt{x}) \qquad \kappa = 2\sqrt{\frac{1}{L}\frac{t_r}{T_R}}. \tag{5.4.18}$$

We see that already the linear part of coherent amplification is essentially different from the usual exponential law for incoherent amplification by stimulated emission. If the phase relaxation of the polarization is weak, then the signal—after passing through a distance x in the inverted medium—will be amplified as $\exp(\kappa\sqrt{x})$ instead of the usual Lambert–Beer law, $\exp(\alpha x)$. The inverse effect, anomalous linear absorption, was considered first by Crisp [C70]. For amplifying media it was discussed by Hopf *et al* [HMM76] in connection with swept gain amplifiers. The first experimental observation of this effect was reported by Chung, Lee and De Temple [CLDT81] in the far-infrared domain. In this latter work the slow amplification process connected with the square root in the exponential was termed lethargic gain. The process was observed experimentally also in ruby and in YAG:Nd at low temperatures [VGKMMBT86], where pulses of area $\pi/200$ were amplified, so that the output pulses still had a small area: $\pi/20$ (see section 2.2.2 for the details).

Figure 5.2 shows the calculated amplification of undamped pulses ($\tau_2^{-1} = 0$) of various durations over a fixed amplification distance, along with the amplification of a single pulse over various amplification distances. The input

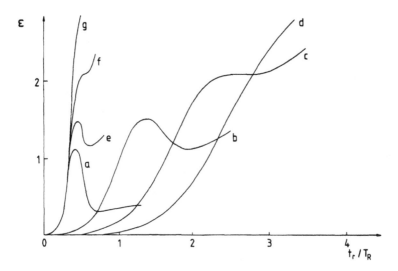

Figure 5.2.   Lethargic amplification ($\tau_2 = \infty$) of a Gaussian incoming pulse: $\varepsilon = \varepsilon_0 \exp[-(t - 4\sigma)^2/2\sigma^2]$ as a function of the retarded time: $t_r/T_R = (t - x/c)/T_R$ for a fixed amplification length $L = cT_R$, with different durations: (a) $\sigma = 0.1T_R$; (b) $\sigma = 0.3T_R$; (c) $\sigma = 0.5T_R$; (d) $\sigma = 0.7T_R$; and for a fixed duration: $\sigma = 0.1T_R$; for different amplification lengths: (e) $L = 3cT_R$; (f) $L = 5cT_R$; (g) $L = 7cT_R$.

pulse here is a Gaussian, $\varepsilon_0(t) = \varepsilon_0 \exp[-(t - 4\sigma)^2/2\sigma^2]$, for $|t - 4\sigma| < 4\sigma$, with a duration of $T_p = 2\sqrt{2\ln 2}\sigma$ at half maximum. The shape of the amplified pulse varies only slightly if $T_p \ll T_R$. With increasing the amplification distance the shape of the pulse becomes smoother, and over a certain distance L, or input pulse duration, σ, the peak in the output pulse changes to a point of inflection.

This means that the amplified signal transforms into the radiation that originates from the polarization generated by the pulse itself. Evaluating the integral in (5.4.16) with the above Gaussian input pulse, we find numerically that this limit is approximately at $\sigma L = L T_R/2$, which does not depend upon L. This is a particular case of a more general scaling property: we see from the solution (5.4.16) that if we scale the input pulse in accordance with $\varepsilon_0(t) \to \varepsilon_0(\beta t)$ and scale the relaxation time $\tau_2 \to \beta\tau_2$, the output pulse scales as

$$\varepsilon(x, t_r) \to \varepsilon(x/\beta, \beta t_r) \qquad (5.4.19)$$

which is the manifestation of the scaling already discussed in section 1.5. Figure 5.2 illustrates this scaling property: all the pulses shown there are similar, and, for instance, the value of the field amplitude at the point of inflection in the output pulse is the same (about $2.06\varepsilon_0$), independent of L and σ.

5.4.3 Transition from lethargic to ordinary gain

Let us now consider the effects of relaxation on the amplification process. According to (5.4.9), a Lorentzian inhomogeneous line shape has the same effect upon the output field as homogeneous broadening. We would expect a Gaussian inhomogeneous line to have a weaker damping effect, because in that case the number of atoms being further out of resonance is less than that for a Lorentzian. From (5.4.16) we see that if $\tau_2^{-1} \neq 0$ then a pulse of finite duration will not increase infinitely, as formally happens for an infinite τ_2, but after a sufficiently long time it will tend towards zero, since in the limit $t \to \infty$ the exponential function in the integrand decreases more rapidly than the function I_1 increases.

In fact the linear approximation retains its validity until $Z \approx 1$ in equations (5.1.8) and (5.1.9). If $T_2^{-1} = 0$ it then follows from those equations that $Z^2 + |R|^2 = 1$, thus $Z \approx 1$ is equivalent to $|R| \ll 1$. Solving equation (5.4.2) we can write the polarization as

$$R(x, t, v) = \int_0^t \varepsilon(x, t') \exp\left[\left(-\mathrm{i}v - \frac{1}{T_2}\right)(t - t')\right] \mathrm{d}t'. \qquad (5.4.20)$$

In spite of relaxation, represented by the constant $1/T_2$, during the amplification process $R(x, t)$ grows with x. The domain of validity of the linear approximation can be estimated from the area theorem (5.3.10). For the linearized equations the theorem states that

$$\frac{\mathrm{d}\mathcal{A}}{\mathrm{d}x} = \tfrac{1}{2}\alpha\mathcal{A} \qquad (5.4.21)$$

and consequently $\mathcal{A} = \mathcal{A}_0 \exp(\alpha x/2)$. So if the area of the outcoming pulse $\mathcal{A}_0 \exp(\alpha x/2) \ll 1$, then the linear approximation remains valid during the whole process.

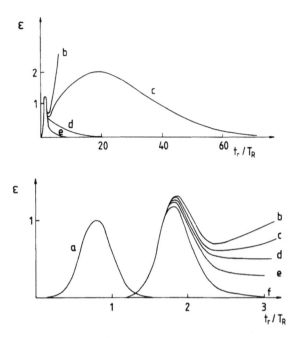

Figure 5.3. The effect of relaxation upon the lethargic amplification of a Gaussian pulse with $\sigma = 0.2T_R$, $L = cT_R$. (a) incoming pulse; (b) $\tau_2 = \infty$; (c) $\tau_2 = 5T_R$; (d) $\tau_2 = 2T_R$; (e) $\tau_2 = T_R$; (f) $\tau_2 = 0.24T_R$. The initial stage of the amplification process is shown in the lower graph.

Let us consider the transition from lethargic to ordinary gain. If the relaxation time τ_2 is finite, it follows from (5.4.16) that a step pulse is amplified as

$$\varepsilon(x, t) = \varepsilon_0 I_0 \left(2\sqrt{x(t - x/c)/LT_R}\right) e^{-(t-x/c)/\tau_2} \theta(t - x/c)$$

$$+ \frac{\varepsilon_0}{\tau_2} \int_{x/c}^{t} I_0 \left(2\sqrt{x(t' - x/c)/LT_R}\right) e^{-(t'-x/c)/\tau_2} \, dt'. \quad (5.4.22)$$

For large values of t, the first term in (5.4.22) tends to zero, whilst the second one approaches a constant value. Setting the upper limit of the integral equal to infinity, we find—after substituting $y = 2\sqrt{x(t - x/c)/LT_R}$ and setting $\alpha L/2 = \tau_2/T_R$—that the integral can be expressed in a closed form [27], and the final result is

$$\varepsilon(x, t = \infty) = \varepsilon_0 \frac{2}{\alpha} \int_0^\infty I_0(y) \exp(-y^2/2\alpha) \frac{y}{x} \, dy = \varepsilon_0 \, e^{\alpha x/2}. \quad (5.4.23)$$

This is the usual Lambert–Beer law for incoherent amplification. From equation (5.4.22) it also follows that the characteristic law of lethargic amplification can be observed only for $t < \tau_2$.

Figure 5.3 shows the amplification of Gaussian pulses with $\sigma = 0.2$, for several values of τ_2. We see that the shape of the amplified pulse depends strongly upon τ_2.

The experimental investigation of the amplification of small area pulses [VGKMMBT86] has shown qualitatively the same form for the amplified pulse as presented in figure 5.3. The transition from lethargic amplification to the more common incoherent gain was also observed in that experimental work. By increasing the pulse duration upwards from below τ_2, the gain gradually achieved its steady-state value corresponding to the Lambert–Beer law.

5.5   Linear régime of super-radiance

Let us consider now the case when super-radiation is initiated by spontaneous fluctuations of the polarization in the medium.   In the semiclassical approximation these fluctuations are represented by a δ-correlated random initial polarization

$$\langle R^*(x', 0)\ R(x, 0)\rangle = \frac{4L}{N}\delta(x - x'). \tag{5.5.1}$$

This initial condition corresponds to the quantum nature of super-radiance. As we have seen in Chapter 4, it allows us to obtain quantum expectation values of the relevant quantities using the semiclassical equations. Here we generalize equation (5.5.1), taking into account the different resonant frequencies of the atoms, i.e. inhomogeneous broadening. We divide the domain of the resonant frequencies into intervals of length Δ_i, and assume that the fluctuations of the polarization are independent in the different intervals. As the number of atoms having a resonance frequency in the interval Δ_i is $Ng(\nu_i)\Delta_i$, the correlation function for these atoms is

$$\langle R^*(x', 0, \nu_i)\ R(x, 0, \nu_j)\rangle = \frac{4L}{Ng(\nu_i)\Delta_i}\delta(x - x')\delta_{ij}. \tag{5.5.2}$$

For $\Delta_i \to 0$, we obtain

$$\langle R^*(x', 0, \nu')R(x, 0, \nu)\rangle = \frac{4L}{Ng(\nu)}\delta(x - x')\delta(\nu - \nu'). \tag{5.5.3}$$

In the present section no external triggering has been assumed, so the other initial condition is

$$\varepsilon(x, t = 0) = 0. \tag{5.5.4}$$

For simplicity's sake, as before we restrict the consideration to unidirectional propagation and omit homogeneous broadening for a moment. Then in the linear approximation, $Z \approx 1$, the system of Maxwell–Bloch equations (5.4.1) and (5.4.2) can be solved using the Laplace transform technique introduced in section 5.4.1. Using the notations introduced in equations (5.4.5) we now obtain

$$\varepsilon(q,s) = \frac{1}{q + s/c - \varphi(s)/L} \frac{1}{L} \int \mathrm{d}\nu \frac{g(\nu)}{s + i\nu} R(q, 0, \nu). \qquad (5.5.5)$$

Taking the inverse Laplace transform of (5.5.5) with respect to q we obtain

$$\varepsilon(x,s) = \frac{1}{L} \int_0^x \mathrm{d}x' \, \mathrm{e}^{-(s-(c/L)\varphi(s))(x-x')/c} \int \mathrm{d}\nu \frac{g(\nu)}{s + i\nu} R(x', 0, \nu). \quad (5.5.6)$$

We again assume a Lorentzian inhomogeneous line shape and use $\varphi_L(s)$ from equation (5.4.8). (For a discussion of other line shapes, see [HHKSG81].) We can then perform the inverse Laplace transformation with respect to s, and after some manipulation we obtain

$$\varepsilon(x,t) = \frac{1}{L} \int_0^x \mathrm{d}x' \int \mathrm{d}\nu \, g(\nu) R(x - x', 0, \nu) K(x', t, \nu) \qquad (5.5.7)$$

where

$$K(x, t, \nu) = \theta\left(t - \frac{x}{c}\right) \mathrm{e}^{-(t-x/c)/T_2^*}$$

$$\times \left\{ I_0\left(2\left[\frac{x(t - x/c)}{LT_R}\right]^{1/2}\right)\right.$$

$$+ \int_0^{t-x/c} \mathrm{d}t' \, I_0\left(2\sqrt{xt'/LT_R}\right)\left(-i\nu + \frac{1}{T_2^*}\right)$$

$$\left. \times \exp\left[\left(-i\nu + \frac{1}{T_2^*}\right)\left(t - t' - \frac{x}{c}\right)\right]\right\}. \qquad (5.5.8)$$

The intensity of the field can now be calculated according to (5.1.6). Using the correlation function for the initial polarization (5.5.3), we obtain at $x = L$ [Ss80, HHKSG81]

$$I(L, t) = \frac{1}{NT_R L} \int_0^L \mathrm{d}x \, \theta\left(t - \frac{x}{c}\right)$$

$$\times \left[\mathrm{e}^{-2(t-x/c)/T_2^*} I_0^2\left(2\sqrt{x(t - x/c)/LT_R}\right)\right.$$

$$\left. + \frac{2}{T_2^*} \int_0^{t-x/c} \mathrm{d}t' \, \mathrm{e}^{-2t'/T_2^*} I_0^2\left(2\sqrt{xt'/LT_R}\right)\right]. \qquad (5.5.9)$$

This result can be simplified if we neglect retardation by setting $c = \infty$. Then in we can perform the x integration (5.5.9). The result can be written as the sum of two terms

$$I(L, t) = I^a(L, t) + I^b(L, t) \qquad (5.5.10)$$

where

$$I^a(L, t) = \frac{1}{NT_R} e^{-2t/T_2^*} \left[I_0^2\left(2\sqrt{t/T_R}\right) - I_1^2\left(2\sqrt{t/T_R}\right) \right] dt \quad (5.5.11)$$

and

$$I^b(L, t) = \frac{2}{NT_R T_2^*} \int_0^t dt' \, e^{-2t'/T_2^*} \left[I_0^2\left(2\sqrt{t'/T_R}\right) - I_1^2\left(2\sqrt{t'/T_R}\right) \right]$$

$$= \frac{2}{T_2^*} \int_0^t dt' \, I^a(t'). \quad (5.5.12)$$

The first term in (5.5.10), $I^a(L, t)$, describes a decaying pulse, whilst the second, $I^b(L, t)$ approaches a constant value when $t \to \infty$

$$I^b(L, \infty) = \frac{2}{NT_2^* T_R} \int_0^\infty dt \, e^{-2t/T_2^*} \left[I_0^2\left(2\sqrt{t/T_R}\right) - I_1^2\left(2\sqrt{t/T_R}\right) \right]. \quad (5.5.13)$$

Let us start by considering the first term $I^a(L, t)$ in (5.5.10). As we shall see below, in the case of high gain ($\alpha L = 2T_2^*/T_R \gg 1$) the duration of the pulse described by (5.5.11) is significantly longer than T_R and T_2^*. Therefore in the analysis of the results we may use the asymptotic form of the modified Bessel functions: $I_n(z) = (1/\sqrt{2\pi z}) \exp(z)$. Then for $I^a(L, t)$ we obtain

$$I^a(L, t) = \frac{e^{\alpha L}}{8\pi N t} \exp\left[-2\left(\sqrt{\frac{t}{T_2^*}} - \sqrt{\tfrac{1}{2}\alpha L} \right)^2 \right]. \quad (5.5.14)$$

This expression describes a pulse with full width at half maximum

$$\Delta t_{1/2} = 2T_2^* \sqrt{\alpha L \ln 2}. \quad (5.5.15)$$

Its maximum is delayed by

$$T_D = \tfrac{1}{2} T_2^* \alpha L \quad (5.5.16)$$

and the value of the maximal intensity is

$$I_{max}^a = \frac{e^{\alpha L}}{2\pi N T_R (\alpha L)^2}. \quad (5.5.17)$$

As has been already mentioned, in the limit $\alpha L \gg 1$ the duration of this pulse, $\Delta t_{1/2}$, is much longer than T_R and T_2^*.

In the derivation of (5.5.15)–(5.5.17) the factor $1/t$ multiplying the exponential in (5.5.14) was substituted by $1/T_D$. This approximation is valid for large gain, $\alpha L \gg 1$.

Applying analogous approximations to the second term in (5.5.10), $I^b(L, t)$, we obtain the following result

$$I^b(L, t) = \frac{e^{\alpha L}}{2\sqrt{\pi} N(\alpha L)^{3/2}} \left[1 + \mathrm{erf}\left(\sqrt{2t/T_2^*} - \sqrt{\alpha L}\right)\right] \quad (5.5.18)$$

where $\mathrm{erf}(z) = (2/\sqrt{\pi}) \int_0^z \mathrm{d}x\, e^{-x^2}$. As follows from equations (5.5.17)–(5.5.18)

$$\frac{I_{\max}^a}{I_{\max}^b} = \frac{1}{2\sqrt{\pi \alpha L}} \ll 1. \quad (5.5.19)$$

Therefore the dynamics of the radiation process is dominated in this model by the term $I^b(t)$, except for a short initial period. For $t \to \infty$ we have

$$I(L, \infty) = I^b(L, \infty) = \frac{1}{NT_R\sqrt{\pi}} \frac{e^{\alpha L}}{(\alpha L)^{3/2}}. \quad (5.5.20)$$

As the value of $I^b(L, t)$ is the time integral of $I^a(L, t)$ (see (5.5.12)), the characteristic time while $I(L, t)$ achieves $\frac{1}{2}I^b(L, \infty)$, is equal to $\Delta t_{1/2}$. The probability of achieving $I(L, \infty)$ in the linear régime is determined by the relation $I(L, \infty)\Delta t_{\frac{1}{2}}(L) < N$, which leads to the condition $\alpha L < \ln N$.

Let us now consider the case where the line broadening is predominantly homogeneous. As we have already mentioned in section 5.1, the correct treatment of homogeneous broadening via the term $-R/T_2$ demands the introduction of a source term for the polarization. Indeed, the relaxation damps the initial polarization, but actually there is a source, which is always present while the system is in an inverted state. To imitate this effect, we add a random noise term, Λ, to the equation determining the dynamics of the polarization, and we modify equation (5.4.2) in the following way

$$\frac{\partial R}{\partial t} = -\frac{1}{T_2}R + \frac{1}{T_R}\varepsilon + \Lambda(x, t). \quad (5.5.21)$$

(For the sake of simplicity we have here omitted the inhomogeneous broadening.)

The correlation function for the noise term can be obtained utilizing equation (5.5.1) in a standard way [28]

$$\langle \Lambda^*(x, t)\Lambda(x', t')\rangle = \frac{8L}{NT_2}\delta(x - x')\delta(t - t'). \quad (5.5.22)$$

Using the initial conditions

$$R(x, t = 0) = R_0(x) \qquad \varepsilon(x, t = 0) = 0 \quad (5.5.23)$$

and applying again the Laplace transform technique, we obtain

$$\varepsilon(x,t) = \frac{1}{L}\int_0^x dx'\, R_0(x-x')\, e^{-(t-x'/c)/T_2}$$

$$\times I_0\left(2\sqrt{x'(t-x'/c)/LT_R}\right)\theta(t-x'/c)$$

$$+\frac{1}{L}\int_0^t dt'\int_0^x dx'\, I_0\left(2\sqrt{x'(t'-x'/c)/LT_R}\right) e^{-(t-x'/c)/T_2}$$

$$\times\theta(t'-x'/c)\Lambda(x-x',t-t'). \tag{5.5.24}$$

Using the correlation functions (5.5.1) for the polarization and (5.5.22) for the source term we obtain for the intensity the same equation (5.5.9) as for the case of inhomogeneous broadening, but with T_2^* replaced by T_2.

We now derive the limit of the intensity for $t > L/c$, which is valid also for small values of αL. We integrate the second term in (5.5.9) (where T_2^* is replaced by T_2) by parts with respect to t' and find

$$I(L,t) = \frac{1}{NT_R}\left(1+\frac{1}{L}\int_0^L dx\int_0^{t-x/c} dt'\, e^{-2t'/T_2}\right.$$

$$\left.\times\frac{d}{dt'}I_0^2\left(2\sqrt{xt'/LT_R}\right)\right). \tag{5.5.25}$$

Using the formula [2]

$$\frac{dI_0^2(y)}{dy} = 2I_0(y)I_1(y) = \frac{dI_1^2(y)}{dy} + \frac{2I_1^2(y)}{y} \tag{5.5.26}$$

we can recast the integral into the form

$$I(L,t) = \frac{1}{NT_R}\left[1+\frac{1}{L}\int_0^L dx\int_0^{t-x/c} dt'\, e^{-2t'/T_2}\right.$$

$$\times\left(\frac{d}{dt'}I_1^2\left(2\sqrt{xt'/LT_R}\right)\right.$$

$$\left.\left.+\frac{1}{t'}I_1^2\left(2\sqrt{xt'/LT_R}\right)\right)\right]. \tag{5.5.27}$$

If the retardation is neglected, i.e. $t-x/c$ is replaced by t, then after integrating by parts with respect to x we arrive at the relatively simple result

$$I(L,t) = \frac{1}{NT_R}\left(1+\int_0^t dt'\,\frac{e^{-2t'/T_2}}{t'}I_1^2\left(2\sqrt{t'/T_R}\right)\right). \tag{5.5.28}$$

Figure 5.4 shows the dependence of the output intensity on time for several values of T_2.

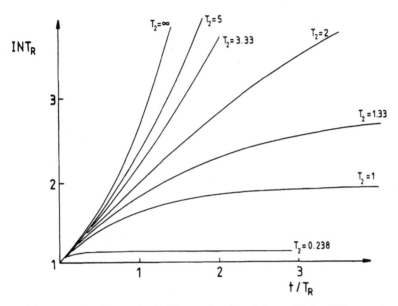

Figure 5.4. The radiated intensity $I N T_R$ as a function of time t/T_R for different relaxation times, as calculated from (5.5.28).

The integral in (5.5.28) is convergent for $t \rightarrow \infty$. Introducing $y = 2\sqrt{t'/T_R}$, we can express the integral in terms of the generalized hypergeometric function $_2F_2$ [27] as follows [BT86a]

$$I(L, \infty) = \frac{1}{N T_R}\left(1 + \frac{T_2}{2T_R}\,_2F_2\left(\tfrac{3}{2}, 1; 2, 3; \frac{2T_2}{T_R}\right)\right). \qquad (5.5.29)$$

We shall use this result in the next section to formulate a threshold condition for super-radiance in the presence of homogeneous or inhomogeneous broadening.

5.6 Threshold condition and induced super-radiance

Super-radiation is initiated by spontaneous fluctuations of the polarization in the medium. As has been stated earlier, in a system that consists of N inverted atoms the root mean square of the initial polarization is $\langle R_0^2 \rangle^{1/2} = 2/\sqrt{N}$. The smaller this quantity, the slower the evolution of the super-radiant pulse. As has been shown in Chapter 1, the delay time of super-radiation is $T_D = (T_R/4)(\ln \sqrt{\pi N})^2$.

If the relaxation times T_2 and T_2^* are less than T_D, then the super-radiant pulse does not have enough time to evolve. In such cases we observe either ordinary spontaneous emission or incoherent amplification. In activated crystals, where, compared with gases, relatively high inversion densities can be achieved, T_2 and T_2^* are of the order of a few picoseconds or less. Therefore the condition $T_R < T_2, T_2^*$ is satisfied, but the other condition, $T_D < T_2, T_2^*$ is not, because in solids the typical value of N is 10^{18}–10^{19}. Nevertheless it is still possible to obtain super-radiant pulses even in such systems, if the initial polarization is generated by a short pulse of given area. This initial polarization can prevail over the spontaneous fluctuations of R_0, which leads to a significant shortening of the delay time T_D. This process, which was investigated in section 5.4, is very close in principle to the process of coherent amplification of small-area pulses. We shall define the threshold condition for super-radiance as a condition by which the super-radiance intensity becomes equal to the intensity of ordinary spontaneous emission into a complete solid angle 4π. Let us recall that for a system with Fresnel number $F = 1$ the radiation is concentrated in the diffraction solid angle. So the threshold condition can be written in the form

$$I(L, \infty) = \gamma. \qquad (5.6.1)$$

Using the result (5.5.29) of the previous section and the definition (5.2.7) of the gain, αL, we have [BT86a]

$$I(L, \infty) = \frac{1}{NT_R} \left(1 + \tfrac{1}{4}\alpha L \, _2F_2(\tfrac{3}{2}, 1; 2, 3; \alpha L)\right). \qquad (5.6.2)$$

Figure 5.5 shows the dependence of $\log_{10}[I(\infty)NT_R]$ upon αL. Note that for large arguments the function $_2F_2$ has an exponential asymptotic behaviour.

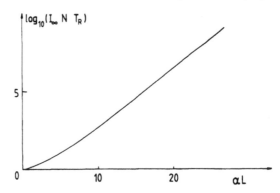

Figure 5.5. The stationary value of the super-radiant intensity as a function of the amplification exponent αL.

Using this graph one can estimate the value of the gain needed to observe super-radiance. Assuming, for example, $\gamma = 10^4$ s^{-1}, $T_2^* = 10$ ps, $N = 10^{16}$ (as in garnet and ruby), we find that $I(\infty) \geq \gamma$ is fulfilled for $\alpha L \geq 20$. If the gain is below this value, then usual spontaneous emission rather than super-radiance will determine the evolution of the atomic population.

Thus both the inhomogeneous and the homogeneous broadening can be responsible for the fact that an inverted system, even with high gain, may retain the inversion for a fairly long time (comparable with the time of radiative decay) and hence the inversion can be produced by relatively slow pumping. The weak intensity of the amplified radiation is owed to the smallness of the initial fluctuations of the field and polarization, which are proportional to $N^{-1/2}$. At the same time, nonlinear coherent amplification can be observed in such a system. If the duration of the input pulse is comparable with T_R, then the pulse will pass through the system without significant changes; it will create, however, a polarization which considerably exceeds the fluctuational value. Then the evolution of this externally created polarization generates super-radiation, which may be called *induced super-radiation* or *triggered super-radiance* [MT84, BT85, BT86a, ZMT88, MMV93]. Induced super-radiation was observed in Cs vapour in [VS79] and [CJSGH80]. This effect is of particular interest in solids, where pure super-radiance proves to be inhibited as a result of the non-negligible inhomogeneous broadening. The effect has been confirmed in the experiments [VKLMMT84] and [VGKMMBT86], where the active material was YAG:Nd and ruby. For a more detailed description of these experiments, see Chapter 2.

5.7 Solution of the nonlinear problem

In order to treat the full nonlinear evolution of the system we have to use numerical methods. Equations (5.1.7)–(5.1.9) have been solved with the spatially homogeneous initial condition $R(x, 0, v) = R_0$. The results of the calculations where the de-phasing processes and the retardation effects have been neglected ($T_2 = T_2^* = \infty$ and $L/c < T_D$) are shown in figure 5.6. The time dependence of the pulse at the end of the sample, figure 5.6(top), shows oscillations in time. As was discussed in Chapter 1, the reason for these oscillations is the reabsorption and coherent amplification of super-radiation during propagation through the sample. Correspondingly, as is seen in figure 5.6(bottom), the spectrum has a characteristic doublet structure, brought about by the optical nutations in the emitted field [MMT80, MMT81, MMT82]. The delay time, T_D, and the peak time of the first maximum, T_p, are in good agreement with the estimates given earlier in [MGF76, MGF81a]

$$T_D = \tfrac{1}{4} T_R \ln^2 \left(\frac{2\pi}{|R_0|} \right) \tag{5.7.1}$$

$$T_p = T_R \ln \left(\frac{2\pi}{|R_0|} \right). \tag{5.7.2}$$

We can see that both T_p and T_D are sufficiently larger than T_R. This is because of the smallness of the initial polarization $|R_0|$.

Now let us consider the effects of inhomogeneous broadening upon the nonlinear stage of super-radiance. The emitting atoms radiate at different frequencies, and this may have a considerable effect upon the process. Figure 5.7 shows the results of numerical solutions of the Maxwell–Bloch equations

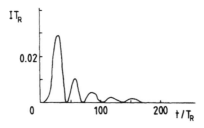

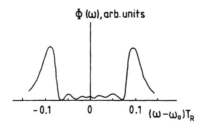

Figure 5.6. Time dependence of the intensity of a super-radiant pulse, and its spectrum Φ, without retardation, $R_0 = 4 \times 10^{-4}$.

for a system with inhomogeneous broadening in the absence of homogeneous broadening ($T_2^{-1} = 0$). In these calculations it has been assumed that the initial polarization is the same along the inhomogeneous contour and also along the length of the medium, that is, $|R(x, 0, v)| = R_0$. For the line shape $g(v)$ a Gaussian distribution (5.1.2) has been assumed. In the case of small inhomogeneous broadening ($T_2^* = 50T_R$: top part of figure 5.7) the emitted pulse differs only slightly from the sharp line result. Its spectrum is broader than the inhomogeneous line and the frequency distribution of the atoms remains approximately constant during the emission process. In this case atoms having different transition frequencies, but localized in a given domain of the system, radiate synchronously. As T_2^* decreases ($T_2^* = 20T_R$, and $T_2^* = 12.5T_R$: middle and bottom parts of figure 5.7) the spectrum of the pulse becomes narrower than the inhomogeneous line, and the number of excited atoms is depleted primarily near the centre of the inhomogeneous contour. A further increase of the inhomogeneous linewidth causes a sharp decrease in radiation intensity, and the number of cooperatively radiating atoms becomes small compared with the total number of excited atoms. Thus the radiation process changes towards a linear régime in which the population inversion may be assumed practically constant.

In order to demonstrate the effect of the random distribution of the initial polarization along the inhomogeneous contour, we present the results of a numerical calculation summarized in figure 5.8. The calculation was performed for a constant initial polarization as well as for two random realizations of initial

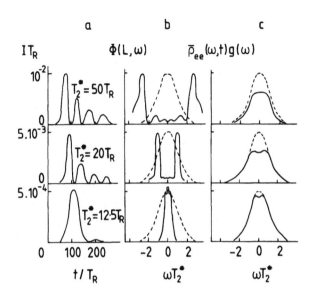

Figure 5.7. Effect of inhomogeneous broadening upon super-radiance. (a) super-radiance pulse; (b) super-radiance spectrum; (c) distribution of excited atoms over the inhomogeneous contour at time $t = 400T_R$ (broken curves denote the inhomogeneous contour).

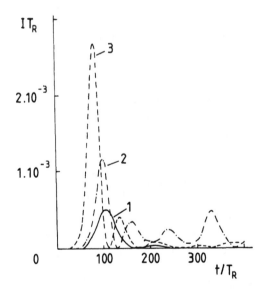

Figure 5.8. Realization of super-radiant pulses under different initial conditions, $T_2^* = 12.5T_R$, $L = cT_R$. (1) with uniform polarization amplitude, $R_0 = 2 \times 10^{-4}$; (2), (3) random distribution of the initial polarization both in frequency and in space.

polarization in the case of moderate inhomogeneous broadening, $T_2^* = 12.5T_R$, $L = cT_R$. The random initial polarization was approximated by a function of the form

$$R(x, 0, \nu) = \frac{2e^{i\phi}}{\sqrt{(\Delta x/L)Ng(\nu)\Delta\nu}} \qquad (5.7.3)$$

where Δx and $\Delta\nu$ are the intervals related to the length and to the inhomogeneous contour respectively, and ϕ is a random number uniformly distributed over the interval $[0, 2\pi]$.

It is evident from figure 5.8 that the stochasticity of the initial polarization (which simulates the quantum character of the emission process) appreciably affects the super-radiance pulse. As the inhomogeneous broadening increases, the realization of the pulse shows an increasing scatter. This means that the realization with a spatially homogeneous polarization can no longer be used as the characteristic one. This suggests that, in estimating the effects of inhomogeneous broadening on super-radiance, the spectrally homogeneous condition for the polarization can only be used when the broadening is not too large.

We shall now discuss homogeneous broadening of the super-radiant emission. We present the results of numerical calculations performed in order to follow the changes over a broad range of de-phasing times T_2. These calculations have been carried out for a system of length $L = 400cT_R$ with the following initial and boundary conditions: $Z(x, 0) = 1$, $R(x, 0) = R_0 = 4 \times 10^{-2}$, $\varepsilon(x, 0) = 0$, $\varepsilon(0, t) = 0$. The inhomogeneous broadening was neglected here. Figure 5.9 illustrates the change in the shape of the emission pulse as a function of T_2. Figure 5.9(a) for $T_2 = \infty$ shows a series of 2π pulses emerging unless $t < L/c$. They correspond to spatially homogeneous solutions of equations (5.1.7)–(5.1.9) and can be modelled in the following way. As was shown in section 1.4, in the absence of relaxation terms equations (5.1.7)–(5.1.9) can be recast in the simplified form of equation (1.4.24). In that equation we can set $\kappa = 0$ for a long system and for $t > L/c$. There exists an analytic solution of equation (1.4.24) for $\kappa = 0$, yielding

$$\varepsilon = -2i\sqrt{\frac{cT_R}{L}} \, \text{sech} \left[\sqrt{\frac{c}{LT_R}} (t - t_D') \right]. \qquad (5.7.4)$$

Note that $\sqrt{c/LT_R} = \Omega_0$, where Ω_0 is given by equation (1.4.14). These 2π pulses are seen in figure 5.9(a) if $t_D' < L/c$, where $t_D' = \sqrt{LT_R/c} \, \ln(4/R_0)$.

For sufficiently short de-phasing times (see figure 5.9(d)), the radiated intensity has roughly the form of an amplified spontaneous emission pulse (compare with figure 5.1). The field obtained in this latter calculation still does not have that very steep trailing edge shown in figure 5.1, because of the finiteness of T_2. The observation of the transition from super-radiance to ASE in ruby has been reported by Varnavsky *et al* [VGKMMBT86], and in KCl crystals with O_2^- centres by Malcuit *et al* [MMSB87, MMRB89].

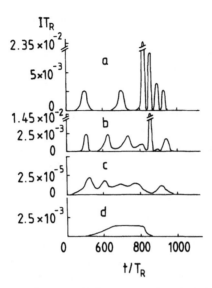

Figure 5.9. Shape modification of the cooperative emission with homogeneous broadening, $R_0 = 4 \times 10^{-2}$, $L = 400cT_R$, $T_2^* = \infty$. (a) $T_2 = \infty$; (b) $T_2 = 400T_R$; (c) $T_2 = 100T_R$; (d) $T_2 = 10T_R$.

5.8   Cooperative Raman scattering

We have discussed so far coherent emission processes from systems of two-level atoms. Several new and interesting phenomena may arise if we take into consideration a third level which is resonantly coupled to one of the other two. In the last two sections of this chapter we shall treat such effects.

One of them is Raman scattering of light incident on molecules. This is a second-order scattering process that results in the scattered light having a frequency which is lower or higher than the frequency of the incident photons, ω_0. The effect is due to partial energy transfer between the incident photons and some internal degrees of freedom of the molecules. The phenomenon is known as Stokes'-type Raman scattering if the frequency of the scattered light is lower than ω_0, and as anti-Stokes'-type Raman scattering if that frequency is higher than ω_0. In ordinary circumstances, e.g. under stationary excitation, the molecules interact with the light field independently. Then, even in the case of a strong incident field when a stimulated Raman effect may take place [15], one actually has an incoherent amplification process, where the intensity of the scattered field is proportional to the number of scattering centres.

We are going to discuss below spontaneous cooperative Raman scattering (CRS) which is different from the stimulated effect and also different from the so-called CARS (coherent anti-Stokes' Raman scattering) where an additional incident field at the scattering frequency is injected into the sample [15]. In spontaneous CRS we have only one incident field, but the atomic dipoles that oscillate at the scattering frequency, retain their phase memory and therefore they radiate cooperatively. This can be realized if the excitation and the scattering processes are fast enough to prevent phase relaxation destroying atomic coherence. If a level population is changed significantly during the interaction, then the intensity of Raman scattering can be proportional to the square of the number of the scattering centres, N^2, which is the characteristic feature of all super-radiant effects.

Spontaneous CRS has been observed by Pivtsov *et al* [PRSFC79] in molecular hydrogen (see also [RSC86]). Its properties have been investigated theoretically in several publications [C80, ES79a, ES79b, MS84a, RC77, RC79, RC80, Ru88, Sv80, Sv84a, Sv84b, Sl77]. In these works different régimes of spontaneous CRS and various approximations for describing them have been examined.

We shall consider some features of spontaneous CRS in extended systems where the propagation effects become significant. The semiclassical approach is particularly useful in this case. The main results for spontaneous CRS to be discussed below have been obtained by Trifonov *et al* [TTS80a, TTS80b, TTS83].

We use the three-level model for the atoms shown in figure 5.10, and relate

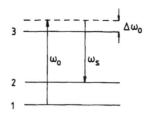

Figure 5.10. Raman scattering by a three-level model atom.

the transition $1 \to 3 \to 2$ to an elementary event of Stokes' Raman scattering $\hbar\omega_0 \to \hbar\omega_s$. Let us assume the following inequalities

$$|\omega_0 - \omega_{31}| \ll \omega_{21}$$
$$|d_{32}E_0|, \ |d_{31}E_s| \ll \hbar\omega_{21} \qquad\qquad (5.8.1)$$
$$|d_{21}E_0| \ll \hbar|\omega_0 - \omega_{21}| \qquad |d_{21}E_s| \ll \hbar|\omega_s - \omega_{21}|.$$

Here ω_0 is the frequency of the incident field, $\omega_s = \omega_0 - \omega_{21}$ is the scattered Stokes' frequency, d_{ik} are the transition dipole moments, E_0 is the field amplitude at the incident frequency and E_s is the Stokes' wave amplitude. The imposed conditions (5.8.1) allow us to neglect the effect of the incident field $\mathcal{E}_0$ on the transition $2 \leftrightarrow 3$ and the effect of the Stokes' field $\mathcal{E}_s$ on the transition $1 \leftrightarrow 3$. The transition $1 \leftrightarrow 2$ can be neglected, as well.

The equation of motion for a three-level atom is

$$\dot{\rho} = -\frac{\mathrm{i}}{\hbar}[\widehat{H}_0 + \widehat{V}, \rho] \tag{5.8.2}$$

where $\hat{\rho}$ is the density matrix of the atom with the matrix elements ρ_{11}, ρ_{22}, ρ_{33}, ρ_{21}, ρ_{31}, ρ_{32} and their complex conjugates, $\widehat{H}_0$ is the single-atom Hamiltonian, and $\widehat{V} = -\widehat{d}\mathcal{E}$ is the atom–field interaction in the dipole approximation. The electric field obeys the wave equation (1.4.6) in which the electric polarization $\mathcal{P}$ is expressed by the formula

$$\mathcal{P} = N_0 \operatorname{Tr}(\widehat{d}\hat{\rho}). \tag{5.8.3}$$

Below we shall use the scalar one-dimensional model, assuming, as before, the same directions for both the field polarization vector and all transition dipole moments of the atoms. Therefore the vector notation will be omitted in further considerations. We shall seek a solution of the wave equation (1.4.6) and of (5.8.2) in the following form

$$\begin{aligned}
\mathcal{E} &= \tfrac{1}{2}\{E_0 \exp[-\mathrm{i}(\omega_0 t - k_0 x)] \\
&\quad + E_s \exp[-\mathrm{i}(\omega_s t - k_s x)] + \text{c.c.}\}
\end{aligned} \tag{5.8.4a}$$

$$\rho_{21} = \tfrac{1}{2} R_{21} \exp\{-\mathrm{i}[\omega_{21} t - (k_0 - k_s)x]\} \tag{5.8.4b}$$

$$\rho_{31} = \tfrac{1}{2} R_{31} \exp[-\mathrm{i}(\omega_0 t - k_0 x)] \tag{5.8.4c}$$

$$\rho_{32} = \tfrac{1}{2} R_{32} \exp[-\mathrm{i}(\omega_s t - k_s x)]. \tag{5.8.4d}$$

Here $k_0 = \omega_0/c$, $k_s = \omega_s/c$, and R_{ik} are the slowly varying amplitudes of the matrix elements. Then the electric polarization $\mathcal{P}$ is given by the following expression

$$\begin{aligned}
\mathcal{P} &= \tfrac{1}{2} N_0 \{d_{13} R_{31} \exp[-\mathrm{i}(\omega_0 t - k_0 x)] \\
&\quad + d_{23} R_{32} \exp[-\mathrm{i}(\omega_s t - k_s x)] + \text{c.c.}\}.
\end{aligned} \tag{5.8.5}$$

Now, let us write down the fundamental semiclassical equations for the three-level system eliminating all fast (spatial and temporal) variations in the same way as we have done in section 1.4. Assuming that SVEA and RWA are valid we arrive at a system of equations of the type (1.4.11). The fields resonant with the transitions $1 \leftrightarrow 3$ and $3 \leftrightarrow 2$ are driven by the corresponding dipole moments, and the wave equation (1.4.6) takes the form

$$\frac{\partial E_0}{\partial x} + \frac{1}{c}\frac{\partial E_0}{\partial t} = \mathrm{i} 2\pi d_{31} k_0 N_0 R_{31} \tag{5.8.6a}$$

$$\frac{\partial E_s}{\partial x} + \frac{1}{c}\frac{\partial E_s}{\partial t} = \mathrm{i} 2\pi d_{32} k_s N_0 R_{32}. \tag{5.8.6b}$$

Introducing the (time-dependent) Rabi frequencies

$$D_0 = d_{31} E_0/\hbar \qquad D_s = d_{32} E_s/\hbar \qquad (5.8.7)$$

(5.8.2) can be written as

$$\frac{\partial \rho_{11}}{\partial t} = -\frac{i}{4}(D_0 R_{31}^* - D_0^* R_{31}) \qquad (5.8.8a)$$

$$\frac{\partial \rho_{22}}{\partial t} = -\frac{i}{4}(D_s R_{32}^* - D_s^* R_{32}) \qquad (5.8.8b)$$

$$\frac{\partial \rho_{33}}{\partial t} = \frac{i}{4}(D_0 R_{31}^* - D_0^* R_{31}) + \frac{i}{4}(D_s R_{32}^* - D_s^* R_{32}) \qquad (5.8.8c)$$

$$\frac{\partial R_{31}}{\partial t} = i\Delta\omega_0 + iD_0(\rho_{11} - \rho_{33}) + \frac{i}{2}D_s R_{21} \qquad (5.8.8d)$$

$$\frac{\partial R_{32}}{\partial t} = i\Delta\omega_0 + iD_s(\rho_{22} - \rho_{33}) + \frac{i}{2}D_0 R_{21}^* \qquad (5.8.8e)$$

$$\frac{\partial R_{21}}{\partial t} = \frac{i}{2}(D_s^* R_{31} - D_0 R_{32}^*). \qquad (5.8.8f)$$

Here $\Delta\omega_0 = \omega_0 - \omega_{31}$. There are no relaxation terms in these equations. By introducing a damping term of the form $-\Gamma_{21} R_{21}$ into the right-hand side of (5.8.8f) and looking for the stationary solution of the resulting system, we would obtain the rate equations describing the ordinary Raman effect [TTS80b]. In order to derive that result it is not necessary to introduce relaxation terms into the equations for R_{32} and R_{31} provided that the de-tuning $\Delta\omega_0$ is considerably larger than the transverse damping constants of these transitions. We shall not give here this derivation which follows the same route as the one described in section 5.2 where we have discussed the transition of coherent amplification into ordinary ASE. In the next section we shall see that the influence of both population and polarization relaxation related to level 3 may lead to novel effects.

In order to solve the above system numerically it is convenient to introduce the dimensionless variables $\tau = \Omega_0 t$ and $\xi = \Omega_0 x/c$, and dimensionless amplitudes $\epsilon_0 = -id_{31} E_0/\hbar\Omega_0$ and $\epsilon_s = -id_{31} E_s/\hbar\Omega_0$, where $\Omega_0 = (2\pi d_{31}^2 \omega_0 N_0/\hbar)^{1/2}$. In terms of these quantities, we then obtain the following equations for the slowly varying amplitudes

$$\frac{\partial \epsilon_0}{\partial \xi} + \frac{\partial \epsilon_0}{\partial \tau} = R_{31} \qquad (5.8.9a)$$

$$\frac{\partial \epsilon_s}{\partial \xi} + \frac{\partial \epsilon_s}{\partial \tau} = g_s \mu R_{32} \qquad (5.8.9b)$$

$$\frac{\partial R_{21}}{\partial \tau} = \tfrac{1}{2}(\mu R_{31}\epsilon_s^* + R_{32}^*\epsilon_0) \qquad (5.8.9c)$$

$$\frac{\partial R_{31}}{\partial \tau} = iq R_{31} + B\epsilon_0 - \tfrac{1}{2}\mu^* R_{21}\epsilon_s \qquad (5.8.9d)$$

$$\frac{\partial R_{32}}{\partial \tau} = iq R_{32} + \mu^* C\epsilon_s - \tfrac{1}{2}\epsilon_0 R_{21}^* \qquad (5.8.9e)$$

$$\frac{\partial B}{\partial \tau} = -\operatorname{Re}\left(\epsilon_0^* R_{31} + \tfrac{1}{2}\mu\epsilon_s^* R_{32}\right) \tag{5.8.9f}$$

$$\frac{\partial C}{\partial \tau} = -\operatorname{Re}\left(\tfrac{1}{2}\epsilon_0^* R_{31} + \mu\epsilon_s^* R_{32}\right) \tag{5.8.9g}$$

Here the following notation has been used

$$B = \rho_{33} - \rho_{11} \qquad C = \rho_{33} - \rho_{22}$$

$$\mu = \frac{d_{23}}{d_{13}} \qquad g_s = \frac{\omega_s}{\omega_0} \tag{5.8.10}$$

$$q = \frac{\Delta\omega_0}{.\Omega} \qquad \Delta\omega_0 = \omega_0 - \omega_{31}.$$

The initial and boundary conditions must guarantee the start of the evolution of the CRS because, as we know, the semiclassical approach does not describe spontaneous emission and spontaneous scattering at the Raman frequencies.

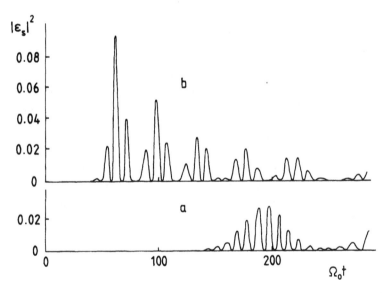

Figure 5.11. The Stokes' component (intensity at the output of the sample per number of atoms in arbitrary units): de-tuning $\Delta\omega_0 = 0$, initial polarization amplitude $R_{32}^0 = 0.002$, $\mu = g_s = 1$. Sample length $L = 0.5c/\Omega_0$ and input field $E_0^{(0)} = 0.5\hbar\Omega_0/|d_{31}|$ are fixed. Concentration of atoms: (a) N_0; (b) $4N_0$. According to (5.8.12), the quantity $|\epsilon_s|^2$ is proportional to I_s/N_0. The first maximum of pulse (b) is about 3.2 times higher than that of pulse (a), thus demonstrating the nonlinear dependence of the intensity on the number of active atoms. This is somewhat lower than a quadratic dependence on N_0, which would have given a factor of four.

The results of a numerical solution of the system (5.8.9a)–(5.8.9g) are presented in figure 5.11. It was assumed there that $\mu = g_s = 1$, and the following initial and boundary conditions were used

$$\epsilon_0(\xi, 0) = \begin{cases} \epsilon_0^{(0)} & \xi = 0 \\ 0 & 0 < \xi < l = \Omega_0 L/c \end{cases} \tag{5.8.11a}$$

$$\epsilon_s(\xi, 0) = 0 \tag{5.8.11b}$$

$$R_{21}(\xi, 0) = R_{31}(\xi, 0) = C(\xi, 0) = 0 \tag{5.8.11c}$$

$$R_{32}(\xi, 0) = R_{32}^{(0)} = 2 \times 10^{-3} \tag{5.8.11d}$$

$$B(\xi, 0) = 1 \tag{5.8.11e}$$

$$\epsilon_0(0, t) = \begin{cases} 0 & \tau < 0 \\ \epsilon_0^{(0)} & \tau > 0 \end{cases} \tag{5.8.11f}$$

$$\epsilon_s(0, \tau) = 0. \tag{5.8.11g}$$

The intensity of the CRS can be calculated from the expression

$$I_s(L, t) = \frac{c|E_s(L, t)|^2}{8\pi\hbar\omega_s N_0 L} = \frac{c}{4L}|\epsilon_s(l, \tau)|^2. \tag{5.8.12}$$

In the initial part of the process the exciting pulse creates inversion on the transition $1 \leftrightarrow 3$ and prepares the atomic system for a super-radiant transition between levels 3 and 2. During this period there is no scattered wave. After a certain delay, the scattered wave appears and, as shown in figure 5.11, it exhibits a typical doubly modulated signal. The modulation with the lower frequency, i.e. the pulse envelope, is similar to the series of super-radiant pulses seen in figure 5.6. The additional modulation at the higher frequency is generated by oscillations of the matrix element R_{32} with the frequency $|d_{31}E_0/\hbar|$. The shape of the pulses and the nonlinear dependence of their intensities on concentration shown in figure 5.11, allow us to interpret CRS as super-radiance observed in a transient régime with resonant quasi-stationary excitation.

More detailed calculations, with different values of the parameters, can be found in [TTS80a, TTS80b, TTS83].

5.9 Lasing without inversion

In this final section we describe briefly lasing without inversion, a phenomenon that has been widely discussed in the course of the last few years. Lasing without inversion may take place in three-level resonant atomic systems and, like super-radiance, results from atomic coherence. This effect was pointed out first by Kocharovskaya and Khanin [KK88] and by Harris [H89]. These authors paid attention to the fact that even without a positive population difference of the levels in question one may obtain a positive gain provided that an additional intermediate level coupled coherently to the upper level, is taken into account. The corresponding 'Λ' level scheme is depicted in figure 5.12. The field $\mathcal{E}_0$

affects the transition $1 \leftrightarrow 3$ whilst the field $\mathcal{E}_c$ produces a 'trapped' (see below) state of the transition $1 \leftrightarrow 2$. The latter is crucial for the effect in question. We shall consider the case where both fields are resonant with the corresponding transitions. The level scheme is actually the same as for Raman scattering, but here $\mathcal{E}_c$ is an external field which couples levels 3 and 2. The damping constant Γ describes the decay of the upper level to others, and it is assumed to be much larger than spontaneous emission rates of all other actual transitions which will be neglected further on.

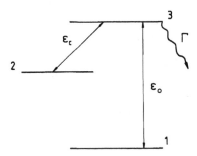

Figure 5.12. A scheme of lasing without inversion in the channel $1 \leftrightarrow 3$. $\mathcal{E}_0$ is the amplified field, $\mathcal{E}_c$ is the field producing a 'trapped' state of the transition $1 \leftrightarrow 2$, and Γ is the damping constant of the uppermost level 3. There is a positive stationary gain on the transition $1 \leftrightarrow 3$ if the atomic system starts from the intermediate level $|2\rangle$, while there is no amplification if the atom is initially in the upper state $|3\rangle$.

We shall calculate the gain for the transition $1 \leftrightarrow 3$ if the atoms in the interaction region are prepared in the intermediate state $|2\rangle$. The equations describing the dynamics of the field are essentially the same as (5.8.6a) and (5.8.6b), with E_s being replaced by E_c. Accordingly, the corresponding Rabi frequency (5.8.7) of the transition $3 \leftrightarrow 2$ will be denoted by D_c.

Assuming exact resonance, i.e. setting $\Delta\omega_0 = 0$, and introducing the damping of level 3, equations (5.8.8a)–(5.8.8f) are modified in the following way

$$\frac{\partial \rho_{11}}{\partial t} = -\frac{i}{4}(D_0 R_{31}^* - D_0^* R_{31}) \tag{5.9.1a}$$

$$\frac{\partial \rho_{22}}{\partial t} = -\frac{i}{4}(D_c R_{32}^* - D_c^* R_{32}) \tag{5.9.1b}$$

$$\frac{\partial \rho_{33}}{\partial t} = \frac{i}{4}(D_0 R_{31}^* - D_0^* R_{31}) + \frac{i}{4}(D_c R_{32}^* - D_c^* R_{32}) - \Gamma\rho_{33} \tag{5.9.1c}$$

$$\frac{\partial R_{31}}{\partial t} = iD_0(\rho_{11} - \rho_{33}) + \frac{i}{2}D_c R_{21} - \frac{\Gamma}{2}R_{31} \tag{5.9.1d}$$

$$\frac{\partial R_{32}}{\partial t} = iD_c(\rho_{22} - \rho_{33}) + \frac{i}{2}D_0 R_{21}^* - \frac{\Gamma}{2}R_{32} \tag{5.9.1e}$$

$$\frac{\partial R_{21}}{\partial t} = \frac{i}{2}(D_c^* R_{31} - D_0 R_{32}^*). \tag{5.9.1f}$$

It has been assumed here that the relation between the phase and population relaxation constants for level 3 is the same as for individual atoms [SSL74]. The solution of the system (5.9.1a)–(5.9.1f) with given field strengths, i.e. with given time-independent Rabi frequencies D_0 and D_c, has been obtained by Scully [Sy94] in the general case where the atoms are initially prepared in a coherent mixture: $a_0|1\rangle + b_0|2\rangle + c_0|3\rangle$. We quote here the result for the initial condition $b_0 = 1$ and $a_0 = c_0 = 0$, which corresponds to the initial values: $\rho_{11} = \rho_{33} = R_{12} = R_{23} = R_{31} = 0$, $\rho_{22} = 1$. Then

$$R_{31} = -i\frac{D_c^2 D_0}{(D_0^2 + D_c^2)D}\exp\left(-\frac{\Gamma}{2}t\right)\sin(Dt)$$
$$\times\left(-1 + \exp\left(-\frac{\Gamma}{2}t\right)\left[\cos(Dt) + \frac{\Gamma}{2D}\sin(Dt)\right]\right) \tag{5.9.2}$$

where $D^2 = D_0^2 + D_c^2 - \Gamma^2/4$.

We are going to look for stationary gain. Let us suppose that the atoms are injected into the interaction region with some injection rate r. Therefore we integrate R_{31} over time and multiply it by r. The result is

$$R_{31} = -i\frac{D_c^2 D_0}{2(D_0^2 + D_c^2)^2}r. \tag{5.9.3}$$

Substituting this expression into the equation for the field E_0 (5.8.6a), we see that the coefficient of E_0 on the right-hand side is $(\pi d_{31}^2 k_0 N_0/\hbar)(D_c^2/(D_0^2 + D_c^2)^2)r$. This is the gain coefficient for E_0, and it is always positive, i.e. the atoms contribute to the amplification of the emerging field. If the atoms are injected into the interaction region on the upper level there will be no stationary gain because of saturation of the transition $1 \leftrightarrow 3$. If, however, they begin their evolution from state $|2\rangle$ they can be 'trapped' in a time-independent (in the rotating frame) superposition state $|a\rangle = (D_c|1\rangle - D_0|2\rangle)/(D_0^2 + D_c^2)^{1/2}$, and are not destined to leave via decay from the state $|3\rangle$ with zero net stimulated emission [Sy94].

If we now calculate the stationary population of the upper level starting from the solution for ρ_{33} [Sy94]

$$\rho_{33} = \frac{D_c^2}{D^2}\exp(-\Gamma t)\sin^2(Dt) \tag{5.9.4}$$

we obtain

$$\rho_{33}^{st} = \frac{r}{2\Gamma}\frac{D_c^2}{D_0^2 + D_c^2}. \tag{5.9.5}$$

The latter equation shows that in spite of having amplification, we do not have more than half of the atoms on level 3. Thus amplification (a laser) without inversion is possible provided we include the effects of quantum coherence and population trapping. We refer for more details to the special issue of the journal *Quantum Optics* [75] with material from the Crested Butte Conference on Atomic Coherence Effects, where several other papers devoted to the current progress in these and related topics can be found. We discuss somewhat further the possibility of a super-radiant laser on a three-level system in Chapter 11.

5.10 Concluding remarks

The semiclassical method is a powerful tool that is useful for the treatment of propagation effects in super-radiance as well as in such coherent emission processes as self-induced transparency, photon echo and others. The use of the Maxwell–Bloch equations allows one to consider these phenomena in elongated systems. In Chapter 8 we shall extend the semiclassical treatment to more than one spatial dimension, and in Chapters 9 and 10 to the case where the boundary of the medium is explicitly taken into account. A treatment of the problem of propagation within the quantum theory of these phenomena would be a much more difficult task.

Fortunately, the semiclassical approach is readily applicable to low-density systems, i.e. to gases. However, one has to exercise greater care when applying it to coherent emission in solids. In high-density systems there is a strong mixing of the field and atomic states in the vicinity of resonance. Such a mixing results, as is well known (see, for example, [14]), in polaritonic states. These collective states decay in a distinctly different manner as compared with the original atomic states. So, we expect that the usual way of introducing relaxation into the material equations, by means of the two constants T_1 and T_2, would break down for systems of high atomic density. The correct approach in this case requires quantum theory, at least for handling the relaxation phenomena.

Chapter 6

The influence of dipole–dipole inter-atomic coupling upon super-radiance

The influence of the dipole–dipole inter-atomic coupling on collective spontaneous emission has been considered by many authors using both semiclassical [SELM72, ZMT83] and quantum [FHM72, M74, CF78, NS75, NS76, NW80, Sl80, Sl85, R83, AZM85, ASS87] theories. Specific results have been obtained for simple cases of regular atomic systems with a finite number of atoms [CF78, Ss80, Sl85]. For disordered systems it has been found that the dipole–dipole coupling causes relaxation of the polarization phase (the effect of de-phasing), thus suppressing super-radiance. In the present chapter the main object of study will be a regular linear chain of identical two-level atoms fully excited in the initial moment of time. Apart from being amenable to a rigorous mathematical treatment, this model is an important limiting case of a multi-atomic system with a small Fresnel number, $F \ll 1$. In our treatment of the dipole–dipole coupling for the case of a linear chain we shall follow [ZMT83] and [AZM85].

6.1 Preliminary remarks

In previous chapters we have considered the atom–field interaction by taking into account only the transverse component of the electromagnetic field. The latter increases the phase correlation of atomic dipoles during the emission process and gives rise to the emission of the intensity which is proportional to the square of the density of the inverted atomic levels. For sufficiently dense systems with average atomic spacing, a, in a sample smaller than the wavelength of the radiation, the dipole–dipole interaction between atoms, $\mathcal{E}_D$, can compete with the interaction with the radiation field $\mathcal{E}_R$. The electric dipole near-zone field $\mathcal{E}_D$ is of the order of d/a^3 at a distance a from the atomic dipole, whereas the radiation field $\mathcal{E}_R$ is of the order of $\hbar d / T_R$, where T_R is the super-radiance time, which depends upon the sample geometry.

Let us consider a system of two-level atoms in a volume with linear dimension L which is less than the wavelength of the radiation ($L \ll \lambda$). For such a system the super-radiance time $T_R \sim 1/\gamma N$, where $\gamma^{-1} = 3\hbar\bar{\lambda}^3/4d^2$ is the spontaneous emission time of a single atom, $\bar{\lambda} = \lambda/2\pi$, and N is the number of atoms in the system (see Chapter 1). Then for the ratio $\mathcal{E}_D/\mathcal{E}_R$ we obtain

$$\frac{\mathcal{E}_D}{\mathcal{E}_R} \sim \frac{3\bar{\lambda}^3}{4Na^3} \sim \left(\frac{\bar{\lambda}}{L}\right)^3.$$

Thus, for a point system ($L \ll \bar{\lambda}$) the field $\mathcal{E}_D$ is much larger than the field $\mathcal{E}_R$, and it should be taken into account.

If the system extends predominantly in one direction only, i.e. if $L \gg \bar{\lambda}$, then it is characterized by a Fresnel number $F = D^2/L\lambda < 1$. For such a system the super-radiance time $T_R \sim a/\gamma\bar{\lambda}$ (see Chapter 3) and $\mathcal{E}_D/\mathcal{E}_R \sim (\bar{\lambda}/a)^2 \gg 1$. Thus the near-zone field $\mathcal{E}_D$ is stronger than the radiation field $\mathcal{E}_R$, similarly to the case of a point system.

Another limit of a multi-atomic system is the case of a large Fresnel number $F = D^2/L\lambda > 1$. Then $T_R \sim 1/\gamma N_0\bar{\lambda}^2L$, where N_0 is the density of the population inversion and $\mathcal{E}_D/\mathcal{E}_R \sim N_0a^3\bar{\lambda}/L$. Taking into account that $N_0a^3 \sim 1$, we obtain $\mathcal{E}_D/\mathcal{E}_R \sim \bar{\lambda}/L$. This estimate shows that for a system with large Fresnel number the radiative field $\mathcal{E}_R$ is stronger than the near-zone field $\mathcal{E}_D$ if $L \gg \bar{\lambda}$ and it is smaller than the latter field if $L \ll \bar{\lambda}$.

To summarize, we can say that the dipole–dipole inter-atomic coupling will influence super-radiance emission if at least one of the dimensions of the atomic system is smaller than the wavelength of the emitted light.

Let us consider a system of two identical two-level atoms separated by a distance $a \ll \lambda$, which are perturbed by the dipole–dipole interaction $\widehat{V}$.

For the unperturbed system, $\widehat{V} = 0$, the energy levels and corresponding wavefunctions are given by equations (1.1.3) and (1.1.4). Let us now switch on the interaction, making $\widehat{V} \neq 0$. This will cause splitting of the two-fold degenerate intermediate energy level $E_g + E_e$ into a doublet, whilst the bottom and top energy levels will remain unchanged. To the first order the perturbed energy levels are

$$2E_g \qquad E_g + E_e - |\langle\widehat{V}\rangle| \qquad E_g + E_e + |\langle\widehat{V}\rangle| \qquad 2E_e \quad (6.1.1)$$

where

$$\langle\widehat{V}\rangle = \langle 1g, 2e|\widehat{V}|1e, 2g\rangle \qquad (6.1.2)$$

whilst the eigenstates and the dipole moment of the system remain the same as those for the unperturbed system. Therefore radiative transitions are possible between symmetric states only (see Chapter 1). The level structure and the allowed radiative transitions in the system of two coupled identical two-level

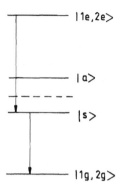

Figure 6.1. Collective energy levels of a close pair of two-level atoms with dipole–dipole inter-atomic coupling.

atoms are shown in figure 6.1. The relative position of the 'a' and 's' levels depends upon the sign of the matrix element of the dipole–dipole operator $\widehat{V}$. The frequencies of the allowed transitions differ by the value of the energy splitting, $2|\langle\widehat{V}\rangle|$. As in the case of the unperturbed system considered in Chapter 1, the dipole moments of the allowed transitions are enhanced by a factor of $\sqrt{2}$. Thus the dipole–dipole interaction does not change the transition characteristic of the two-atom system, but may considerably change the spectrum.

It will be shown below that the main conclusions derived from the model of two two-level atoms with dipole–dipole interaction are valid in the more general case of a linear regular system of atoms with equidistant separation. In the latter case the time dependence of the super-radiance intensity remains unchanged, but the super-radiance spectrum changes considerably compared with the case where the dipole–dipole interaction is neglected.

6.2   Dipole–dipole interaction in the semiclassical theory

Let us consider a system of N identical two-level atoms whose arbitrary positions in space are given by the position vectors r_k, $k = 1, \ldots, N$. In our treatment of this system we shall apply the semiclassical theory. We recall that in this theory the two-level atom is described by a single-particle density matrix ρ_{ab}, where the indices a and b label possible states of the atom. For the density matrix $\rho_{ab}^{(k)}$ of the kth atom, the semiclassical equations are as follows (see equations (1.4.8))

$$\dot{\rho}_{gg}^{(k)} = -i\frac{d_k \mathcal{E}_k}{\hbar}[\rho_{ge}^{(k)} - \rho_{eg}^{(k)}]$$

$$\dot{\rho}_{ee}^{(k)} = i\frac{d_k \mathcal{E}_k}{\hbar}[\rho_{ge}^{(k)} - \rho_{eg}^{(k)}]$$

$$(6.2.1)$$

$$\dot{\rho}_{eg}^{(k)} = -i\omega_0\rho_{eg}^{(k)} + i\frac{d_k\mathcal{E}_k}{\hbar}[\rho_{gg}^{(k)} - \rho_{ee}^{(k)}]$$

$$\dot{\rho}_{ge}^{(k)} = \rho_{eg}^{(k)*}.$$

where d_k is the transition dipole moment of the kth atom, and $\mathcal{E}_k$ is the electric field acting on the kth atom at the position r_k. This field is a superposition of the microscopic fields $\mathcal{E}_{lk}$ produced at the point r_k by all the remaining lth atoms, plus the self-action field $\mathcal{E}_{kk}$ (see later on)

$$\mathcal{E}_k = \sum_l \mathcal{E}_{lk}. \tag{6.2.2}$$

The electromagnetic field emitted by the lth atom is treated as the field of a classical dipole with moment $D_l(t)$ equal to the mean quantum mechanical dipole moment $D_l(t) = d_l[\rho_{12}^{(l)}(t) + \rho_{21}^{(l)}(t)]$. The electric field $\mathcal{E}_{lk}$ produced by the lth atom at the position $k \neq l$ may be written as follows (see e.g. [46])

$$\mathcal{E}_{lk}(t) = \left[\frac{3D_l(t')}{r_{lk}^3} + \frac{3\dot{D}_l(t')}{cr_{lk}^2} + \frac{\ddot{D}_l(t')}{c^2r_{kl}}\right](m_l n_{lk})n_{lk}$$
$$- \left[\frac{D_l(t')}{r_{lk}^3} + \frac{\dot{D}_l(t')}{cr_{lk}^2} + \frac{\ddot{D}_l(t')}{c^2r_{kl}}\right]m_l \tag{6.2.3}$$

where

$$t' = t - \frac{r_{lk}}{c} \qquad m_l = \frac{d_l}{d_l} \qquad r_{lk} = r_l - r_k \qquad n_{lk} = \frac{r_{lk}}{r_{lk}}.$$

Allowance for the self-action field $\mathcal{E}_{kk}$ in equation (6.2.2) can be made using the energy balance (as in the classical theory of spontaneous emission)

$$\frac{dU_a}{dt} = -I \qquad I = \int\int S\left(t - \frac{r}{c}\right)d\sigma. \tag{6.2.4}$$

Here U_a is the energy stored in the atomic subsystem, I is the total radiation intensity, and S is the Poynting vector averaged over the optical period. The integration in equation (6.2.4) is performed over a sphere of radius $r \gg L$. We note that for systems containing a large number of atoms, the effect of the self-action field on the development of super-radiance turns out to be insignificant, and we shall neglect it in what follows.

6.2.1 The slowly varying envelope approximation in time

Let us assume that the characteristic time of the variation of the amplitudes of both the electromagnetic fields $\mathcal{E}_k$ and $\mathcal{E}_{lk}$, and the off-diagonal element of the density matrix $\rho_{ge}^{(k)}$ are significantly longer than the optical period $2\pi/\omega_0$. Then we separate the fast dependences in the atomic and field characteristics as shown

$$\rho_{eg}^{(k)}(t) = \frac{1}{2}R^{(k)}(t)e^{-i\omega_0 t}. \tag{6.2.5}$$

Substituting equation (6.2.5) into equations (6.2.2) and (6.2.3), we obtain for $\mathcal{E}_{lk}$ and $\mathcal{E}_k$ expressions similar to equation (6.2.5)

$$\mathcal{E}_{lk}(t) = \tfrac{1}{2} E_{lk}(t)\, e^{-i\omega_0 t} + \text{c.c.}$$

$$\mathcal{E}_k(t) = \tfrac{1}{2} E_k(t)\, e^{-i\omega_0 t} + \text{c.c.} \qquad (6.2.6)$$

where the amplitudes $E_{lk}(t)$ and $E_k(t)$ are given by the formulae

$$
E_{lk} = \left\{ \left[\frac{3}{r_{lk}^3} - \frac{3ik_0}{r_{lk}^2} - \frac{k_0^2}{r_{lk}} \right] (d_l n_{lk}) n_{lk} \right.
$$
$$
\left. - \left[\frac{1}{r_{lk}^3} - \frac{ik_0}{r_{lk}^2} - \frac{k_0^2}{r_{lk}} \right] d_l \right\} R^{(l)}(t')\, e^{ik_0 r_{lk}} \qquad (6.2.7)
$$

$$E_k(t) = \sum_{l \neq k} E_{lk}(t). \qquad (6.2.8)$$

Retardation of the interaction manifests itself in the space-dependent oscillating exponential factors in equation (6.2.7) as well as in the argument of the amplitude of the off-diagonal element of the density matrix, $R^{(l)}$. We shall neglect the retardation in the amplitudes, assuming that the time for light to propagate through the system, L/c, is shorter than the characteristic super-radiance time T_R. Note that this condition imposes a limitation upon the number of atoms in the system.

Substituting equations (6.2.5) into equations (6.2.1) and neglecting the rapidly oscillating terms (i.e. using the SVEA in time and the RWA) we obtain the following system of equations for the slowly varying amplitudes

$$\dot{R}^{(k)} = -i \frac{d_k E_k}{\hbar} Z^{(k)}$$

$$\dot{Z}^{(k)} = i \frac{d_k E_k}{\hbar} R^{(k)*} + \text{c.c.} \qquad (6.2.9)$$

$$Z^{(k)} = \rho_{ee}^{(k)} - \rho_{gg}^{(k)}.$$

Using (6.2.9) it is not difficult to verify that the square of the modulus of the Bloch vector, $Z^{(k)2} + |R^{(k)}|^2$, is conserved for each atom.

The initial conditions for equations (6.2.9) are chosen in the usual way for the semiclassical approach: all atoms are assumed to be excited, and a small initial polarization R_0 stimulating the spontaneous fluctuations is introduced at the initial instant of time

$$Z^{(k)}(0) = 1 \qquad R^{(k)}(0) = R_0. \qquad (6.2.10)$$

In the derivation of the reduced system (6.2.9) we did not extract the spatial factor $\exp(ik_0 r)$, as is often done in the semiclassical formulation. This allows us to follow the variations that occur in the behaviour of the super-radiance as we go from the Dicke model ($L \ll \tilde{\lambda}$) to the extended system ($L \gg \tilde{\lambda}$).

Equations (6.2.9) can be rewritten in the following form

$$\dot{R}^{(k)} = \sum_{l \neq k} (i\Delta_{lk} + \gamma_{lk}) R^{(l)} Z^{(k)}$$

$$\dot{Z}^{(k)} = -\tfrac{1}{2} i \sum_{l \neq k} \Delta_{lk} (R^{(l)} R^{(k)*} - R^{(l)*} R^{(k)})$$

$$\qquad\qquad -\tfrac{1}{2} \sum_{l \neq k} \gamma_{lk} (R^{(l)} R^{(k)*} + R^{(l)*} R^{(k)}) \qquad (6.2.11)$$

where the matrices Δ_{lk} and γ_{lk} are given by the formulae

$$\Delta_{lk} = \frac{1}{\hbar} \left[\left(\frac{\chi_{lk}}{r_{lk}^3} - k_0^2 \frac{\kappa_{lk}}{r_{lk}} \right) \cos(k_0 r_{lk}) + k_0 \frac{\chi_{lk}}{r_{lk}^2} \sin(k_0 r_{lk}) \right] \qquad (6.2.12)$$

$$\gamma_{lk} = \frac{1}{\hbar} \left[\left(k_0^2 \frac{\kappa_{lk}}{r_{lk}} - \frac{\chi_{lk}}{r_{lk}^3} \right) \sin(k_0 r_{lk}) + k_0 \frac{\chi_{lk}}{r_{lk}^2} \cos(k_0 r_{lk}) \right] \qquad (6.2.13)$$

$$\chi_{lk} = (d_l d_k) - 3(d_l n_{lk})(d_k n_{lk})$$
$$\kappa_{lk} = (d_l d_k) - (d_k n_{lk})(d_l n_{lk}). \qquad (6.2.14)$$

The matrix γ_{lk} is identical to the relaxation matrix calculated in quantum theory using the Hamiltonian which contains the interaction between the atoms and the transverse field only. It describes the decrease, resulting from emission, of the energy stored in the atomic subsystem. This can easily be verified by summing the second equation of the system (6.2.11) over the atoms

$$\sum_k Z^{(k)} = \sum_{l,k \neq l} \gamma_{lk} R^{(l)} R^{(k)*}. \qquad (6.2.15)$$

The matrix Δ_{lk} describes the frequency shifts caused by the interaction of the atomic system with longitudinal and transverse components of the field. It also determines the re-distribution (migration) of excitation amongst the atoms of the system.

The matrices γ_{lk} and Δ_{lk} have a simpler form in the case of a small system ($L \ll \tilde{\lambda}$). It follows from equations (6.2.12) and (6.2.13), for $k_0 r_{lk} \ll 1$, that

$$\Delta_{lk} = \frac{\chi_{lk}}{\hbar r_{lk}^3} \qquad (6.2.16)$$

$$\gamma_{lk} = \frac{2d^2 k_0^3}{3\hbar} = \tfrac{1}{2}\gamma. \qquad (6.2.17)$$

Here we leave out the dominant contribution to the matrix Δ_{lk} resulting from the dipole–dipole interaction, and the matrix elements γ_{lk} become equal to half of the radiative constant γ of a single atom.

We shall determine the intensity in a given direction of radiation at an observation point as the Poynting vector in the wave zone $r \gg L, \tilde{\lambda}$ averaged over the optical period $2\pi/\omega_0$

$$S(r, t) = \frac{c}{4\pi}\overline{[\mathcal{E} \times \mathcal{H}]} \qquad (6.2.18)$$

where $\mathcal{E}$ and $\mathcal{H}$ are the electric and magnetic fields produced by the system at the point r. When $r \gg L, \tilde{\lambda}$ the vectors $\mathcal{E}$ and $\mathcal{H}$ are of equal magnitude, and, making use of equations (6.2.7) and (6.2.8), we obtain the following expression for the field $\mathcal{E}$

$$\mathcal{E}(r, t) = k_0^2 \sum_l \frac{d_l - (d_l n_l)n_l}{|r - r_l|}|R^{(l)}(t')| \cos(\omega_0 t' + \varphi_l(t')). \qquad (6.2.19)$$

Here $|r - r_l|$ is the distance from the lth dipole to the point of observation, which can be assumed to be r, and $n_l = r_l/r_l$, $t' = t - r_l/c$, and φ_l is the phase of $R^{(l)}$. The time scale of the variation of φ_l, as well as that of $|R^{(l)}|$, is significantly greater than $2\pi/\omega_0$. Therefore we shall assume them to be constants in the averaging over the optical period in equation (6.2.18). Hence

$$S(r, t) = n\frac{ck_0^4}{8\pi r^2} \sum_{l,l'}[(d_l d_{l'}) - (d_l n)(d_{l'}n)]R^{(l)*}(t')R^{(l')}(t')\, e^{ik_0(r_l - r_{l'})}.$$

$$(6.2.20)$$

where n is the unit vector along r.

6.2.2   Super-radiance of a regular linear chain of two-level atoms

The expressions obtained above are valid for any spatial configuration of atoms. Below we shall consider the super-radiance of a regular linear chain of atoms. Such a system is the limiting case of a multi-atomic system characterized by a small Fresnel number. It includes the point system ($L \ll \tilde{\lambda}$) as a particular case.

We shall assume that dipole matrix elements d_l of individual atoms are directed normally with respect to the chain. In this case the expression for the Poynting vector takes a simplified form

$$S(r, t) = n\frac{\hbar\omega_0}{4\pi r^2}\left[1 - \frac{(dn)^2}{d^2}\right]F(t) \qquad (6.2.21)$$

$$F(t) = \frac{3\gamma}{8} \sum_{l,l'} R^{(l)*}(t')R^{l'}(t')e^{ik_0 r_{ll'} \cos\theta_0}. \qquad (6.2.22)$$

where θ_0 is the angle between the axis of the system and the direction of observation, n. The factor in the square brackets in equation (6.2.21) is the directivity pattern of a single dipole, and $F(t)$ is a factor arising from the interference of the fields of the atomic dipoles.

The spatially homogeneous model. We start our discussion of a regular linear chain of N atoms by considering a case in which the density matrices of all atoms are assumed to be identical. Strictly speaking, such a model may be justified only for an infinite chain. The usefulness of this model consists in allowing us to obtain an analytical solution of the problem in which dipole–dipole interaction is included.

Let $R^{(k)} = R$ and $Z^{(k)} = Z$ in equations (6.2.11). Then we have

$$\dot{R} = \left(i\Delta + \frac{1}{T_R} \right) RZ \qquad \dot{Z} = -\frac{1}{T_R}|R|^2. \qquad (6.2.23)$$

The expressions for Δ and T_R follow from equations (6.2.12), (6.2.13) and (6.2.17), with allowance made for the condition stipulated above

$$\Delta = \frac{3\gamma}{2(k_0 a)^3} \sum_{l=1}^{N/2} \left\{ \left[\frac{1}{l^3} - \frac{(k_0 a)^2}{l} \right] \cos(k_0 a l) + \frac{k_0 a}{l^2} \sin(k_0 a l) \right\} \qquad (6.2.24)$$

$$\frac{1}{T_R} = \frac{3\gamma}{2(k_0 a)^3} \sum_{l=1}^{N/2} \left\{ \left[\frac{(k_0 a)^2}{l} - \frac{1}{l^3} \right] \sin(k_0 a l) + \frac{k_0 a}{l^2} \cos(k_0 a l) \right\} \qquad (6.2.25)$$

where a is the lattice constant.

Summing the series in equations (6.2.24) and (6.2.25) in the limit $k_0 a \ll 1$, we obtain

$$\Delta = \tfrac{3}{2}\zeta(3)\gamma \left(\frac{\tilde{\lambda}}{a} \right)^3 = \tfrac{3}{2}\zeta(3)\frac{\gamma}{(k_0 a)^3} \qquad (6.2.26)$$

$$\frac{1}{T_R} = \tfrac{1}{2}\gamma N \qquad L \ll \tilde{\lambda} \qquad (6.2.27)$$

$$\frac{1}{T_R} = \frac{3\pi}{4}\frac{\gamma\tilde{\lambda}}{a} = \frac{3\pi}{4}\frac{\gamma}{k_0 a} \qquad L \gg \tilde{\lambda} \qquad (6.2.28)$$

where $\zeta(x)$ is the Riemann zeta function. As can be seen from equation (6.2.28), for an extended chain ($L \gg \tilde{\lambda}$) the super-radiance constant T_R^{-1} is determined by the number $N_a = \tilde{\lambda}/a$ of atoms located at a distance equal to the wavelength $\tilde{\lambda}$, rather than by the total number of atoms in the chain. It follows from equations (6.2.26) and (6.2.28) that $T_R \Delta \gg 1$, a result which is independent of the ratio $L/\tilde{\lambda}$. Therefore the process of collective emission is significantly slower than the rate of migration of the population inversion.

Dynamics of the super-radiance pulse. Using the conservation law for the Bloch vector, $Z^2 + |R|^2 = 1 + |R_0|^2 \approx 1$, we integrate equation (6.2.23) for the population inversion to obtain

$$Z(t) = -\tanh\left(\frac{t - t_D}{T_R} \right). \qquad (6.2.29)$$

Here t_D is the delay time of the super-radiance pulse, which is given by

$$t_D = -T_R \ln \tfrac{1}{2}|R_0|. \tag{6.2.30}$$

Having obtained $Z(t)$, equation (6.2.23) for R yields

$$R(t) = \frac{R_0}{|R_0|} \operatorname{sech}\left(\frac{t - t_D}{T_R}\right) e^{i\varphi(t)} \tag{6.2.31}$$

where

$$\varphi(t) = \Delta \int_0^t Z(t')\,dt'$$
$$= \Delta T_R \ln\left[\cosh\left(\frac{t_D}{T_R}\right) \operatorname{sech}\left(\frac{t - t_D}{T_R}\right)\right]. \tag{6.2.32}$$

The radiation intensity is then given by

$$I(t) = -N\hbar\omega_0 \dot{Z} = \frac{N\hbar\omega_0}{T_R} \operatorname{sech}^2\left(\frac{t - t_D}{T_R}\right). \tag{6.2.33}$$

Notice that the parameter Δ, which is the inverse of the oscillation frequency for neighbouring atoms, does not enter into the expressions for Z and I. Thus, in the case of a *spatially homogeneous* system the dipole–dipole interaction between atoms has no effect upon the super-radiance dynamics.

Self-phase modulation of the super-radiant pulse. Expression (6.2.31) for R differs from that corresponding to the case without dipole–dipole coupling (i.e. when $\Delta = 0$) in that it has a time-dependent phase factor describing the self-phase modulation.

At the initial stage of the process, when $t < t_D$

$$R(t) = \frac{R_0}{|R_0|} \operatorname{sech}\left(\frac{t - t_D}{T_R}\right) e^{i\Delta t}. \tag{6.2.34}$$

Later, at the end of the process, when $t > t_D$

$$R(t) = \frac{R_0}{|R_0|} \operatorname{sech}\left(\frac{t - t_D}{T_R}\right) e^{-i\Delta(t - 2t_D)}. \tag{6.2.35}$$

A comparison of equations (6.2.34) and (6.2.35) shows that the phase of the amplitude of the off-diagonal density matrix element R changes sign in the course of the decay of the super-radiance pulse. This means that there is an increase of the super-radiance frequency from the value $\omega_0 - \Delta$ to $\omega_0 + \Delta$ in accordance with the following law

$$\omega(t) = \omega_0 + \dot\varphi(t) = \omega_0 + \Delta \tanh\left(\frac{t - t_D}{T_R}\right). \tag{6.2.36}$$

Correspondingly, the super-radiance spectrum becomes broadened by an amount 2Δ, since $\Delta \gg 1/T_R$, and this broadening will determine the width of the super-radiance spectrum.

Let us obtain the spectrum of the super-radiance pulse in the wave zone $r \gg L, \tilde{\lambda}$. Defining the spectrum as $\Phi(\omega) = |\tilde{\mathcal{E}}(\omega)|^2/T_R$ where $\tilde{\mathcal{E}}(\omega)$ is the Fourier transform of the electric field in the wave zone, we have

$$\Phi(\omega) = \frac{k_0^4 d^2}{r^2 T_R} \left[1 - \frac{(nd)^2}{d^2} \right] \{ |\tilde{R}(\omega - \omega_0)|^2 + |\tilde{R}(\omega + \omega_0)|^2 \}$$
$$\times \sum_{l,l'} e^{ik_0 r_{ll'} \cos \theta}. \tag{6.2.37}$$

Here $\tilde{R}(\omega)$ is the Fourier transform of the amplitude of the off-diagonal density matrix element $R(t)$.

Note that should we neglect the dipole–dipole coupling between atoms, the polarization spectrum would then be of the form

$$|\tilde{R}(\omega)|^2 = \tfrac{1}{4}\pi^2 T_R^2 \operatorname{sech}^2(\tfrac{1}{2}\pi \omega T_R). \tag{6.2.38}$$

Therefore the spectral width is determined in this case by the inverse super-radiance time T_R^{-1}.

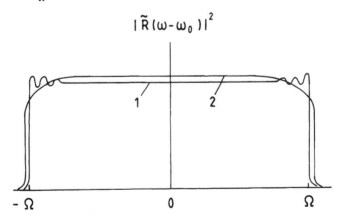

Figure 6.2. Super-radiance spectrum of a point system ($L \ll \tilde{\lambda}$) with dipole–dipole coupling ($N = 20$, $R_0 = 0.02$): (1) spatially homogeneous approximation; (2) calculation using equations (6.2.11).

The super-radiance spectrum with allowance for the dipole–dipole coupling, calculated for the spatial homogeneous approximation, is shown in figure 6.2 (curve (1)). It is symmetric relative to the frequency ω_0 and it has a more or less rectangular shape, with width $2\Delta \gg T_R^{-1}$.

The small-scale structure of the super-radiance spectrum is caused by the finiteness of the super-radiance delay time, t_D, and it decreases in amplitude as t_D grows.

Field patterns. In order to obtain the field pattern for a super-radiance pulse we sum the series in equation (6.2.22), thus deriving the following expression for the collective part of the Poynting vector F

$$F(t) = \frac{3\gamma}{8} \operatorname{sech}^2 \left(\frac{t - t_D}{T_R} \right) \frac{\sin^2(\frac{1}{2} k_0 a N \cos \theta_0)}{\sin^2(\frac{1}{2} k_0 a \cos \theta_0)}. \qquad (6.2.39)$$

If the chain is short (i.e. if $k_0 a N \ll 1$), the collective factor F does not depend upon the angle θ_0

$$F(t) = \frac{3\gamma}{8} N^2 \operatorname{sech}^2 \left(\frac{t - t_D}{T_R} \right) \qquad L \ll \tilde{\lambda} \qquad (6.2.40)$$

and the field pattern for the super-radiance pulse coincides, according to equation (6.2.21), with that for the radiation emitted by a single dipole. For a long chain (i.e. for $L \gg \tilde{\lambda}$) the collective factor F has a principal maximum at $\theta_0 = \pi/2$ (in accordance with (6.2.40)) with width of the order of $1/k_0 a N$, and a series of side lobes the positions of which are determined by the equation

$$\cos \theta_0 = (n + \tfrac{1}{2}) \frac{\lambda}{L} \qquad n = 0, 1, 2, \dots \qquad (6.2.41)$$

It follows from (6.2.41) that the number of side lobes is determined by the number of wavelengths that fit into the length of the chain. According to condition (6.2.41), the quantity $L \cos \theta_0$ is the difference between the paths of the two waves emitted by the extreme dipoles of the chain in the direction θ_0. The value of the nth lobe is given by

$$F^{(n)}(t) = \frac{3}{8\pi^2} \gamma \left(\frac{N}{n + \frac{1}{2}} \right)^2 \operatorname{sech}^2 \left(\frac{t - t_D}{T_R} \right) \qquad (6.2.42)$$

which decreases rapidly as n increases.

Thus, in the spatially homogeneous approximation the field pattern for super-radiance generated by an extended linear chain of atoms has a sharp maximum in a direction normal to both the axis of the chain and the direction of the dipole moments. The pattern is similar to that of a linear chain of classical dipole emitters oscillating in phase.

Super-radiance of a short linear chain ($L \ll \tilde{\lambda}$). If we neglect the dipole–dipole inter-atomic coupling by setting $\Delta_{lk} = 0$ in equation (6.2.11) for the case $L \ll \tilde{\lambda}$, this will automatically lead to the spatially homogeneous model considered in the previous section. Indeed, according to equation (6.2.17), in the limit $L \ll \tilde{\lambda}$ the relaxation matrix γ_{lk} is equal to $\gamma/2$, and does not depend upon the position of the atom in the chain.

The dipole–dipole inter-atomic coupling destroys the spatial homogeneity. This is a consequence of the fact that the positions of the atoms in the chain are not equivalent, owing to dipole–dipole coupling. In the quantum mechanical description the dipole–dipole operator does not commute with the operator for the square of the total spin energy of the system.

To estimate the effect of the dipole–dipole inter-atomic coupling upon the super-radiance pulses emitted by a short linear chain ($L \ll \tilde{\lambda}$), equations (6.2.11) are solved numerically. Figure 6.3 shows the distribution of the population

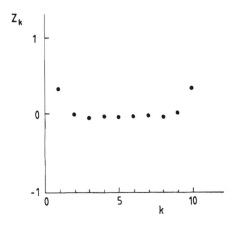

Figure 6.3. The inversion profile for a short linear chain ($N = 10$) at the instant t_D, calculated with allowance for near-field dipole–dipole inter-atomic coupling.

inversion along a chain of ten atoms at the instant $t = t_D$ when the super-radiance intensity is maximal. The initial value of the polarization amplitude was chosen to be $R_0 = 0.02$. A significant variation in the inversion distribution along the linear chain occurs only for the atoms at the ends of the chain. For the inner atoms, Z depends weakly upon the site number. Consequently, the resulting super-radiance pulse (see curve (1) in figure 6.4) differs from that generated either in the spatially homogeneous model or in the absence of the dipole–dipole interaction (curve (2) in figure 6.4) by an increase in the delay time of the pulse. The shape and the amplitude of the pulses differ only slightly. The increase of the super-radiance delay time is caused by the spatially inhomogeneous phase modulation of the polarization amplitude, which takes place at the initial stage of the super-radiance emission. This leads to the slowing down of the growth of the polarization, hence to an increase in the super-radiance delay time in the model where dipole–dipole coupling is included.

Super-radiance of a long linear chain ($L \gg \tilde{\lambda}$). The neglect of the dipole–dipole inter-atomic coupling in the description of the super-radiance dynamics

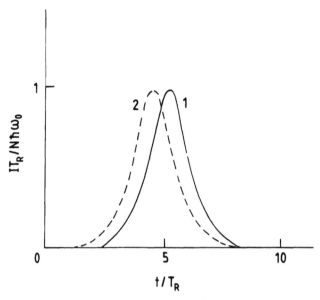

Figure 6.4. Super-radiant pulses for a short linear chain: (1) with dipole–dipole interaction ($\Delta T_R = 360$); and (2) for the spatially homogeneous model.

of a long linear chain ($L \gg \tilde{\lambda}$) does not lead to a spatially homogeneous model as it does for a short chain ($L \ll \tilde{\lambda}$). The collective relaxation constants

$$\tilde{\gamma}_k = \sum_{l \neq k} \gamma_{lk} \tag{6.2.43}$$

are functions of the site number, for $L \gg \tilde{\lambda}$. Consequently all the atomic characteristics will also depend upon the position of an atom in the chain.

The super-radiance dynamics calculations without dipole–dipole coupling, but with uniform initial polarization amplitude $R_0 = 0.02$ reveal the existence of two emission régimes—single-pulse and two-pulse (see figure 6.5)—which alternate with each other as the length of the chain is increased. We can see the reason for this behaviour in the spatial dependence of the field acting at the initial instant of time. This field is determined by the relaxation matrix γ_{lk}: $E_k = \sum_{l \neq k} \gamma_{lk} R^l$. At the initial instant of time $E_k(0) = \sum_{l \neq k} \gamma_{lk} R^l(0) = \tilde{\gamma}_k R_0$.

Taking into account the condition $k_0 a \ll 1$, the summation above can be replaced by integration. Thus

$$\tilde{\gamma}(x) = \frac{1}{k_0 a} \left[\int_0^{k_0 a(L/2+x)} \gamma_1(x') \, dx' + \int_0^{k_0 a(L/2-x)} \gamma_1(x') \, dx' \right] \tag{6.2.44}$$

where

$$\gamma_1(x) = \tfrac{3}{2}\gamma \left(\frac{\sin x}{x} + \frac{\cos x}{x^2} - \frac{\sin x}{x^3} \right). \tag{6.2.45}$$

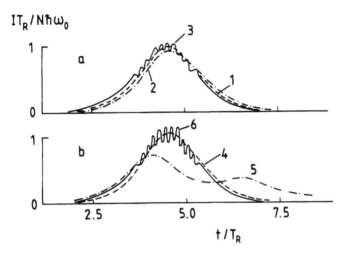

Figure 6.5. Super-radiance dynamics for a linear chain: (a) $N = 100$, $a/\bar{\lambda} = 0.1$; (b) $N = 120$, $a/\bar{\lambda} = 0.1$; (1, 4) spatially homogeneous model; (2, 5) without dipole–dipole coupling; (3, 6) with dipole–dipole coupling.

The functions $\gamma_1(x)$ and $\bar{\gamma}(x)$ are shown in figures 6.6 and 6.7. For $N = 100$ we have the single-pulse super-radiance régime (curve (2) in figure 6.5); for $N = 120$, the two-pulse régime (curve (5) in figure 6.5). Figure 6.7 demonstrates the existence in the sample of regions differing from each other in the de-excitation rates $\bar{\gamma}(x)$. The centres of the regions are determined by the extremal points of $\bar{\gamma}(x)$

$$\gamma_1(k_0(\tfrac{1}{2}L + x)) - \gamma_1(k_0(\tfrac{1}{2}L - x)) = 0. \tag{6.2.46}$$

The region with a higher de-excitation rate $\bar{\gamma}(x)$ begins to develop faster than the region with a lower de-excitation rate. The subsequent evolution significantly depends upon the sign of the coupling between these regions. The coupling will be taken as positive (respectively negative) if $\gamma_1(\delta x) > 0$ (respectively $\gamma_1(\delta x) < 0$), where δx is the distance between the centres of the neighbouring regions.

In systems with positive coupling (which is true for $N = 100$) the regions with a lower de-excitation rate $\bar{\gamma}(x)$ follow in their development the regions with a higher rate (see figure 6.8, curves (1–4)). This is ensured by the property that, because $\gamma_1(\delta x) > 0$, the electric fields of the indicated regions are in phase. As a result, a super-radiance pulse is formed that is close in shape to the pulse obtained in the spatially homogeneous model (figure 6.5, curve (2)).

In systems with negative coupling (such a situation occurs for $N = 120$) the fields of the indicated regions are in opposite phase. Then the regions have the development at the start where the de-excitation rates $\bar{\gamma}(x)$ are the highest (see figure 6.8, curves (5–8)). After these regions have depleted their inversion

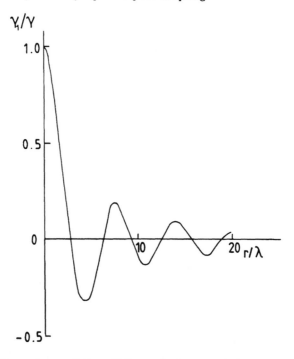

Figure 6.6. Dependence of the off-diagonal element of the relaxation matrix γ_1, equation (6.2.45), upon the atomic spacing.

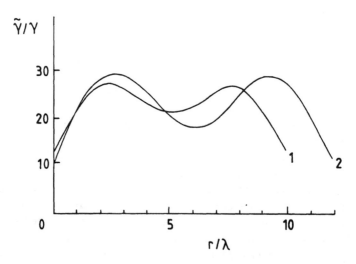

Figure 6.7. Dependence of the constant $\tilde{\gamma}$ of collective relaxation upon the position of the atom: (1) $N = 100$; (2) $N = 120$.

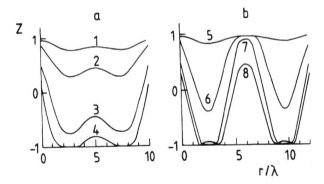

Figure 6.8. Inversion profile in a linear chain without dipole–dipole coupling, at different instants of time: (a) $N = 100$, $a/\bar{\lambda} = 0.1$; (1) $t = 3.6T_R$; (2) $t = 4.3T_R$; (3) $t = 5.3T_R$; (4) $t = 6.4T_R$; (b) $N = 120$, $a/\bar{\lambda} = 0.1$; (5) $t = 2.9T_R$; (6) $t = 4.3T_R$; (7) $t = 5.7T_R$; (8) $t = 6.4T_R$.

and formed one peak in the super-radiance pulse, the region with the small de-excitation rate begins to develop, and it eventually forms a second peak of the super-radiance pulse (see figure 6.5, curve (5)).

These features of the super-radiance dynamics for the case when the dipole–dipole inter-atomic coupling is neglected are closely tied to the appearance of a large-scale (in comparison with $\tilde{\lambda}$) inversion gradient during the super-radiance process (see figure 6.8).

As has already been noted, the dipole–dipole interaction between atoms leads to the transfer of excitation from one atom to another, i.e. to the homogeneity of spatial inversion. This process occurs significantly faster than the super-radiance emission. It is therefore natural to expect that the dipole–dipole coupling can have a strong effect upon the super-radiance dynamics of a long linear chain. Figure 6.5 (curves (3, 4)) shows super-radiance pulses obtained by numerical integration of equation (6.2.11). We see that the calculated super-radiance pulses have two features: (i) there is a structure in the pulse; (ii) they are similar to the pulses obtained in the spatially homogeneous approximation. This is explained as being a result of the unification effect of dipole–dipole interaction, an effect which inhibits the appearance of large-scale inversion gradients during the super-radiance emission. The inversion profiles at different moments of time show that there is indeed spatial homogeneity 'on average' (see figure 6.9).

The appearance of the structure in the super-radiance pulse is caused by the exchange of excitation between macroscopic regions with different de-excitation rates $\bar{\gamma}(x)$. The modulation of the de-excitation rate with frequency equal to the rate of redistribution of the inversion amongst these regions is the cause of the structure in the super-radiance pulse. The time scale of the structure is thus determined by the transfer of energy from one region to another. We can estimate the characteristic time T_{tr} of this process by recognizing that the

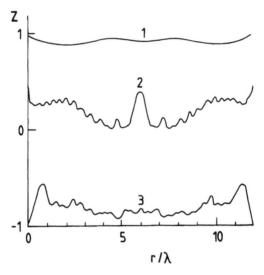

Figure 6.9. Inversion profile in a linear chain with dipole–dipole coupling at different moments of time: $N = 120$, $a/\bar{\lambda} = 0.1$; (1) $t = 2.9T_R$; (2) $t = 4.3T_R$; (3) $t = 5.7T_R$.

velocity of the coherent motion of excitons along a linear chain is of the order of Δa (see e.g. [70]). The distance between the indicated regions is $\delta x \sim \bar{\lambda}$ (see figure 6.7). Therefore $T_{tr} \sim \delta x/(\Delta a) \sim k_0 a T_R$. This value corresponds, in order of magnitude, to the structural scale of the super-radiance pulse (see figure 6.5, curves (3, 6)). The fact that this structure disappears as the length of the system is decreased corroborates the proposed explanation of the structure in the super-radiance pulses. This results from the uniformity of the de-excitation rates for the various regions as L is decreased. (Let us recall that for $L \ll \bar{\lambda}$ the rate $\bar{\gamma}(x) = N\gamma/2$.)

6.3 Quantum theory of super-radiance from a regular short linear chain of two-level atoms

As we have seen above, in the framework of the semiclassical theory for the evolution of super-radiance, both the super-radiance pulse and the population inversion depend only slightly upon the dipole–dipole coupling of atoms. At the same time this coupling influences the super-radiance spectrum very strongly. In order to shed light on these features of the super-radiance of a regular linear chain of two-level atoms, we shall consider below another approach to the problem, which includes, as the first stage, the solution of the quantum eigenvalue problem, for a regular linear chain with dipole–dipole coupling of atoms.

6.3.1   Energy spectrum of a linear chain of N two-level atoms

We shall restrict ourselves to nearest-neighbour dipole–dipole coupling of atoms in a linear chain. The Hamiltonian of such a system, in the nearest-neighbour approximation, has the following form

$$\widehat{H} = \sum_{k=1}^{N} \hbar\omega_0 \widehat{B}_k^{\dagger} \widehat{B}_k + \sum_{k=1}^{N-1} V(\widehat{B}_k^{\dagger}\widehat{B}_{k+1} + \widehat{B}_{k+1}^{\dagger}\widehat{B}_k). \qquad (6.3.1)$$

Here $\widehat{B}_k$ (respectively $\widehat{B}_k^{\dagger}$) is the annihilation (respectively creation) operator of the kth atom's excitation, and V is the resonant matrix element of the dipole–dipole coupling operator. We shall consider the case of $V > 0$, as this corresponds to the model discussed above, in which atomic dipoles are both parallel to each other and normal to the linear axis of the chain.

The Hamiltonian (6.3.1) is diagonalized by applying the Jordan–Wigner transformation [38] (see also [3, 11, 48])

$$\widehat{b}_j = \sum_{k=1}^{N} C_{jk} \widehat{\varepsilon}_k \widehat{B}_k \qquad (6.3.2)$$

$$C_{jk} = \sqrt{\frac{2}{N+1}} \sin\left(\frac{\pi jk}{N+1}\right) \qquad (6.3.3)$$

$$\widehat{\varepsilon}_k = \exp\left(i\pi \sum_{q=1}^{k-1} \widehat{B}_q^{\dagger}\widehat{B}_q\right). \qquad (6.3.4)$$

In the new representation we have

$$\widehat{H} = \sum_{j=1}^{N} E_j \widehat{b}_j^{\dagger}\widehat{b}_j \qquad (6.3.5)$$

$$E_j = \hbar\omega_0 + 2V \cos\left(\frac{\pi j}{N+1}\right). \qquad (6.3.6)$$

The annihilation, $\widehat{b}_j$, and creation, $\widehat{b}_j^{\dagger}$, operators of the collective excitations—excitons—obey the fermionic commutation relation

$$\widehat{b}_j^{\dagger}\widehat{b}_{j'} + \widehat{b}_{j'}\widehat{b}_j^{\dagger} = \delta_{jj'} \qquad (6.3.7)$$

As a result, the total energy of the system takes the values

$$E = \sum_{j=1}^{N} E_j n_j \qquad n_j = 0, 1. \qquad (6.3.8)$$

As follows from equations (6.3.6) and (6.3.8), the dipole–dipole coupling between atoms splits each of the energy levels $E_m = m\hbar\omega_0$ (initially $N!/m!(N-$

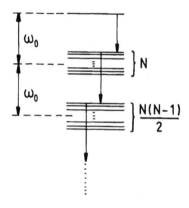

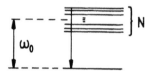

Figure 6.10. Energy-level structure of a short linear chain with dipole–dipole coupling (arrows show allowed radiative transitions).

m)!-fold degenerate) into the exciton band (see figure 6.10). Top $(E_m^{(+)})$ and bottom $(E_m^{(-)})$ energy levels of the mth exciton band are determined according to equation (6.3.8) to be

$$E^{(\pm)} = E_m \pm \tfrac{1}{2}\delta E_m \qquad (6.3.9)$$

where the bandwidth δE_m is given by the relation

$$\delta E_m = 4V \cos\left(\frac{\pi(m+1)}{2(N+1)}\right) \sin\left(\frac{\pi m}{2(N+1)}\right) \operatorname{cosec}\left(\frac{\pi}{2(N+1)}\right) \quad (6.3.10)$$

The bands with one and $N-1$ excitons have the simplest energy structure. The states of these bands are $|1, j\rangle = b_j^\dagger|0\rangle$ and $|N-1, j\rangle = b_j|N\rangle$. Here $|0\rangle$ is the ground state of the system and $|N\rangle$ is the fully excited state corresponding to all the atoms being excited. The corresponding energy spectra are

$$E_j^{(1)} = E_j = \hbar\omega_0 + 2V \cos\left(\frac{\pi j}{N+1}\right)$$

$$\tag{6.3.11}$$

$$E_j^{(N-1)} = \hbar\omega_0 - 2V \cos\left(\frac{\pi j}{N+1}\right).$$

The widths of both exciton bands are equal to $4V$, and their levels are non-degenerate. It is easy to see that the density of states of these bands increases at the edges of the band.

As an example let us consider the density of states of the one-exciton band. Defining this as $\rho(E) = (1/N) \sum_{j=1}^{N} \delta(E - E_j)$, in the limit of large N we shall obtain

$$\rho(E) = \frac{1}{\pi} \frac{1}{\sqrt{4V^2 - (E - \hbar\omega_0)^2}}. \tag{6.3.12}$$

From this it follows that the density of states, $\rho(E)$, of the one-exciton band is inverse square-like near the edges of the band.

Finally, we note that the results of this subsection are valid for a linear chain of arbitrary length.

6.3.2 Radiative transition rates and emission dynamics

Interaction of atoms with the radiative field, in the case of a short linear chain $(L \ll \tilde{\lambda})$, is described by the Hamiltonian

$$\widehat{H}_{int} = -\widehat{d}\widehat{\mathcal{E}} \tag{6.3.13}$$

where $\mathcal{E}$ is the operator of the transverse electrical field, and $\widehat{d}$ is the total dipole-momentum operator of the atoms

$$\widehat{d} = \sum_{k=1}^{N} \widehat{d}_k = \widehat{d}^{(+)} + \widehat{d}^{(-)}. \tag{6.3.14}$$

Here

$$\widehat{d}^{(+)} = d \sum_{k=1}^{N} \widehat{B}_k^\dagger = d \sqrt{\frac{2}{N+1}} \sum_{j=1}^{N} \widehat{b}_j^\dagger \sum_{k=1}^{N} \widehat{\varepsilon}_k \sin\left(\frac{\pi jk}{N+1}\right) \tag{6.3.15}$$

$$\widehat{d}^{(-)} = \widehat{d}^{(+)\dagger}.$$

The atom–field interaction, described by the operator $\widehat{H}_{int}$, stimulates radiative transitions between the states of neighbouring excitonic bands $(\Delta m = \pm 1)$. In the dipole approximation the radiative transition rate between two states 'i' and 'j' is given by the relation (cf [WW30])

$$w_{ij} = \frac{|(\widehat{d})_{ij}|^2}{d^2} \gamma. \tag{6.3.16}$$

Let us calculate now the matrix elements $(\widehat{d})_{ij}$ between the ground state $|0\rangle$ and states of the one-exciton band $|1, j\rangle$, and between the ground state and the fully excited state $|N\rangle$. Since $\widehat{\varepsilon}_k|0\rangle = |0\rangle$ and $\widehat{\varepsilon}_k|N\rangle = (-1)^k|N\rangle$, we have

$$\langle 1, j|\widehat{d}|0\rangle = d\sqrt{\frac{2}{N+1}} \sum_{k=1}^{N} \sin\left(\frac{\pi jk}{N+1}\right)$$

$$= d\sqrt{\frac{2}{N+1}}\frac{1-(-1)^j}{2}\cot\left(\frac{\pi j}{2(N+1)}\right) \qquad (6.3.17)$$

$$\langle N-1, j|\hat{d}|N\rangle = d\sqrt{\frac{2}{N+1}}\sum_{k=1}^{N}(-1)^k\sin\left(\frac{\pi jk}{N+1}\right)$$

$$= d\sqrt{\frac{2}{N+1}}\frac{1-(-1)^{N+1-j}}{2}\cot\left(\frac{\pi(N+1-j)}{2(N+1)}\right). \qquad (6.3.18)$$

Here, the standard time-dependent factors describing oscillations of the matrix elements at the transition frequency $E_j/\hbar$ are omitted. Substituting equations (6.3.17) and (6.3.18) into equation (6.3.16) we obtain the following formulae for the transition rates

$$w(0 \to 1, j) = \begin{cases} 0 & j\ \text{even} \\ \gamma\dfrac{2}{N+1}\cot^2\left(\dfrac{\pi j}{2(N+1)}\right) & j\ \text{odd} \end{cases} \qquad (6.3.19)$$

$$w(0 \to N-1, j) = \begin{cases} 0 & N+1-j\ \text{even} \\ \gamma\dfrac{2}{N+1}\cot^2\left(\dfrac{\pi(N+1-j)}{2(N+1)}\right) & N+1-j\ \text{odd}. \end{cases}$$

$$(6.3.20)$$

It is easy to show that the total transition rates from both the ground state $w_0 = \sum_{j=1}^{N} w(0 \to 1, j)$ and the fully excited state $w_N = \sum_{j=1}^{N} w(N \to N-1, j)$ are equal to the Dicke spontaneous decay rate $N\gamma_0$ (the sum rule for the oscillator strengths of transitions).

Let us consider transitions to the states situated near the band edges, i.e. for $j \ll N$ and $N+1-j \gg N$. Within these limits, making use of equations (6.3.19) and (6.3.20) we then obtain

$$w(0 \to 1, j) = \frac{8}{\pi^2 j^2}(N+1)\gamma = \frac{0.81}{j^2}(N+1)\gamma \qquad (6.3.21)$$

$$w(0 \to N-1, j) = \frac{8}{\pi^2(N+1-j)^2}(N+1)\gamma \qquad (6.3.22)$$

Thus, transitions to the states situated near the top of the bands are more probable. For example, the transition rates to the states with $j = 1$ and $j = N - 1$ constitute 81% of the total transition rate $N\gamma$, and those to the states with $j = 3$ and $j = N - 3$ only 9%. Taking into account the dominant transition rate from the ground state to the states of the top one-exciton band, and the increase of the density of the exciton states near the band edge, it confirms that the transition into the one-exciton band predominates with the frequency $\omega_0 + 2V/\hbar$. The transition from the fully excited state to the $N - 1$ exciton band occurs with frequency $\omega_0 - 2V/\hbar$.

Calculation of the transition rates between two arbitrary exciton bands in general is very demanding owing to the complicated structure of the dipole momentum operator in the collective state representation (6.3.15). But the calculations that were carried out for the three- and four-atom linear chains showed that the transitions between the top exciton states are dominant, and their rates are approximately equal to those for the Dicke states. At the same time, the transition rates from the top state to the others decrease rapidly as the state number increases.

The decrease in the transition rates is explained by the oscillation of the amplitudes C_{ij}, which describe the contribution of the site states to the exciton states (see figure 6.11). It is clear that a similar selection rule will in general

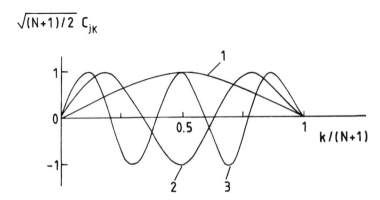

Figure 6.11. Site amplitudes C_{jk} of collective states $|j\rangle$ versus site number k: (1) $j = 1$; (2) $j = 3$; (3) $j = 5$.

operate in the case of transitions between any nearest-neighbour pair of the exciton band, since the collective states corresponding to the top exciton bands comprise the linear combinations of the site states that have extremely smooth amplitudes C_{1k} along a linear chain. The approximate selection rule for a finite chain described above is the trace of the conservation of quasi-momentum for an infinite medium. In our case the number $\pi j/(N + 1)$ plays the rôle of the quasi-momentum.

Thus, according to the quantum mechanical treatment above, the super-radiance decay of a short system ($L \ll \bar{\lambda}$) results from the transition between the top states of the nearest-neighbour excitonic bands produced by the dipole–dipole inter-atomic coupling from initially degenerate collective atomic levels of a many-atom system (see figure 6.10). As the corresponding transition rates differ only slightly from the same values for the Dicke states $w_m = \gamma m(N - m + 1)$, the super-radiance dynamics is expected to be similar to that without the dipole–dipole inter-atomic coupling, in accordance with the conclusion drawn from the semiclassical treatment.

6.3.3   Self-phase modulation and the super-radiance spectrum

We shall now apply the model developed in subsection 6.3.2 to treat self-phase modulation in the super-radiance spectrum. According to this model, radiative transitions occur mainly between the top states of the exciton bands. The transition frequency between the tops of the $(m + 1)$th and mth bands is equal to

$$\omega_m = \omega_0 + \frac{2V}{\hbar} \cos\left(\frac{\pi(m+1)}{N+1}\right). \qquad (6.3.23)$$

The frequency difference of nearest-neighbour exciton bands is then

$$\delta\omega_m = \omega_{m-1} - \omega_m = \frac{2\pi V}{\hbar(N+1)} \sin\left(\frac{\pi(m+\frac{1}{2})}{N+1}\right) \qquad (6.3.24)$$

which is much larger than the characteristic radiation width of the exciton states that is equal to $N\gamma$ if $m \sim 1, N$, and $\sim N^2\gamma$ if $1 \ll m \ll N$. Indeed, $\delta\omega_m \sim V/\hbar(N+1)^2$ for $m \sim 1, N$, and $\delta\omega_m \sim V/\hbar(N+1)$ for $1 \ll m \ll N$, whilst $V \sim \hbar\gamma(\tilde{\lambda}/a)^3 = \hbar\gamma N^3(\tilde{\lambda}/L)^3$. For $L \ll \tilde{\lambda}$ we have

$$\delta\omega_m \sim \begin{cases} \dfrac{V}{\hbar(N+1)^2} \sim \gamma N \left(\dfrac{\tilde{\lambda}}{L}\right)^3 \gg \gamma N & m \sim 1, N \\[4mm] \dfrac{V}{\hbar(N+1)} \sim \gamma N^2 \left(\dfrac{\tilde{\lambda}}{L}\right)^3 \gg \gamma N^2 & 1 \ll m \ll N. \end{cases} \qquad (6.3.25)$$

By virtue of the assumptions made above, it follows from (6.3.25) that the structure of the super-radiance spectrum for a short system consists of the isolated spectral components corresponding to the transitions between nearest-neighbour exciton bands. The super-radiance spectrum can be obtained by taking into account that there is one photon per frequency interval $\delta\omega_m$

$$\Phi(\omega_m) = \frac{1}{N} \frac{1}{\delta\omega_m}. \qquad (6.3.26)$$

Substituting expression (6.3.24) for $\delta\omega_m$ into equation (6.3.26), and using equation (6.3.23) to express m via ω_m, we obtain

$$\Phi(\omega) = \frac{1}{\pi} \frac{1}{\sqrt{(2V/\hbar)^2 - (\omega - \omega_0)^2}} \qquad |\omega - \omega_0| < \frac{2V}{\hbar}. \qquad (6.3.27)$$

Note that there is some difference between the quantum mechanical and semiclassical shapes of the super-radiance spectrum. The quantum mechanical formula (6.3.27) for the super-radiance spectrum has a singularity at $|\omega - \omega_0| = 2V/\hbar$ (similarly to the expression (6.3.12) for the exciton density), whereas the semiclassical super-radiance spectrum is continuous in the full range of frequencies. We note, however, that the latter property can be formally obtained by imposing the condition $\hbar \to 0$ in equations (6.3.27) (see figure 6.2).

6.4 Concluding remarks

In conclusion, we note that for short regular systems ($L \ll \tilde{\lambda}$) the dipole–dipole inter-atomic coupling modifies the super-radiance dynamics only slightly, but has a strong effect on the shape and width of the spectrum. Without the dipole–dipole interaction, the width of the spectrum is determined by the reciprocal time $1/T_R \sim N/\tau_0$. If this interaction is present, the spectrum has a width $d^2/\hbar a^3 \gg T_R^{-1}$. Dipole–dipole coupling in the super-radiance dynamics also increases the delay time of the super-radiance pulse.

For a long regular linear chain of atoms ($L \gg \tilde{\lambda}$), the near-field coupling produces the same effect on the spectrum as in the case of a short system, and there is no radical change in the super-radiance dynamics. This coupling causes transfer of coherent excitation, which leads to the approximate spatial homogeneity of the population inversion along the linear chain. Therefore, the super-radiance pulse turns out to be close to the one calculated using the spatially homogeneous model. This justifies, to some extent, the application of the latter model to systems with small Fresnel numbers.

On the other hand, if atoms are placed randomly, then dipole–dipole interaction between them leads to the de-phasing of the atomic dipoles, i.e. plays the rôle of a destructive factor [FHM72, CF78, S180, S185, R83].

Finally, we note that, despite its simplicity, a linear chain of two-level atoms is a useful model and it can be applied e.g. for treating some real chemical structures. There are known J-aggregates in concentrated dye solutions (for example, pseudoisocyanine) which are linear chains of dye molecules bound together with the dipole–dipole inter-molecular interaction [37, 65]. Related materials are polysilanes, a class of polymers that are very promising for many technological applications. Recently these objects were intensively studied because of their unique spectroscopic and nonlinear optical properties (see e.g. [SM89a, SM89b, TTZH90, TOZH89, M91, M93] and [18, 19, 66]).

Chapter 7

Super-radiance of multi-spin systems

The spin magnetic moment in a static magnetic field is a perfect realization of the 'two-level' atomic model. In fact, this physical system was the origin of this model, and was extended to a wide class of resonant phenomena, including super-radiance (see e.g. the book by Allen and Eberly [AE75]).

This chapter will be devoted mainly to the description and theoretical interpretation of super-radiance experiments in a multi-proton spin system [KPSY88, BKP88, BBKPTTS89, BBZKMT90].

7.1 Preliminary remarks

It is well known that ordinary spontaneous emission from electron or nuclear spins in a static magnetic field is impossible to observe, owing to the very long radiation decay time τ_0. The expression for τ_0 has the same form as that for the optical transition (1.1.2), but with the electric transition dipole moment d being replaced by the magnetic dipole moment μ

$$\tau_0^{-1} = \frac{4}{3} \frac{\omega_0^3 \mu^2}{\hbar c^3}. \tag{7.1.1}$$

Here ω_0 is the resonant frequency of the transition between Zeeman levels, i.e. the Larmor frequency of the magnetic spin moment in a static magnetic field H_0

$$\omega_0 = \gamma H_0 \tag{7.1.2}$$

where γ is the gyromagnetic ratio for the spin, $\gamma_e = |e|/mc = 1.76 \times 10^7 \, \text{G}^{-1} \, \text{s}^{-1}$ for the electron, $\gamma_p = |e|/m_p c = 9.58 \times 10^3 \, \text{G}^{-1} \, \text{s}^{-1}$ for the proton. Thus, for a field $H_0 \sim 10^4$ Oe the lifetimes τ_0 would have the astronomically large values of 10^{18} s and 10^{25} s for the electron and proton respectively.

To estimate the super-radiance time for such a system we can take into account that this is just the case in which the wavelength of the transition is of the same order, $\lambda \approx 1$ cm (for electrons), or more, $\lambda \approx 10$ m (for protons), as the dimension of the sample. Therefore $T_R = \tau_0/N$, where N, the total number of spins in the sample, can be of the order of 10^{24}.

Nevertheless, there are difficulties in observing super-radiance, too. The reason for this is the homogeneous broadening of resonance spectral lines as a result of spin–spin interaction [1, 4]. The homogeneous width of the line is of the same order of magnitude as the interaction between two spins, separated by an average distance of $\mu N_0^{-1/3}$, where N_0 is the spin density number. This implies an estimated de-phasing time $T_2 \simeq \hbar/\mu^2 N_0$. On the other hand, the super-radiance time T_R for the system is

$$T_R = \frac{\tau_0}{N} = \frac{3\hbar\lambda^3}{32\pi^3\mu^2 N}. \tag{7.1.3}$$

Taking this into account, and that $N = N_0 V$, we find $T_2/T_R \simeq V/\lambda^3$, where V is the volume of the sample. Thus, for a small volume ($V \leq \lambda^3$) the de-phasing caused by dipole–dipole coupling out-paces the emission of super-radiance, and the latter will be suppressed.

The possibility of the observation of super-radiance in a multi-spin system was discussed at almost the same time as Dicke's original paper [D54] appeared. Bloembergen and Pound [BP54] showed that the super-radiance effect can be observed if one puts the sample into a resonant cavity. Indeed, the density of resonant radiation inside a cavity of volume V_c, and quality factor Q, is larger than that in free space by a factor of $3Q\lambda^3/(8\pi^2 V_c)$. The collective radiation rate becomes

$$T_{R \text{ (in cavity)}}^{-1} = \tau_0^{-1}\frac{3Q\lambda^3}{8\pi^2 V_c}N_0 V$$
$$= 2\pi\eta Q\gamma M_0 \tag{7.1.4}$$

where $\eta = V/V_c$ is the so-called filling factor of the cavity, and $M_0 = \mu N_0$ is the magnetization of the sample. The additional factor ηQ in (7.1.4) can be of the order of 10^3, and thus the inequality $T_R < T_2$ required for super-radiance can be satisfied.

7.2 Experimental observations of super-radiance of multi-spin systems in a cavity

7.2.1 Electron-spin super-radiance

Super-radiance of electron-spin systems in a cavity was observed in the microwave domain for the first time by Feher *et al* [FGBGT58] and Chester *et al* [CWC58]. In Feher's experiments phosphorus donors in silicon were used. The electronic spin resonance, which is inhomogeneously broadened by hyperfine interactions of the donor electrons with the ^{29}Si nuclei, was narrowed from 2.7 Oe in width to 0.22 Oe by using a crystal of isotopically purified silicon (of which the estimated final isotopic purity was $(99.88 \pm 0.08)\%$ ^{28}Si). The sample used in this experiment had a volume of about 0.3 cm^3 and a phosphorus concentration of 4×10^{16} atom cm^{-3}. The cavity was resonant at ~ 9 GHz, and its Q factor at 1.2 K was ~ 2000. Population inversion of the electron spin levels was produced by a fast adiabatic change. Its relaxation time at the operating temperature of 1.2 K was one minute. Under these conditions, spontaneous and coherent radiation of the spin system in a resonant cavity, at the Larmor precession frequency of the spins, was observed in the form of oscillatory modulation of the microwave signal envelope, with a duration of about 10 μs on back-sweeping through resonance. The measured power of the signal was about 1 μW.

In Chester's experiments the materials used were single crystals of quartz and magnesium oxide, each containing defects introduced by neutron irradiation. Samples were mounted in a reflection cavity, resonant at 9 GHz, with a loaded Q of ~ 6000. Inversion of the electron populations was brought about by rapid adiabatic change, in which the magnetic field swept through resonance. With a quartz sample containing $\sim 10^{18}$ spins, the inverted state persisted for 2 ms at 4.2 K. Coherent microwave oscillation was observed on the return through resonance. The peak power emitted during oscillation was 12 mW in a pulse of about 10 μs duration. For a theoretical discussion of the experiments of Chester *et al* see the work of Yariv [Y60].

7.2.2 Nuclear-spin super-radiance

The first attempt to observe super-radiance in the radiofrequency (RF) domain was reported by Bosiger *et al* [BPM77, BBM78], using a nuclear subsystem of ^{27}Al atoms in ruby crystals.

Recently super-radiance has been observed in a system of proton spins (Kiselev *et al* [KPSY88] and Bazhanov *et al* [BBZKMT90, BKP88, BBKPTTS89]). The experimental set-up used in these experiments was, in essence, an RF circuit, with a characteristic frequency $f = 54$ MHz, placed in a static magnetic field H_0 oriented perpendicularly to the axis of the coil (see figure 7.1). The sample under investigation (in this case propandiol C$_3$H$_8$O$_2$) was located inside the coil, with a proton-spin concentration of $N_0 = 4.5 \times 10^{22}$ cm^{-3}. Using the method of dynamic nuclear polarization (see e.g. Jeffries [36] and Abragam [1]) a population inversion was created in the proton Zeeman levels produced by the external field H_0. The paramagnetic impurity used to bring this about was $^{5+}$Cr at a concentration of 1.8×10^{20} cm^{-3}. After creating the population inversion in a field $H_0 = 2.45 \times 10^4$ Oe (the proton resonance frequency was $f_0 = 104.3$ MHz, and the uniformity of the field where the sample was located was $\Delta H_0 / H_0 \sim 10^{-4}$) the system of proton spins was cooled to a temperature of 60 mK. The magnetic field was then scanned at a rate $dH_0 / dt \leq 50$ Oe s^{-1}. As the value of the proton resonance frequency approached f, a powerful RF pulse (or a series of pulses) was recorded in the

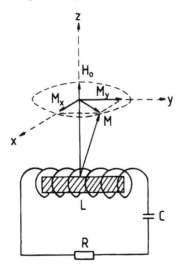

Figure 7.1. Experimental geometry for observing super-radiance in the NMR region (the sample is shown cross-hatched).

circuit. The magnetization in this case had decreased and even reversed.

The experimentally observed basic, functional relationships which characterize the phenomenon are shown in figure 7.2. This figure shows the proton system polarization P_f after the emission process, the amplitude U of the detected RF voltage pulse and the increased time of the pulse τ, all plotted against the value of the initial polarization P_i. We use the term 'polarization' here to describe the fraction of proton spins oriented parallel to the external magnetic field H_0. The polarization vector P is related to the magnetization M by the formula

$$M = \tfrac{1}{2}\hbar\gamma_p N_0 P. \tag{7.2.1}$$

The generation begins at a certain threshold value of negative initial polarization $P_{i1} \simeq -7\%$, in which case a single weak pulse is observed. As the absolute value of P_i increases, the value of the final polarization P_f remains constant (negative) at -7% until P_i reaches the value $P_{i2} \simeq -25\%$. For $P_{i1} > P_i > P_{i2}$ a series of pulses is detected, unequally spaced in time and decreasing in amplitude (figure 7.3(a)). As P_i approaches P_{i2} the amplitude of the first pulse grows. In the region $P_i < P_{i2}$ of values of initial polarization, again a single (but already powerful) pulse is observed (figure 7.3(b,c)). The final value of the nuclear polarization P_f then increases linearly with $|P_i|$. Particularly noteworthy are the sharp changes observed in the amplitude of the voltage U and the increased time τ as a function of P_i near the point P_{i2}.

Below we shall consider the theory of nuclear-spin super-radiance, and then discuss the observed peculiarities of the phenomenon in detail.

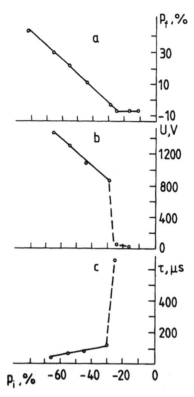

Figure 7.2. Experimental dependence upon the initial polarization P_i [BBZKMT90] of:
(a) the final polarization P_f; (b) the amplitudes of the voltage U of the RF pulses; (c)
their rise times τ.

7.3   Theory of super-radiance of proton multi-spin systems in a cavity

For the experiment under consideration (see figure 7.1) the electric current I,
induced in the coil, obeys Kirchhoff's law

$$\frac{L}{c^2}\dot{I} + RI + \frac{1}{C}\int I\,dt = -\frac{4\pi}{c}n_0\sigma\eta\dot{M}_x \qquad (7.3.1)$$

where M_x is the x component of the magnetization, n_0 is the number of windings
on the coil, σ is its transverse cross-section, $\eta = \sigma_0/\sigma$, σ_0 is the sample's cross-
section, and L, R, C are respectively the inductance, resistance and capacitance
of the circuit.

The current I, in turn, gives rise to a magnetic field directed along the x
axis

$$H_x = \frac{L}{n_0\sigma}I \qquad (7.3.2)$$

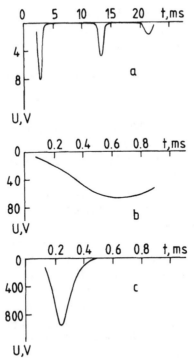

Figure 7.3. Oscilloscope traces of the pulses obtained by slowly scanning with an external field H_0 [BBZKMT90]: (a) the initial polarization $P_i \approx -18\%$; (b) $P_i \approx -25\%$; (c) $P_i \approx -30\%$.

which, together with the static magnetic field H_0 directed along the z axis, controls the magnetization according to

$$\dot{M} = \gamma_p [M \times H].$$ (7.3.3)

The vector H has components $(H_x, 0, H_0)$.

To obtain the classical Bloch equations of motion for the magnetization M we complete (7.3.3) by adding relaxation terms, and then obtain

$$\dot{M}_x = \omega_0 M_y - \frac{M_x}{T_2}$$

$$\dot{M}_y = -\omega_0 M_x + \gamma_p M_z H_x - \frac{M_y}{T_2}$$ (7.3.4)

$$\dot{M}_z = -\gamma_p M_y H_x - \frac{M_z - M_0}{T_1}$$

where $\omega_0 = \gamma_p H_0$ is the Larmor frequency, T_1 is the relaxation time for the longitudinal component of magnetization, T_2 is the transversal relaxation time and M_0 is the stationary value of the magnetization M_z. Differentiating both sides of equation (7.3.1) with respect to time and expressing I in terms of H_x with the help of (7.3.2), we obtain

$$\ddot{H}_x + \frac{\omega}{Q}\dot{H}_x + \omega^2 H_x = -4\pi\eta\ddot{M}_x \qquad (7.3.5)$$

where $\omega = c/\sqrt{LC}$ is the resonant frequency of the circuit and $Q = \omega L/c^2 R$ is its quality factor. Equations (7.3.4) together with equation (7.3.5) form a closed system of equations for H_x and M.

We do not take into account the spread in Larmor frequencies of individual spins, assuming that the width of the NMR line results solely from dipole interactions between these spins. The effect of this interaction is treated by including the relaxation time T_2 in the equations for the components of M transverse to the field H_0.

The magnitude of the relaxation time T_1 for the longitudinal component of the magnetization is determined by two factors: spontaneous emission and spin–lattice relaxation. The first of these processes for nuclear spins is negligible, as we have seen above. The spin–lattice relaxation time, at an operating temperature of ~ 0.1 K, and with a value of the external magnetic field $H_0 \simeq 2 \times 10^4$ Oe, is of the order of hundreds of hours. The super-radiation time in experiments [KPSY88, BBZKMT90] is much shorter than the relaxation time T_1, we therefore set $T_1 = \infty$.

Let us factor the basic time variation at the frequency ω_0 out of the quantities H_x and M_x

$$H_x = ih(t)\exp(-i\omega_0 t) + \text{c.c.} \qquad (7.3.6a)$$
$$M_x = \tfrac{1}{2}m(t)\exp(-i\omega_0 t) + \text{c.c.} \qquad (7.3.6b)$$

where we assume that the amplitudes $h(t)$ and $m(t)$ vary slowly over a time interval of the order of the Larmor precession period $2\pi/\omega_0$: $|\dot{h}| \ll \omega_0|h|$, $|\dot{m}| \ll \omega_0|m|$. The factor i ($\pi/2$ phase shift) has been introduced in H_x for later convenience. In this approximation M_y can be expressed with the help of the first of equations (7.3.4) in the form

$$M_y = \frac{1}{\omega_0}\left(\dot{M}_x + \frac{M_x}{T_2}\right) \qquad (7.3.7)$$

The conditions that correspond to the experiment are $|\omega - \omega_0| \ll \omega_0$ (quasi-resonance), $\gamma_p|M_z| \ll \omega_0$ (a negligible value of the magnetization, M_z, relative to the direct external magnetic field, H_0) and $Q \gg 1$. Taking these into account we obtain the following equations for the amplitudes h, m, and the longitudinal component of magnetization M_z

$$\dot{h} + \left(\frac{\omega}{2Q} - i\Delta\right)h = \pi\eta\omega_0 m \qquad \Delta = \omega_0 - \omega \qquad (7.3.8a)$$

$$\dot{m} = -\gamma_p M_z h - \frac{m}{T_2} \qquad (7.3.8b)$$

$$\dot{M_z} = \tfrac{1}{2}\gamma_p(mh^* + m^*h). \qquad (7.3.8c)$$

7.3.1 Resonant case in the absence of de-phasing

First, let us consider the simplest case of multi-spin super-radiance in a cavity. We assume both that the Larmor precession frequency is equal to the resonant frequency of the RF circuit ($\Delta = 0$) and that the de-phasing is very slow ($T_2^{-1} = 0$). Taking these conditions into account, we can rewrite equations (7.3.8a)–(7.3.8c) in the following form

$$\dot{h} + \frac{\omega}{2Q}h = \pi\eta\omega_0 m \qquad (7.3.9a)$$

$$\dot{m} = -\gamma_p M_z h \qquad (7.3.9b)$$

$$\dot{M_z} = \gamma_p mh. \qquad (7.3.9c)$$

Here all quantities can be considered as real, without loss of generality.

Taking into account the conservation law $m^2 + M_z^2 = \text{constant}$ which follows from equations (7.3.9b) and (7.3.9c), we can put

$$m(t) = B\sin\theta(t) \qquad (7.3.10a)$$

$$M_z(t) = -B\cos\theta(t) \qquad (7.3.10b)$$

where $B = [M_z^2(0) + m^2(0)]^{1/2}$ is the modulus of the Bloch vector.

Substituting equations (7.3.10a) and (7.3.10b) into equation (7.3.9a) we find the relationship between the magnetic field amplitude, h, and the polar angle of the Bloch vector, θ

$$h = \frac{1}{\gamma_p}\dot{\theta}. \qquad (7.3.11)$$

Then by using equations (7.3.10a), (7.3.10b) and (7.3.9a) we obtain the equation for the polar angle θ

$$\ddot{\theta} + \frac{\omega}{2Q}\dot{\theta} - \Omega^2\sin\theta = 0 \qquad (7.3.12)$$

where the quantity Ω is defined as

$$\Omega^2 = \pi\eta\gamma_p\omega_0 B = \pi\eta\gamma_p\omega_0[M_z^2(0) + m^2(0)]^{1/2}. \qquad (7.3.13)$$

The time Ω^{-1} characterizes the time scale of the SR process provided there is no loss in the cavity ($Q = \infty$).

Equation (7.3.12) describes the damped oscillation of a mathematical pendulum near the state of unstable equilibrium.

Let us consider two extreme cases: (1) the high-Q limit when $\omega/2Q \ll \Omega$; (2) the opposite case of a bad cavity with a very large damping $\omega/2Q \gg \Omega$.

High-Q limit. In this case we neglect the term in equation (7.3.12) proportional to the first derivative, and obtain the mathematical pendulum equation

$$\ddot{\theta} - \Omega^2 \sin\theta = 0 \tag{7.3.14}$$

with initial conditions

$$\cos\theta(0) = \cos\theta_0 = \frac{|M_z(0)|}{[M_z^2(0) + m^2(0)]^{1/2}} \tag{7.3.15a}$$

$$\dot{\theta}(0) = \gamma_p h(0) = 0. \tag{7.3.15b}$$

If the spin system is prepared almost in the fully inverted state, so that $|m(0)| \ll |M_z(0)|$, then we have

$$\theta_0 \cong \left|\frac{m(0)}{M_z(0)}\right| \ll 1.$$

We have already discussed the solution of equation (7.3.14) in section 1.4. It describes the periodic exchange of energy between the spin system and the magnetic field in a cavity (see figure 7.4(a)) with a period

$$T_0 = \frac{4}{\Omega} \ln\frac{8}{\theta_0}. \tag{7.3.16}$$

By taking into account the damping term ($\sim \dot{\theta}$) while $\omega/2Q \ll \Omega$, the train of electric current pulses obtained will have decreasing amplitudes and repetition times (T_0) (figure 7.4(b)).

Bad cavity limit. In the opposite case of a bad cavity ($\omega/2Q \gg \Omega$) equation (7.3.12) describes the overdamped aperiodic motion of a mathematical pendulum, and the second derivative of the polar angle θ can be neglected. As a result we obtain the first-order equation

$$\dot{\theta} - \frac{1}{T_R} \sin\theta = 0 \tag{7.3.17}$$

where

$$T_R = \frac{2Q}{\omega}\Omega^2$$

$$= 2\pi\eta\gamma_p Q[M_z^2(0) + m^2(0)]^{1/2}. \tag{7.3.18}$$

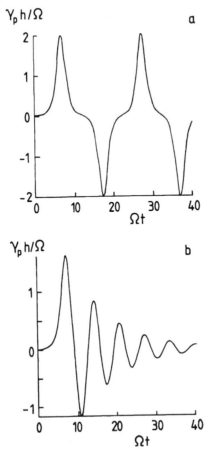

Figure 7.4. Time dependence of the magnetic field amplitude as calculated with the help of equations (7.3.11) and (7.3.12) (high-Q limit).

This value of T_R is just the same as was determined in (7.1.4), because

$$\sqrt{M_z^2(0) + m^2(0)} \approx M_z(0) = \mu_p N_0 = \tfrac{1}{2}\hbar \gamma_p N_0. \tag{7.3.19}$$

The time constant T_R defines the time scale of the super-radiance process in the case of a bad cavity. It depends inversely upon the quality factor Q of the cavity.

The solution of equation (7.3.17) with the initial condition $\theta(0) = \theta_0$ is

$$\theta = 2 \tan^{-1} \exp\left(\frac{t - t_D}{T_R}\right) \tag{7.3.20}$$

$$t_D = T_R \ln \frac{2}{\theta_0}. \tag{7.3.21}$$

Substituting equation (7.3.20) in equation (7.3.11) we obtain

$$h = \frac{1}{\gamma_p T_R} \operatorname{sech} \frac{t - t_D}{T_R}. \tag{7.3.22}$$

So in the case of a bad cavity the super-radiance pulse has a hyperbolic secant shape of characteristic duration T_R, and the peak of the pulse coincides with t_D (see figure 7.5).

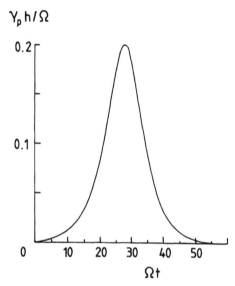

Figure 7.5. Time dependence of the magnetic field in the bad-cavity limit.

In the experiments of Bazhanov *et al* [BKP88, BBKPTTS89, BBZKMT90] the values of the frequency and the quality factor were $\omega = 2\pi f \cong 3 \times 10^8$ s^{-1} and $Q \approx 100$, so the damping time $2Q/\omega$ was of the order of magnitude of one microsecond, which was considerably shorter than the super-radiance time observed ($T_R \sim 100 \ \mu s$). Thus the case of a bad cavity has been realized here. Nevertheless, the solution (7.3.22) for the magnetic field amplitude cannot be used for describing the experimental data, because in that neither the de-tuning $\Delta = \omega_0 - \omega$ nor the relaxation time T_2 are taken into account. We are going to include these effects in the next subsection.

7.3.2   Non-resonant case with de-phasing: bad cavity

Let us now consider the more general case of a bad cavity with $\Delta \neq 0$ and $T_2 \neq \infty$. The fact that the cavity lifetime $2Q/\omega$ is the shortest time of all allows us to simplify equation (7.3.8a) for the amplitude of the RF field h. Using the inequality $|\dot{h}| \ll |(\omega/2Q - i\Delta)h|$ we omit the derivative in the left-hand side of equation (7.3.8a). Under this approximation the field amplitude h follows the magnetization adiabatically, that is

$$h = \beta m \tag{7.3.23}$$

where

$$\beta = \frac{\pi \eta \omega_0}{\omega/2Q - i\Delta}. \tag{7.3.24}$$

Eliminating the magnetic field amplitude from equations (7.3.8b) and (7.3.8c) by using equation (7.3.23), we obtain

$$\dot{m} = -\beta \gamma_p M_z m - \frac{m}{T_2} \tag{7.3.25a}$$

$$\dot{M}_z = \mathrm{Re}\,(\beta)\gamma_p|m|^2. \tag{7.3.25b}$$

It follows from equation (7.3.25a) that for $M_z(0) > -[\mathrm{Re}\,(\beta)\gamma_p T_2]^{-1}$ the initial transverse magnetization will decay to zero, and this case is not interesting. In the opposite case, $M_z(0) < -[\mathrm{Re}\,(\beta)\gamma_p T_2]^{-1}$, an avalanche-like rise in the transverse and longitudinal magnetization begins. Therefore the inequality

$$M_z(0) > -[\mathrm{Re}\,(\beta)\gamma_p T_2]^{-1} \tag{7.3.26}$$

is a threshold condition for this process. From (7.3.26) it follows that the threshold value of the initial longitudinal magnetization is negative, and its absolute value is equal to

$$[M_z(0)]_{\mathrm{tr}} = [\,\mathrm{Re}\,(\beta)\gamma_p T_2]^{-1} = \frac{(2Q\Delta/\omega)^2 + 1}{2\pi\eta Q\gamma_p T_2}. \tag{7.3.27}$$

Note that this depends upon the amount of de-tuning from resonance, $\Delta = \omega_0 - \omega$.

In order to solve equations (7.3.25a) and (7.3.25b) we make the substitution

$$m = \tilde{m}\exp\left\{-i\,\mathrm{Im}\,(\beta)\gamma_p \int_0^t M_z(t')\,dt'\right\} \tag{7.3.28}$$

which explicitly introduces the phase modulation of the magnetization. Then in place of equation (7.3.25a) and (7.3.25b) we have

$$\dot{\tilde{m}} = -\,\mathrm{Re}\,(\beta)\gamma_p M_z\tilde{m} - \frac{\tilde{m}}{T_2} \tag{7.3.29a}$$

$$\dot{M}_z = \mathrm{Re}\,(\beta)\gamma_p\tilde{m}^2. \tag{7.3.29b}$$

Without loss of generality, the function $\tilde{m}(t)$ can be taken to be real.

The solution of equations (7.3.29a) and (7.3.29b) in the region above the threshold (7.3.26) has the form

$$\tilde{m} = |M_z(0)| \left(1 - \frac{T_R}{T_2}\right) \operatorname{sech} \frac{t - t_D}{T} \qquad (7.3.30a)$$

$$M_z(t) = |M_z(0)| \left\{ -\frac{T_R}{T_2} + \left(1 - \frac{T_R}{T_2}\right) \tanh \left[\frac{t - t_D}{T}\right] \right\} \qquad (7.3.30b)$$

where

$$T_R^{-1} = \operatorname{Re}(\beta)\gamma_p |M_z(0)| = \frac{2\pi \eta Q \gamma_p |M_z(0)|}{(2Q\Delta/\omega)^2 + 1} \qquad (7.3.31)$$

$$T = T_R \frac{T_2}{T_2 - T_R} \qquad (7.3.32)$$

$$t_D = -T \ln \left[\frac{\frac{1}{2}\tilde{m}(0)T}{|M_z(0)|T_R}\right] \qquad (7.3.33)$$

the parameter T characterizes the duration ($T_p = 2.4T$) of the pulse at half maximum, whilst t_D is the delay time. The threshold condition (7.3.26) for the NMR generation can now be cast in the form $T_R/T_2 < 1$.

It is clear from equation (7.3.30b) that for $2T_R/T_2 < 1$, or, according to equation (7.3.31), for $|M_z(0)| > 2[\operatorname{Re}(\beta)\gamma_p T_2]^{-1}$, there occurs a change in the sign of the magnetization after a sufficiently long time. As we see, the threshold value of the initial magnetization for this sign reversal is twice as large as the threshold for generation (see equation (7.3.27)). Note that at the sign reversal threshold ($T_2 = 2T_R$) the equality $T_2 = T$ is fulfilled, allowing us to obtain an estimate of T_2 based upon the experimental value of the pulse length.

We conclude this section by presenting expressions for the transverse magnetization components M_x and M_y and the RF magnetic field H_x via $\tilde{m}(t)$ and $M_z(t)$. Using equations (7.3.28), (7.3.30a) and (7.3.30b), we find

$$M_x(t) = \tilde{m} \cos(\omega_0 t - \varphi) \qquad (7.3.34a)$$

$$M_y(t) = -\tilde{m} \sin(\omega_0 t - \varphi) \qquad (7.3.34b)$$

$$H_x(t) = \frac{4\pi \eta Q}{[(2Q\Delta/\omega)^2 + 1]^{1/2}} \tilde{m} \sin(\omega_0 t - \varphi + \varphi_0) \qquad (7.3.34c)$$

where

$$\varphi_0 = \tan^{-1} \frac{2Q\Delta}{\omega} \qquad (7.3.35a)$$

$$\varphi = \frac{2Q\Delta}{\omega T_R} \int_0^t dt' \frac{M_z(t')}{M_z(0)}$$

$$= \frac{2Q\Delta}{\omega} \left[-\frac{t}{T_2} + \ln \frac{\cosh[(t - t_D)/T]}{\cosh(t_D/T)} \right]. \qquad (7.3.35b)$$

7.4   Comparison with NMR super-radiance experiments

The theory presented in the previous section contains all the essential features of the super-radiance generation observed experimentally [KPSY88, BKP88, BBKPTTS89, BBZKMT90]. It predicts the existence of thresholds for both NMR generation and reversal of the magnetization. In addition, the dependence of the generation threshold on de-tuning from resonance allows us to explain qualitatively the origin of the series of pulses occurring when the Larmor frequency is scanned slowly. Thus, if the initial value of the spin polarization P is small (but larger, of course, than the absolute value of the threshold polarization (for $\Delta = 0$)), then the threshold can be reached only close to resonance, i.e. for $\Delta \approx 0$. In this case a single long pulse of rather small amplitude appears.

Let us now assume that the value of the initial polarization is such that the threshold condition $T_R(\Delta_1, P_i) = T_2$ is fulfilled for some $\Delta_1 \neq 0$. We will also assume that after the creation of the pulse there is no reversal of the polarization (i.e. it remains negative). It is not difficult to convince oneself, by using equations (7.3.30*b*) and (7.3.31), that the value of the final negative polarization, $P_f = P(\infty)$, corresponds to the below-threshold region $T_R(\Delta_1, P_f) > T_2$. Thus, if the de-tuning Δ is fixed, then the repeated generation of RF pulses is impossible. However, in the process of scanning Δ the threshold condition $T_R(\Delta, P_f) = T_2$ can be realized again as soon as $\Delta = \Delta_2 < \Delta_1$. Then a second pulse appears, then a third, etc., until the final value of the negative polarization P_f is equal to the threshold at resonance ($\Delta = 0$).

All the phenomena listed above were observed in the experiments of Bazhanov *et al* [BKP88, BBKPTTS89, BBZKMT90]: after the initial polarization reached the threshold value $P_{i1} = -7\%$ (which corresponds to exact resonance, $\Delta = 0$), a single small-amplitude pulse was observed. For values of the initial polarization P_i corresponding to the horizontal portion of the plot of the function P_f (P_i) (see figure 7.2(a)), a series of pulses appeared (see figure 7.3(a)). In this case, as the absolute value of P_i increased, the number of pulses also increased, at first. However, because of the increasing power of the first pulses, the conditions for generating subsequent pulses became worse. As a result, eventually a single pulse was generated once more, but one of large amplitude (the region of linear increasing of P_f (P_i) in figure 7.2(a)). For values of the initial polarization $P_i \leq -33\%$ reversal of the polarization was observed. In fact, it is just this régime that corresponds to super-radiance.

The experimentally observed values of the threshold for generation ($P_{i1} = -7\%$) and for reversal ($P_{ir} = -33\%$) do not fit the theoretical ratio (for fixed value of Δ) of 1:2. This can be explained (in agreement with what was said earlier) by the fact that the generation threshold corresponds to exact resonance, whilst the reversal threshold appears at a certain finite de-tuning Δ_r. Quantitative agreement with experiment is obtained if we set $\Delta_r \approx 1.2\omega/2Q$.

In order to obtain a detailed picture of pulsed SR generation with a slowly

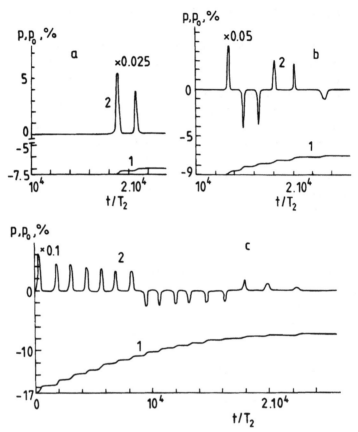

Figure 7.6. Curves showing the evolution of the nuclear polarization for different values of the initial nuclear polarization P_i: (1) the longitudinal component P; (2) the amplitude of the transverse component P_0.

scanned Larmor frequency, a series of calculations were carried out to follow the time evolution of the longitudinal (M_z) and transverse ($\tilde{m}$) components of the magnetization. It was assumed in these calculations that the quantity β in equations (7.3.29a) and (7.3.29b) was a slowly varying function of time (over durations $t \sim T_R$). The time dependence of the off-resonance de-tuning was assumed to be of the form: $\Delta = \Delta_r - vt$, where $\Delta_r = 1.2\omega/2Q$ was the de-tuning that matched the experimental value of the sign-reversal threshold $P_{ir} = -33\%$ when the generation threshold was $P_{i1} = -7\%$. The parameter v was taken to be the experimental value of the scanning rate: 1.6×10^6 s^{-1}. The relaxation time T_2 was determined from the duration of the SR pulse at the sign-reversal threshold. It was assumed that the pulse had a hyperbolic-secant shape, with $T_2 = T_p/2.4 = 50$ μs. The Q factor of the RF circuit was taken

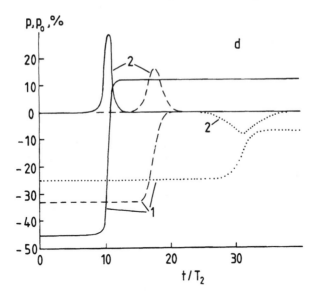

Figure 7.6. (*Continued.*)

to be $Q = 100$. In order to imitate the presence of the noise that initiates SR generation, a random force $F(t)$ with a δ correlation

$$\langle F(t)F(t')\rangle = F_0^2\delta(t - t') \tag{7.4.1}$$

was added to the right-hand side of equation (7.3.29a). This models the noise that originates from both the quantum fluctuations of the transversal component of the magnetization and current fluctuations in the loop of the RF circuit.

The results of these calculations are presented in figure 7.6 The time-dependent longitudinal (P) and transverse (P_0) amplitudes of nuclear polarization were obtained from the calculated magnetization with the help of equation (7.2.1). These quantities are presented as percentages in figure 7.6, at different values of the initial nuclear polarization P_i. It is clear that the theoretical picture of the formation of SR pulses for slow scanning of the Larmor frequency coincides almost completely with experiment. The theory reproduces both the threshold values, $P_i = -7\%$ (figure 7.6(a)) and $P_{ir} = -33\%$ (figure 7.6(d)), and the multi-pulse structure of SR generation when the polarization lies in the interval $-25\% < P_i < -7\%$ (figures 7.6(a)–(d)). In this latter case the theory also predicts the decrease of the nuclear polarization to the threshold value $P_i = -7\%$ within the time required for the Larmor frequency to change from its initial ($\Delta = \Delta_r$) to the resonance ($\Delta = 0$) value. However, the calculated pulses have different polarities, depending on how the fluctuations in the magnetization $\tilde{m}$ (positive or negative) develop, a feature which was not seen in the experiment.

The single-pulse régime of SR generation (figure 7.6(d)) can be described by equations (7.3.30a), (7.3.30b)–(7.3.35a), (7.3.35b) with fixed off-resonance de-tuning. This is possible because the Larmor frequency is practically unchanged during the SR pulse, owing to the relatively slow scanning rate. In particular, one then obtains from equation (7.3.30b) a linear dependence of P_f on P_i. This is in accord with the experiment, as seen in figure 7.2(a).

7.5   Concluding remarks

We note that a good agreement between theory and experiment has been obtained only for a specific choice of values of the relaxation time and initial de-tuning from resonance. From our estimate, T_2 must be about 50 μs, while from NMR experiments it follows that for the material studied the relaxation time, determined by the inverse of the spectral width at a temperature of 50 mK, is of one order of magnitude less (about 4 μs). Another discrepancy is that in the experiment the initial value of the de-tuning is not Δ_r, but is of the order of the resonant frequency. In the approach presented above this gives a *multi-pulse régime* for an arbitrary value of the initial polarization, unless it is below the threshold, but not a *single-pulse régime* with polarization inversion.

This discrepancy between theory and experiment is a major one. It shows the limitations of the method of Bloch equations for describing NMR effects in solids, as mentioned earlier in the literature [26, 58, 59]. Therefore the values of T_2 and Δ_r in our treatment can be considered as certain effective values in a simplified theory.

A refined description of super-radiance of magnetic spins in solids will probably have to take into account the dependence of the relaxation rate on the magnitude of the magnetization, using as a basis the concept of spin temperature [26, 58, 59].

Chapter 8

Effects of diffraction upon super-radiance

The previous chapters have been mainly concerned with the one-dimensional model of super-radiance, where the dependence of both the emission field and the characteristics of the medium upon the transversal coordinates are ignored. These effects are of considerable interest, as they determine the polar diagram of super-radiance. Even in those cases where the geometry of the system specifies a preferred direction for the propagation of the radiation, the divergence and diffraction of the field have to be taken into account because they may substantially influence the kinetics of the processes. The theoretical aspects of diffraction in super-radiance have been discussed in a number of works listed within the bibliography at the end of the book.

In this chapter both diffraction and fluctuation effects on super-radiance will be investigated within the framework of two models. Firstly, the two-dimensional theory of super-radiance of an extended system will be developed. Secondly, we shall consider the super-radiance of a 'sheet of paper' volume. Such a system was realized in $KCl:O_2^-$ by Schiller *et al* [SSS87, SSS88], who observed the diffraction of $KCl:O_2^-$ super-radiance (see Chapter 2). Our main interest is in studying transverse effects and the polar diagram of the collective emission for systems with different Fresnel numbers. Following [AZMT89, AZMT91], the semiclassical approach, which assumes either a uniform or a random initial polarization of the medium, will be applied to calculate characteristics of super-radiance. The results will be compared with the experiments [SSS87, SSS88].

8.1 Two-dimensional super-radiance

8.1.1 The model and basic equations

We start with a two-dimensional model that assumes the field and atomic characteristics of the system to be dependent upon the longitudinal coordinate x $(0 < x < L)$ as well as on one transverse coordinate y $(-D/2 < y < D/2)$.

The optical centres are modelled by two-level atoms with equal frequencies ω_0 and dipole transition momenta, d, oriented along the z axis. In this model the z component of the polarization vector $\mathcal{P}_z$ of the medium, as well as three components of the electromagnetic field (one component of the electric vector $\mathcal{E}_z = \mathcal{E}$, and two components of the magnetic vector $\mathcal{B}_x$ and $\mathcal{B}_y$), are non-zero.

Maxwell's equations for the fields $\mathcal{E}$ and $\mathcal{B}$ are written as follows

$$\frac{\partial \mathcal{E}}{\partial y} = -\frac{1}{c}\frac{\partial \mathcal{B}_x}{\partial t} \qquad \frac{\partial \mathcal{E}}{\partial x} = \frac{1}{c}\frac{\partial \mathcal{B}_y}{\partial t} \tag{8.1.1a}$$

$$\frac{\partial \mathcal{B}_y}{\partial x} - \frac{\partial \mathcal{B}_x}{\partial y} = \frac{4\pi}{c}\frac{\partial \mathcal{P}}{\partial t} + \frac{1}{c}\frac{\partial \mathcal{E}}{\partial t} \tag{8.1.1b}$$

where the polarization of the medium is defined as in (1.4.7)

$$\mathcal{P} = N_0 d(\rho_{ge} + \rho_{eg}). \tag{8.1.2}$$

As earlier, N_0 is the concentration of optical centres, and $\rho_{ge} = \rho_{eg}^*$ is the off-diagonal element of the atomic density matrix which together with ρ_{ee} and ρ_{gg} satisfy the first three of equations (1.4.8) of Chapter 1

$$\dot{\rho}_{ee} = \frac{i}{\hbar}d\mathcal{E}(\rho_{ge} - \rho_{eg}) \tag{8.1.3a}$$

$$\dot{\rho}_{gg} = -\frac{i}{\hbar}d\mathcal{E}(\rho_{ge} - \rho_{eg}) \tag{8.1.3b}$$

$$\dot{\rho}_{eg} = -i\omega_0\rho_{eg} + \frac{i}{\hbar}d\mathcal{E}(\rho_{gg} - \rho_{ee}). \tag{8.1.3c}$$

Assuming that the fields $\mathcal{E}$ and $\mathcal{B}$ arising in the system are not very large, i.e. $d\mathcal{E}, d\mathcal{B} \ll \hbar\omega_0$, we can pick out, as before, the rapidly varying factors of the electromagnetic field and of the off-diagonal elements of the density matrix, connected with the optical frequency and wavelength. In the present chapter, however, we are going to take into account that the super-radiant pulse can propagate in both directions of the x axis and accordingly we set

$$\rho_{eg}(x, y, t) = \tfrac{1}{2}R_1(x, y, t)\exp[-i(\omega_0 t - k_0 x)]$$
$$+ \tfrac{1}{2}R_2(x, y, t)\exp[-i(\omega_0 t + k_0 x)] \tag{8.1.4a}$$

$$\begin{pmatrix} \mathcal{E}(x, y, t) \\ \mathcal{B}_x(x, y, t) \\ \mathcal{B}_y(x, y, t) \end{pmatrix} = \frac{\hbar}{2id T_R}\left[\begin{pmatrix} \varepsilon_1(x, y, t) \\ b_{1x}(x, y, t) \\ b_{1y}(x, y, t) \end{pmatrix}\exp[-i(\omega_0 t - k_0 x)]\right.$$

$$\left. + \begin{pmatrix} \varepsilon_2 \\ b_{2x} \\ b_{2y} \end{pmatrix}\exp[-i(\omega_0 t + k_0 x)]\right] + \text{c.c.} \tag{8.1.4b}$$

Here $k_0 = \omega_0/c$, $T_R = \hbar/(2\pi k_0 d^2 N_0 L)$ is the super-radiance characteristic time, R_j, ε_j, b_{jx} and b_{jy} are dimensionless polarization amplitudes and components of the electromagnetic field. The index $j = 1$, (2) describes a wave propagating in the positive (negative) direction along the x axis. We assume the SVEA for the slow amplitudes R, ε, b (in the longitudinal coordinate and time) and after substitution of (8.1.4a), (8.1.4b) into (8.1.1a), (8.1.1b) and (8.1.3a)–(8.1.3c) and neglecting the rapidly oscillating terms obtain the equations

$$\frac{1}{c}\frac{\partial}{\partial t}\begin{pmatrix}\varepsilon_1\\\varepsilon_2\end{pmatrix} + \frac{\partial}{\partial x}\begin{pmatrix}\varepsilon_1\\-\varepsilon_2\end{pmatrix} - \frac{i}{2k_0}\frac{\partial^2}{\partial y^2}\begin{pmatrix}\varepsilon_1\\\varepsilon_2\end{pmatrix} = \frac{1}{L}\begin{pmatrix}R_1\\R_2\end{pmatrix} \quad (8.1.5a)$$

$$\frac{\partial}{\partial t}\begin{pmatrix}R_1\\R_2\end{pmatrix} = \frac{1}{T_R}Z\begin{pmatrix}\varepsilon_1\\\varepsilon_2\end{pmatrix} \qquad Z = \rho_{ee} - \rho_{gg} \quad (8.1.5b)$$

$$\frac{\partial Z}{\partial t} = -\frac{1}{2T_R}(\varepsilon_1^* R_1 + \varepsilon_2^* R_2 + \text{c.c.}) \quad (8.1.5c)$$

$$b_{jx} = -\frac{i}{k_0}\frac{\partial \varepsilon_j}{\partial y} \qquad b_{jy} = (-1)^j \varepsilon_j. \quad (8.1.5d)$$

Note that in this model we are not making use of the approximation of slow variation of the amplitudes with respect to the transverse coordinate y, so that the corresponding terms in equation (8.1.5a) contain second derivatives in y.

The Poynting vector S, which determines the polar diagram of the radiation, has in this case the following components

$$S = \frac{c}{4\pi}[\mathcal{E} \times \mathcal{B}] = \frac{c}{4\pi}(-\mathcal{E}\mathcal{B}_y, \mathcal{E}\mathcal{B}_x, 0). \quad (8.1.6)$$

From equations (8.1.5a)–(8.1.5d) there follows the local energy conservation law

$$\frac{\partial}{\partial t}(U_a + U_f) + \text{div}\,S = 0 \quad (8.1.7)$$

where U_a and U_f are the atomic and field energy densities respectively

$$U_a = N_0\hbar\omega_0\rho_{ee} \quad (8.1.8a)$$

$$U_f = \frac{N_0\hbar\omega_0 L}{4cT_R}(|\varepsilon_1|^2 + |\varepsilon_2|^2) \quad (8.1.8b)$$

and

$$S_x = \frac{N_0\hbar\omega_0 L}{4T_R}(|\varepsilon_1|^2 - |\varepsilon_2|^2) \quad (8.1.9a)$$

$$S_y = \frac{N_0\hbar\omega_0 L}{4T_R}\frac{i}{2k_0}\left(\varepsilon_1\frac{\partial \varepsilon_1^*}{\partial y} - \varepsilon_2^*\frac{\partial \varepsilon_2}{\partial y}\right) + \text{c.c.} \quad (8.1.9b)$$

are the non-vanishing components of the Poynting vector (8.1.6), averaged over fast temporal and spatial oscillation periods. S may be expressed as the sum $S = S_1 + S_2$, where the vectors

$$S_1 = \frac{N_0 \hbar \omega_0 L}{4 T_R} \left(|\varepsilon_1|^2, \frac{i}{2 k_0} \varepsilon_1 \frac{\partial \varepsilon_1^*}{\partial y} + \text{c.c.}, 0 \right) \tag{8.1.10a}$$

$$S_2 = \frac{N_0 \hbar \omega_0 L}{4 T_R} \left(-|\varepsilon_2|^2, \frac{i}{2 k_0} \varepsilon_2 \frac{\partial \varepsilon_2^*}{\partial y} + \text{c.c.}, 0 \right) \tag{8.1.10b}$$

constitute energy flow densities of waves propagating in the positive and negative directions along the x axis.

It is convenient, in solving equations (8.1.5a)–(8.1.5d), to go over to dimensionless variables $\tau = t/T_R$, $\xi = x/L$ and $\eta = y/D$. In terms of these variables equations (8.1.5a)–(8.1.5d) become

$$\frac{1}{v}\frac{\partial}{\partial \tau}\begin{pmatrix} \varepsilon_1 \\ \varepsilon_2 \end{pmatrix} + \frac{\partial}{\partial \xi}\begin{pmatrix} \varepsilon_1 \\ -\varepsilon_2 \end{pmatrix} - \frac{i}{4\pi F}\frac{\partial^2}{\partial \eta^2}\begin{pmatrix} \varepsilon_1 \\ \varepsilon_2 \end{pmatrix} = \begin{pmatrix} R_1 \\ R_2 \end{pmatrix} \tag{8.1.11a}$$

$$\frac{\partial}{\partial \tau}\begin{pmatrix} R_1 \\ R_2 \end{pmatrix} = Z\begin{pmatrix} \varepsilon_1 \\ \varepsilon_2 \end{pmatrix} \tag{8.1.11b}$$

$$\frac{\partial Z}{\partial \tau} = -\tfrac{1}{2}(\varepsilon_1^* R_1 + \varepsilon_2^* R_2 + \text{c.c.}) \tag{8.1.11c}$$

where we have introduced the dimensionless velocity of light $v = c T_R/L$ and the Fresnel number $F = D^2/L\lambda$.

In the semiclassical theory, the initial values of the quantities in equations (8.1.11a)–(8.1.11c) are specified in the following way

$$\varepsilon_j(\xi, \eta, 0) = 0 \tag{8.1.12a}$$
$$Z(\xi, \eta, 0) = 1 \tag{8.1.12b}$$
$$R_j(\xi, \eta, 0) = R_{j0}(\xi, \eta) \qquad j = 1, 2. \tag{8.1.12c}$$

corresponding to a vanishing initial field, to full inversion and to an initial polarization imitating spontaneous emission. In addition, the field should satisfy the boundary conditions

$$\varepsilon_1(0, \eta, \tau) = \varepsilon_2(1, \eta, \tau) = 0 \tag{8.1.12d}$$
$$\varepsilon_j(\xi, \eta \to \infty, \tau) = 0 \qquad j = 1, 2. \tag{8.1.12e}$$

We shall discuss below two ways of specifying the initial polarization: it will be chosen either to be constant along the sample, or to be a Gaussian random function with a δ correlation.

8.1.2 Integro-differential form of equations

The solution of equation (8.1.11*a*) satisfying the boundary conditions (8.1.12*d*) and (8.1.12*e*) has the form [13]

$$\varepsilon_1(\xi, \eta, \tau) = \int_0^\xi \mathrm{d}\xi' \int_{-1/2}^{1/2} \mathrm{d}\eta' \theta(\tau') R_1(\xi', \eta', \tau') M(\xi - \xi', \eta - \eta') \quad (8.1.13a)$$

$$\varepsilon_2(\xi, \eta, \tau) = \int_\xi^1 \mathrm{d}\xi' \int_{-1/2}^{1/2} \mathrm{d}\eta' \theta(\tau') R_2(\xi', \eta', \tau') M(\xi - \xi', \eta - \eta') \quad (8.1.13b)$$

$$\theta(\tau') = \begin{cases} 1 & \tau' > 0 \\ 0 & \tau' < 0 \end{cases} \quad \tau' = \tau - |\xi - \xi'|/v$$

where the Heaviside step function $\theta(\tau')$ takes into account the retardation effect, and the kernel M, given by the formula

$$M(\xi, \eta) = \left(-\frac{\mathrm{i}F}{|\xi|} \right)^{1/2} \exp(\mathrm{i}\pi F\eta^2/|\xi|) \quad (8.1.14)$$

is the Green's function for the two-dimensional parabolic equation.

In this way the initial system of partial differential equations (8.1.11*a*)–(8.1.11*c*) is transformed into a system of integro-differential equations (8.1.11*b*, *c*) and (8.1.13*a*, *b*).

We confine ourselves to the study of super-radiance for samples where the passage time of light through the system, L/c, is much less than the characteristic super-radiance time, T_R, i.e.

$$\frac{1}{v} = \frac{L}{cT_R} = \frac{2\pi k_0 d^2 N_0 L^2}{\hbar c} \ll 1. \quad (8.1.15)$$

In that case we may ignore retardation effects, i.e. in the equations for the field (8.1.11*a*) neglect the derivatives with respect to time and omit the step function $\theta(\tau')$ in equations (8.1.13*a*) and (8.1.13*b*).

8.1.3 Numerical solution of equations

To solve this system numerically we introduce a rectangular region of integration in the plane of the dimensionless coordinates ξ, η ($0 < \xi < b; -\frac{1}{2}a < \eta < \frac{1}{2}a$). By taking it somewhat larger than the size of the system, it permits the determination of the field both inside and outside the sample, in a rectangular area satisfying $2a > 1$, $b > 1$. The region is divided by a uniform grid into rectangular sub-regions (cells) of dimension $H_\xi \times H_\eta$, numbered by the indices and k, l. It is further assumed that the inversion and polarization are constant inside each cell (k, l), with the field ε_j^{kl} equal to the sum of the contributions

of the fields from cells to the left of the fixed cell (k, l) and acting at its centre
with coordinates $\xi_k = (k + \frac{1}{2})H_\xi$, $\eta_l = l H_\eta$

$$\varepsilon_j^{kl} = \sum_{k' \leq k} \sum_{l'} M_{k-k', l-l'} R_j^{kl} \qquad (8.1.16)$$

$$M_{k-k', l-l'} = \int_{k'H_\xi}^{(k'+1)H_\xi} d\xi' \int_{(l'-\frac{1}{2})H_\eta}^{(l'+\frac{1}{2})H_\eta} d\eta' \left(-i \frac{F}{|\xi_k - \xi'|} \right)^{1/2}$$

$$\times \exp\left[i\pi F \frac{(\eta_l - \eta')^2}{|\xi_k - \xi'|} \right] \theta(\xi_k - \xi'). \qquad (8.1.17)$$

The Heaviside function $\theta(x)$ in equation (8.1.17) is introduced to take correct
account of the contribution of the polarization cells (k', l') to the field acting
upon the cell (k, l).

The expression for the kernel M may be written as

$$M_{kl} = M_{k, -l} = Q(\xi_k, \eta_l + \tfrac{1}{2}H_\eta) - Q(\xi_k, \eta_l - \tfrac{1}{2}H_\eta) \qquad (8.1.18)$$
$$+ Q(\xi_{k-1}, \eta_l - \tfrac{1}{2}H_\eta) - Q(\xi_{k-1}, \eta_l + \tfrac{1}{2}H_\eta)$$

with

$$Q = (-i)^{1/2}\xi[(1 - 2\pi i u^2)\Phi(u) + u \exp(i\pi u^2)] \qquad (8.1.19)$$

where $u = (F/|\xi|)^{1/2}\eta$ and $\Phi(u)$ is the Fresnel integral

$$\Phi(u) = \int_0^u dt \, \exp(i\pi t^2). \qquad (8.1.20)$$

For $k = 0$ the last two terms in equation (8.1.18) are understood to be the
limiting values of the corresponding expressions ($\xi_{-1} \to 0$), which remain finite
for arbitrary values of the index l and for $l = 0$ correspond to the contribution of
the polarization of the given cell to the field calculated for it, i.e. 'self-action'.

The requirement of smooth variation of the kernel M_{kl} from one cell to the
next, which is essential for an accurate representation of the integral relation
by equation (8.1.16), imposes definite restrictions upon the total number of grid
points $N_\xi = 1/H_\xi$, $N_\eta = 1/H_\eta$, to be used in the integration region. For
large values of the Fresnel number the indicated smoothness is ensured for N_ξ,
$N_\eta \gg 1$. For small values of F the corresponding condition is $N_\xi \gg 1/F$,
$N_\eta \gg 1$.

Note that the kernel M does not depend on time. This implies that the
time evolution of the problem is governed by the ordinary differential equations
(8.1.11b) and(8.1.11c) complemented with the system of algebraic relations
(8.1.16).

In specifying the uniform initial polarization, the integration region was
broken up into 300 cells in a scheme of 12×25 ($H_\xi = \frac{1}{12}$, $H_\eta = \frac{1}{25}$) grid
points.

The accuracy of the numerical solution of the Maxwell–Bloch system of equations was assessed by testing the conservation laws for the energy (8.1.7) and for the length of the Bloch vector $|R_1|^2 + |R_2|^2 + Z^2 = $ constant.

Using the solution obtained for equations (8.1.11b), (8.1.11c), (8.1.13a) and (8.1.13b), the components S_x and S_y of the Poynting vector are calculated as follows

$$S_j = n\frac{\hbar\omega_0 N_0 L}{4T_R}|\varepsilon_j|^2 = n\frac{\hbar\omega_0 N_0 L}{4T_R}\frac{FL}{|x|}|f_j(\theta, \tau)|^2 \qquad (8.1.21)$$

where

$$f_j(\theta, \tau) = \int_0^1 d\xi' \int_{1/2}^{1/2} d\eta' R_j(\xi', \eta', \tau) \exp\left[i\pi F\frac{L}{D}\theta\left(\frac{L}{D}\theta\xi' - 2\eta'\right)\right]. \qquad (8.1.22)$$

Here $j = 1, 2$, n is a unit vector directed from the location of the system towards the point of observation and lying in the xy plane, and θ is the angle between n and the x axis. In deriving equation (8.1.22) we have used the approximation $\theta \ll 1$.

The solution allows us to calculate also the output intensity of each of the counter-propagating waves

$$I_j(\tau) = \int S_j \, d\sigma. \qquad (8.1.23)$$

Here S_j ($j = 1, 2$) are given by equations (8.1.21) and (8.1.22) and the integration is taken over the end surfaces of the system.

8.1.4 Results of numerical calcualtions

Spatially homogeneous initial polarization. As we have already remarked, the case of uniform initial polarization corresponds to induced super-radiance. The calculations were performed for the value $R_{01} = 0.04$. The amplitude of the initial polarization for the counter-propagating wave R_{20} was set to zero. In this case that wave does not evolve. The time dependence of super-radiance intensity obtained for several values of the Fresnel number F is shown in figure 8.1 (solid curves). A characteristic feature of the super-radiance pulse for $F > 1$ is that between maxima the radiation intensity approaches some finite value I_{min} and it does not go down to zero, as was the case in the one-dimensional theory (see Chapter 5). Increasing the Fresnel number F reduces this value, whereas decreasing F increases I_{min}, and, at the same time, causes the peak intensity of the second maximum to decrease. They become equal for $F \approx \frac{1}{4}$. As a result, the oscillatory structure of the super-radiance pulse is smeared out, and only one peak is observed in the radiation.

Figures 8.2–8.4 show the distribution of the population inversion throughout the sample and the radiation patterns $|f_1(\theta, \tau)|^2$ for values of the

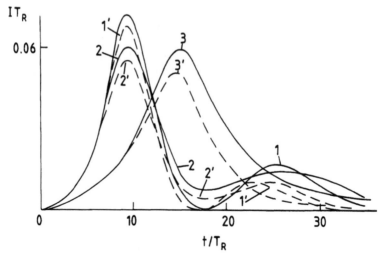

Figure 8.1. The calculated intensity of radiation for samples with different Fresnel numbers. Solid curves—the single-wave approximation; broken curves—the two-wave approximation. Initial polarization $R_0 = 0.04$. Curves: $(1, 1')$ $F = \infty$; $(2, 2')$ $F = 1$; $(3, 3')$ $F = 0.1$.

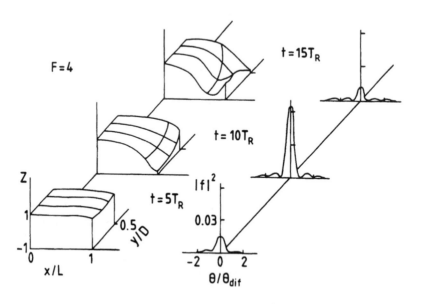

Figure 8.2. Inversion distribution in the sample (left) and the radiation polar diagram (right) at different instants, for Fresnel number $F = 4$. The initial amplitude of polarization is uniform throughout the sample.

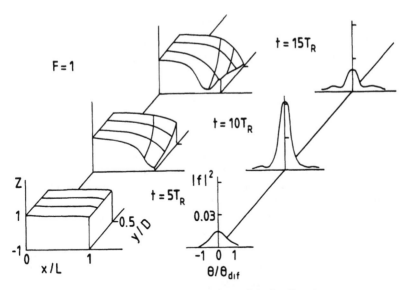

Figure 8.3. The same as figure 8.2, for $F = 1$.

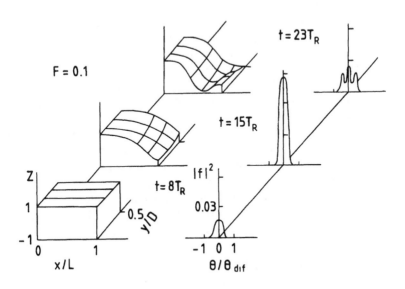

Figure 8.4. The same as figure 8.2, for $F = 0.1$.

Fresnel number $F = 0.1$, 1.0 and 4.0 computed at several instants in the single-wave approximation. Owing to the symmetry of the problem in η, it is sufficient to show the distribution over half of the sample only, for $\eta > 0$. These figures show a common regularity consisting of early development of inversion in the central part of the sample ($\eta \approx 0$) with relatively late evolution at the edges ($\eta = 0.5$). This is most pronounced for $F = 1$. The transverse non-uniformity of the inversion is the reason for the smoothing out of the ringing structure of the super-radiance pulse mentioned above.

As the Fresnel number F increases, the width of the region of the synchronous variation of the inversion grows, encompassing an ever larger part of the cross-section of the sample. At the same time transverse uniformity is established, and the longitudinal directivity of the Poynting vector becomes more pronounced (see figure 8.5) thus ensuring a decrease in the radiation flux that leaves the sample through its side boundaries.

The divergence of the radiation (see figure 8.5) during the major part of the duration of the super-radiance pulse is determined by the angle, equal in order of magnitude to the diffraction angle $\theta_{\text{dif}} = \lambda/D$.

However, at the instants between the maxima the intensity of the radiation is distributed approximately equally over the diffraction side lobes located within the limits of the geometric angle $\theta_{\text{geom}} = D/L$, and it decreases with increasing F.

In this manner, for uniform initial polarization the super-radiance pulse approximately corresponds to a pulse of the one-dimensional model of super-radiance as the Fresnel number increases. Its time scale is determined by the super-radiance time T_R. This can be also deduced from the original equations (8.1.11a)–(8.1.11c). Indeed, by omitting the terms in those equations with both the first-order ($v \gg 1$) and the second-order ($F \gg 1$) time derivatives, we obtain equations which do not contain any physical parameters. Thus the time scale of the super-radiance evolution is determined only by T_R.

For small Fresnel numbers the transverse component of the Poynting vector increases rapidly (see figure 8.5) as the distance from the axis of the sample increases. This indicates that an increasing fraction of the energy leaves the sample through its side boundaries. Thus, the ratio of the total side flux to the longitudinal flux at the maximum of the super-radiance pulse is approximately 3:2 for $F = 0.1$ and it is close to 0.1 for $F = 1$. The intense radiation in the transverse directions is a distinctive feature of the super-radiance pulse for $F \ll 1$: the establishment of transverse uniformity of atom–field characteristics, an increase in the delay time, a decrease in the peak value of the intensity, and an increase in the width of the super-radiance pulse.

Similarly we can determine the time scale of super-radiance in the limit of small Fresnel number ($F \ll 1$) by neglecting the retardation effect ($v \gg 1$) in equations (8.1.11b), (8.1.11c), (8.1.13a) and (8.1.13b). As follows from equation (8.1.14), $M(\xi, \eta) \approx (-iF/|\xi|)^{1/2}$ for $F \ll 1$. This means that the slowly varying amplitudes behave as $\varepsilon_i \sim F^{1/2}$ ($i = 1, 2$). Defining the new

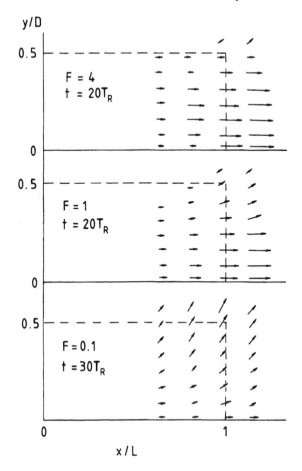

Figure 8.5. The Poynting vector field for maximal intensity of radiation. The initial amplitude of polarization is taken to be uniform throughout the sample.

amplitudes $\varepsilon_i' = \varepsilon_i/F^{1/2}$ and the new time constant $T_R' = T_R/F^{1/2}$, we then obtain equations for the amplitudes ε_i', R_i and for Z in terms of the new time variable $\tau' = t/T_R'$, which do not contain any physical parameters. Therefore the time constant $T_R' = T_R/F^{1/2}$ is the characteristic time of super-radiance in the case of small Fresnel numbers.

The polar diagram of the radiation in the case of small Fresnel numbers (see figure 8.5) shows that the divergence of super-radiance is contained within the limits of the diffraction angle $\theta_{\text{dif}} = \lambda/D$.

The results described above were obtained in the single-wave model. To illustrate the effect of the counter-propagating wave upon the super-radiance characteristics, we show in figure 8.1 the calculations of the super-radiance intensity (broken curves) when the symmetric counter-propagting wave ($R_{20} =$

0.04) is taken into account. According to figure 8.1 the opposing wave affects mainly the later stage of the evolution of super-radiance, resulting in a suppression of the intensity of oscillations in the case $F > 1$. With decreasing Fresnel number ($F < 1$) the rôle of the opposing wave increases, and manifests itself in a decrease in the delay time and in the peak super-radiance intensity, as compared with the single-wave approximation.

Random initial polarization. The influence of random initial polarization on the emission kinetics and the polar diagram of the super-radiance will be discussed below within the single-wave approximation for a specified case of initial polarization of a constant amplitude $|R_{10}| = 0.2$ and with the phase randomly chosen from the interval $[0, 2\pi]$ for each elementary cell.

Super-radiance pulses computed for three random realizations of the initial polarization are shown in figure 8.6. As can be seen, fluctuations in the polarization generate fluctuations of the parameters of the super-radiance pulse:

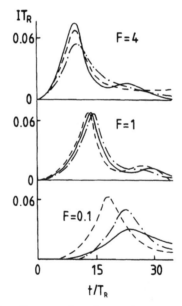

Figure 8.6. Emission pulses for several random realizations of the initial amplitude of polarization and Fresnel numbers F.

its form, delay time and peak intensity. The largest random changes are experienced by the parameters of the super-radiance pulse for systems with small Fresnel number, whilst for $F = 1$ the fluctuations are smallest. With increasing Fresnel number the effect of the fluctuations of the initial polarization upon the delay time is weakened, but the spread in the peak values of super-radiance intensity remains, as before, significant.

Figure 8.7 demonstrates the time evolution of the super-radiance polar

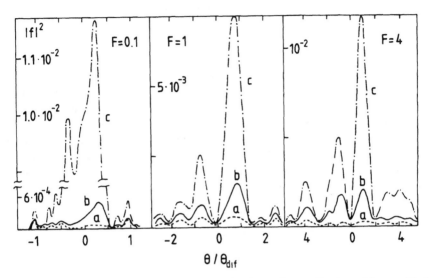

Figure 8.7. The structure of the polar diagram for the random realization of the amplitude of the initial polarization at different instants of time. For $F = 0.1$: (a) $t = 0.5T_R$; (b) $t = 5T_R$; (c) $t = 15T_R$. For $F = 1$ and $F = 4$: (a) $t = 0.5T_R$; (b) $t = 2.5T_R$; (c) $t = 5T_R$.

diagram for a typical realization of the distribution of the initial amplitudes of polarization. The ray structure of the polar diagram is particularly clear, as well as the competition between rays propagating at different angles to the axis of the sample. These angles undergo strong fluctuations from one realization to another. For large Fresnel number ($F > 1$) the fluctuations lie within the limits of the angular dimension D/L of the sample, with the angular dimension of an individual ray having the diffraction scale λ/D. In this way the number of rays is approximately equal to the Fresnel number, $(D/L)/(\lambda/D) = D^2/L\lambda = F$. For small Fresnel numbers ($F < 1$) the total width of the polar diagram is determined by the diffraction angle, and the angular dimensions of the system are manifested in the structure.

8.2 Super-radiance of a 'sheet of paper' volume

Diffraction of super-radiance was observed in $KCl:O_2^-$ crystals by Schiller *et al* [SSS87, SSS88]. From their estimate the active volume had the dimensions $L \times D \times H = 8 \times 1 \times 0.05$ mm^3. Here the dimension D, which is essential for the diffraction pattern, was determined from the absorption depth of the pumping into the sample.

In this section we shall consider the three-dimensional super-radiance problem of the so-called 'sheet of paper' volume when the Fresnel number is small for one of the transversal dimensions. Just such a case has been realized in the experiments of Schiller *et al* [SSS87, SSS88] (see subsection 2.2.1).

8.2.1 Basic equations

Let us examine a thinly stretched three-dimensional sample $(L \gg D \gg H)$, where the Fresnel number corresponding to the least dimension H (along the z axis) will be considered small, $F_H = H^2/L\lambda \ll 1$. Using our usual two-level model for the atomic system, we obtain the following equations for the inversion, $Z = \rho_{ee} - \rho_{gg}$, and for the slowly varying amplitudes of both the dimensionless electric field, ε, and the off-diagonal element of the density matrix, R (for the sake of simplicity we consider only the wave propagating in the positive x direction, the single-wave model)

$$\left[\frac{1}{c} \frac{\partial}{\partial t} + \frac{\partial}{\partial x} - \frac{i}{2k_0} \left(\frac{\partial^2}{\partial y^2} + \frac{\partial^2}{\partial z^2} \right) \right] \varepsilon = \frac{1}{L} R \qquad (8.2.1a)$$

$$\frac{\partial R}{\partial t} = \frac{1}{T_R} Z\varepsilon \qquad (8.2.1b)$$

$$\frac{\partial Z}{\partial t} = -\frac{1}{2T_R} (\varepsilon^* R + \text{c.c.}). \qquad (8.2.1c)$$

The initial and boundary conditions corresponding to the present case are given by

$$\varepsilon(x, y, z, 0) = 0$$
$$Z(x, y, z, 0) = 1 \qquad (8.2.2)$$
$$R(x, y, z, 0) = R_0(x, y, z)$$

$$\varepsilon(0, y, z, t) = \varepsilon(x, +\infty, z, 0)$$
$$= \varepsilon(x, y, +\infty, 0) = 0. \qquad (8.2.3)$$

Using Green's function for the three-dimensional parabolic equation [13] corresponding to the boundary conditions (8.2.3)

$$M(x - x', y - y', z - z')$$
$$= -\frac{i}{L\lambda} \frac{\exp\left[\frac{1}{2} ik_0 \left((y - y')^2 + (z - z')^2 \right) / |x - x'| \right]}{|x - x'|}. \qquad (8.2.4)$$

the electric field amplitude, ε, can be expressed in terms of R

$$\varepsilon(x, y, z, t) = \int_0^x dx' \int_{-D/2}^{D/2} dy' \int_{-H/2}^{H/2} dz' \theta(T) R(x', y', z', T)$$
$$\times M(x - x', y - y', z - z') \qquad (8.2.5)$$

where $T = t - |x - x'|/c$ is the retarded time.

Taking into account that $H \ll D$ and $F \ll 1$ in the kernel of equation (8.2.5), the dependence on $z - z'$ can be neglected for the field inside the sample, that is, for $-D/2 < y, y' < D/2$ and $-H/2 < z, z' < H/2$. Then, averaging ε, R and Z over z, according to the rule

$$\langle A(x, y, z) \rangle = \frac{1}{H} \int_{-H/2}^{H/2} dz\, A(x, y, z)$$

and keeping the previous notation for the average values, we obtain

$$\varepsilon(x, y, t) = \int_0^x dx' \int_{-H/2}^{H/2} dy'\, \theta(T) R(x', y', T) \tag{8.2.6}$$
$$\times M(x - x', y - y')$$

where

$$M(x - x', y - y') = -\frac{iH}{L\lambda} \frac{\exp\left[\frac{1}{2} i k_0 (y - y')^2 / |x - x'|\right]}{|x - x'|}. \tag{8.2.7}$$

We note that although in equations (8.2.6) and (8.2.7) only two spatial variables x and y are present, the model remains three-dimensional, as we have here used the Green's function for the three-dimensional case (cf (8.1.14) and (8.2.7)).

8.2.2 Calculation of the diffraction pattern

In solving the equations obtained above we neglect the retardation effect (i.e. we omit the Heaviside function in (8.2.6)). It is convenient to introduce dimensionless variables $\tau = t/T_R$, $\xi = x/L$ and $\eta = y/D$. Then for ε, R and Z we obtain the following system of equations

$$\frac{\partial R}{\partial \tau} = Z\varepsilon$$
$$\frac{\partial Z}{\partial \tau} = -\tfrac{1}{2}(\varepsilon^* R + \text{c.c.}) \tag{8.2.8}$$
$$\varepsilon = \int_0^\xi d\xi' \int_{-1/2}^{1/2} d\eta'\, R(\xi', \eta', \tau) M(\xi - \xi', \eta - \eta')$$

where

$$M(\xi - \xi', \eta - \eta') = -i\sqrt{FF_H} \frac{\exp\left[i\pi F(\eta - \eta')^2 / |\xi - \xi'|\right]}{|\xi - \xi'|}. \tag{8.2.9}$$

The Poynting vector in the wave zone can now be expressed in the form

$$S = n \frac{\hbar \omega_0 N_0 L}{4 T_R} \frac{F F_H L^2}{x^2} |f(\theta, \tau)|^2 \tag{8.2.10}$$

where the function for $f(\theta, \tau)$ is given by equation (8.1.22) and $F_H = H^2/\lambda L$.

Numerical integration of equations (8.2.8) is carried out using the algorithm described in section 8.1.2. A series of calculations performed for the kinetics and polar diagram of super-radiance for several different Fresnel numbers, F, and for several realizations of the random initial polarization are presented in figures 8.8 and 8.9. Figure 8.8 gives the evolution of the intensity, $I(\tau)$, in

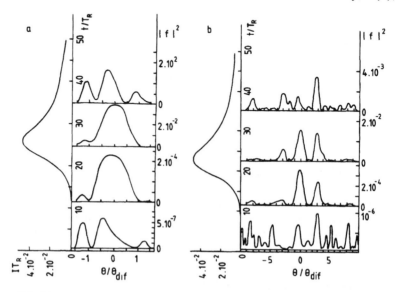

Figure 8.8. Time evolution of the integrated intensity I and angular structure $|f|^2$ of SF for Fresnel numbers: (a) $F = 0.5$ and (b) $F = 4$; random initial polarization $|R_0| = 0.02$.

time, and the polar diagram $|f(\theta, \tau)|^2$ of the super-radiance for Fresnel numbers $F = 0.5$ and 4, similar to that in the case of two-dimensional super-radiance (see subsection 8.1.4). The beam structure of the polar diagram of the super-radiance that develops in the course of competition between rays propagating at different angles to the axis of the sample is clearly seen. The structure undergoes strong fluctuations from one realization to another. For large Fresnel numbers ($F > 1$) these fluctuations remain within the angular size of the sample, D/L, whilst the angular size of the individual beam has the diffraction scale λ/D. The number of rays in this case is approximately equal to the Fresnel number F.

For small Fresnel numbers ($F < 1$) the total width of the polar diagram is determined by the diffraction angle, λ/D, and super-radiance has a single-ray character.

Polar diagrams of super-radiance in the case where the amplitude of the initial polarization is uniform over the sample are presented in figure 8.9. A comparison with figure 8.8 shows that, during the evolution of the polar diagram, competition between the rays and self-organization of polarization play a substantial rôle.

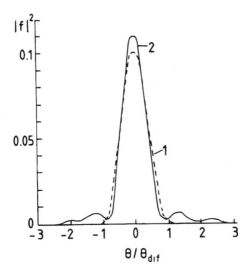

Figure 8.9. Angular structure of SF $|f|^2$ at the maximum of the super-radiance pulse for Fresnel numbers: (1) $F = 0.5$ and (2) $F = 4$. The amplitude of the initial polarization is taken to be uniform over the sample, $|R_0| = 0.02$.

The results obtained above may be compared with the experimental data on super-radiance in a KCl crystal with O_2^- ion centres [SSS87, SSS88] where a diffraction pattern of super-radiance was observed (see subsection 2.2.1). The number of beams increased from one to four with an increase in the pumping intensity. Since the pumping was in the transverse direction, the transverse size D of the active region also changed proportionally to its intensity as that varied. An analysis of the beam pattern and the ratios of the angular sizes of the polar diagram suggests that the experimental situation corresponds to super-radiance with Fresnel numbers $F = 1$–4.

For the wavelength $\lambda = 629.04$ nm used in the experiments, the Fresnel number corresponding to height $H = 0.05$ in the vertical direction is $F_H = H^2/L\lambda \approx 0.5$. This justifies the use of the three-dimensional model of a 'sheet of paper' volume for describing the super-radiance of the experimental sample.

8.3 Concluding remarks

In conclusion, let us briefly recapitulate the basic features of super-radiance in the two-dimensional model and in the three-dimensional 'sheet of paper' volume models, which result from the diffraction of super-radiance and from quantum fluctuations of the initial polarization.

For systems with Fresnel number $F > 1$ the time scale of the super-radiant pulse turns out to be the same as that in the one-dimensional model.

However, quantum fluctuations of the initial polarization give rise to transverse inhomogeneity of atomic and field characteristics, which manifests itself in a change in the shape of the pulse and in the radiation polar diagram. The ringing of the envelope, which is typical of the one-dimensional model, becomes smoother. The super-radiance polar diagram has a ray structure, of which the angular size of individual components (rays) is of the order of the diffraction angle λ/D. Since the full width of the polar diagram is of the order of that for the angular geometry of the sample, D/L, the number of rays approximately equals the Fresnel number F (this conclusion extends to both of the models considered). Owing to competition between the rays, some of them evolve earlier thus causing the stochastic behaviour of the intensity distribution.

With decreasing Fresnel numbers ($F < 1$) the time scale of super-radiance increases as $F^{-1/2}$ for both the two- and three-dimensional models, and the spatial dependence of the inversion as well as that of the slow amplitudes of the field and polarization smooth out. This results in a pulse shape with a single maximum and in a single-ray polar diagram for the super-radiance. The angular size of the ray is close to the diffraction angle λ/D, and its form changes relatively little from one realization of the initial polarization distribution to another.

Chapter 9

Reflection and transmission on the boundary of a resonant medium

It is well known that the reflection and refraction of an electromagnetic wave at the boundary of a material system is a light-scattering phenomenon. The atoms in the medium are brought into oscillation by the incoming wave, and then secondary waves are generated by the induced atomic dipoles. The transmitted wave is the result of the superposition of the scattered wave and the external field, whilst the reflected wave is a result of back-scattering.

The traditional treatment of this process, leading to the Fresnel formulae, is restricted to the linear and stationary régime, when all the transient relaxation processes in the medium have completed their course. One side of this problem, the solution of the quantum mechanical equations of motion of the perturbed atoms (or the forced oscillations in the classical model), is treated in most textbooks in order to calculate the frequency dependence of the polarizability and the dielectric constant. The other side of the problem, however, the actual scattering problem, is usually replaced by phenomenological boundary conditions for the macroscopic fields. This procedure veils the actual physical process, and it is not well suited to describing transient nonlinear processes. A more straightforward way to calculate the laws of transmission and reflection is the method of integral equations. This is known as the Ewald–Oseen extinction theorem and is described in the book by Born and Wolf [9]. In order to treat the transient, and generally nonlinear, effects which we are going to consider in this chapter, we shall use the one-dimensional version of this theory. The method of integral equations has the advantage that phenomenological boundary conditions need not be specified explicitly.

In this chapter we mainly investigate the transmission and reflection properties of a resonant nonlinear medium when the incoming wave is a short pulse, generally shorter than the relaxation times of the atoms in the medium. We will show that if the lifetime of the macroscopic polarization is long enough, and neither the incident field nor the relaxation processes reduce it, then the field originating from this polarization gives rise to effects which may be called *cooperative* reflection and transmission. As is known from the previous chapters, super-radiance takes place only if its characteristic time T_R is shorter than the relaxation time, T_2, of the polarization, otherwise the phase memory of the dipoles will be destroyed. The condition $T_2 > T_R$ will be seen to be substantial for cooperative reflection, too.

As has been shown in Chapter 6, the effect of dipole–dipole interaction may influence the coupled atom–field dynamics, therefore—as we will show in the next chapter—it can affect the reflection and transmission properties, as well. This is especially important in the nonlinear régime, where the inversion of the atoms varies on a sufficiently large scale during the process. For weak fields this effect leads only to a constant shift of the resonance frequency, therefore we shall treat the problem here without this correction, and the necessary modifications will be explained in the next chapter. Nevertheless, we shall also include here the discussion of the weak nonlinear case, where we can neglect the dynamic change of the frequency shift caused by the dipole–dipole interaction. A more detailed explanation of this problem is postponed to Chapter 10.

The problem of non-stationary reflection has been studied so far in only a few works. The discussion of Eilbeck [16] is restricted to the linear case by using a frequency-dependent index of refraction in the Fresnel formulae. In their works, Rupasov and Yudson [RY82, RY87] also use the phenomenological boundary conditions of electrodynamics, and both deduce and solve the equations describing the transmission and reflection of an optically thin layer in the absence of relaxation. We also note some other works related to the problem treated here [VGGS86, ZM87, BM88, EZ80, MBEZMS89, SLT90b, LST91].

In Chapters 1 and 5 dealing with super-radiance in an extended medium we have used the *slowly varying envelope approximation*, SVEA, both in space and in time. In order to treat the problem of reflection we keep this approximation *in the time variable*–and abbreviate it as SVEAT—but we must drop SVEAS, the *slowly varying envelope approximation in the space coordinate*.

9.1   Integral equation for the propagation problem without SVEAS

We shall consider the boundary value problem for the transmission of a light pulse through a plane-parallel resonant layer consisting of two-level atoms. Let the incoming field be a linearly polarized plane wave

$$\mathcal{E}_{in}(x; t) = \tfrac{1}{2} E_{in}\left(t - \frac{x}{c}\right) e^{i(kx - \omega t)} + \text{c.c.} \qquad (9.1.1)$$

with an amplitude $E_{in}(x, t)$ slowly varying in space and time. We restrict ourselves to the case of normal incidence and assume that the transition dipole moments of the atoms are all parallel to the direction of the exciting field. The problem we investigate is one-dimensional in space, thus the macroscopic field obeys the inhomogeneous wave equation

$$\frac{\partial^2 \mathcal{E}}{\partial x^2} - \frac{1}{c^2}\frac{\partial^2 \mathcal{E}}{\partial t^2} = \frac{4\pi}{c^2}\frac{\partial^2 \mathcal{P}}{\partial t^2}. \tag{9.1.2}$$

We shall obtain the solution of equation (9.1.2) with a given $\mathcal{P}$ by using the corresponding Green's function $G(x - x', t - t')$ which obeys the equation

$$\frac{\partial^2 G}{\partial x^2} - \frac{1}{c^2}\frac{\partial^2 G}{\partial t^2} = -\delta(x - x')\delta(t - t'). \tag{9.1.3}$$

As is well known, the retarded solution of this equation is

$$G^{\text{ret}}(\zeta, \tau) = \tfrac{1}{2}c\theta\left(\tau - \frac{|\zeta|}{c}\right) \tag{9.1.4}$$

where $\zeta = x - x'$, $\tau = t - t'$ and θ is the Heaviside step function. Here the domain of θ is the (τ, ζ) plane, and its value is unity if the argument is non-negative, and 0 otherwise. Equation (9.1.4) has an equivalent form

$$G^{\text{ret}}(\zeta, \tau) = \tfrac{1}{2}c\theta\left(\tau - \frac{\zeta}{c}\right)\theta\left(\tau + \frac{\zeta}{c}\right). \tag{9.1.5}$$

The equivalence of (9.1.4) and (9.1.5) follows from the property that in the (τ, ζ) plane the domain $\tau - |\zeta|/c \geq 0$ is valid if and only if both conditions $\tau - \zeta/c \geq 0$ and $\tau + \zeta/c \geq 0$ are fulfilled. We can easily see that (9.1.5) satisfies (9.1.3). Applying the d'Alembertian wave operator to the latter form of G^{ret} and using $d\theta(x)/dx = \delta(x)$, we obtain

$$\left(\frac{\partial^2}{\partial x^2} - \frac{1}{c^2}\frac{\partial^2}{\partial t^2}\right)\frac{c}{2}\theta\left(\tau - \frac{\zeta}{c}\right)\theta\left(\tau + \frac{\zeta}{c}\right)$$

$$= -\frac{2}{c}\delta\left(\tau - \frac{\zeta}{c}\right)\delta\left(\tau + \frac{\zeta}{c}\right) = -\delta(\zeta)\delta(\tau). \tag{9.1.6}$$

In the case considered here, $\mathcal{P}$ is different from zero in a slab placed between $x = 0$ and $x = L$, and the solution of equation (9.1.2) can be written in the form

$$\mathcal{E}(x, t) = \mathcal{E}_{\text{in}}(x, t) - \int_0^L \int_{-\infty}^{\infty} G^{\text{ret}}(x - x', t - t')\frac{4\pi}{c^2}\frac{\partial^2 \mathcal{P}}{\partial t^2}\bigg|_{t=t'} dt'dx' \tag{9.1.7}$$

where $\mathcal{E}_{\text{in}}$ is the solution of the homogeneous equation corresponding to (9.1.2), and is identified with the incoming wave, whereas the second term is a scattered wave which is the superposition of outgoing elementary waves originating in different x' planes. The complete solution is the sum of the incoming wave and the secondary scattered wave. Substituting the form of G^{ret} from (9.1.4) and performing the integration in the t' variable, we have

$$\mathcal{E}(x, t) = \mathcal{E}_{\text{in}}(x, t) - \frac{2\pi}{c}\int_0^L \frac{\partial}{\partial t}\mathcal{P}\left(x', t - \frac{|x - x'|}{c}\right)dx'. \tag{9.1.8}$$

Given an incoming wave, the transmitted wave is determined by the full solution (9.1.8) at $x \geq L$, whilst the reflected wave is described only by the second, integral, term at $x \leq 0$.

In the following we shall use the *slowly varying envelope approximation in time* (SVEAT), but an important point is that the same (slowly varying envelope) approximation in space (SVEAS) will not be exploited. Accordingly we seek the averaged macroscopic field and the polarization inside the medium, in the form

$$\mathcal{E}(x,t) = \tfrac{1}{2}E(x,t)e^{-i\omega t} + \text{c.c.} \tag{9.1.9}$$

$$\mathcal{P}(x,t) = \tfrac{1}{2}P(x,t)e^{-i\omega t} + \text{c.c.} \tag{9.1.10}$$

where the amplitudes E and P are functions of the time varying slowly compared with $e^{-i\omega t}$, but we do not require them to have a similar property in the space variable.

Using the SVEAT, the term $\partial P/\partial t$ can be neglected in the time derivative of P compared with $i\omega P$, and from equations (9.1.8)–(9.1.10) we obtain

$$E(x,t) = E_{\text{in}}\left(t - \frac{x}{c}\right)e^{ikx}$$
$$+ \frac{2\pi i\omega}{c}\int_0^L P\left(x',t - \frac{|x-x'|}{c}\right)e^{ik|x-x'|}\,dx' \tag{9.1.11}$$

where we have introduced the usual notation $k = \omega/c$. We emphasize again that E and P are still rapidly varying functions of x. According to what we said after equation (9.1.8) we can now express the amplitude of the reflected wave $E_{\text{r}}(t)$ at $x = 0$, and that of the transmitted wave $E_{\text{tr}}(t)$ at $x = L$, in the following form

$$E_{\text{r}}(t) = E(0,t) - E_{\text{in}}(t)$$
$$= \frac{2\pi i\omega}{c}\int_0^L P\left(x',t - \frac{x'}{c}\right)e^{ikx'}\,dx' \tag{9.1.12}$$

$$E_{\text{tr}}(t) = E(L,t)$$
$$= \left[E_{\text{in}}\left(t - \frac{L}{c}\right) + \frac{2\pi i\omega}{c}\int_0^L P\left(x',t - \frac{L-x'}{c}\right)e^{-ikx'}\,dx'\right]e^{ikL}. \tag{9.1.13}$$

The dynamics of the polarization P in the two-level atom model will be determined by the optical Bloch equations in the RWA

$$\frac{\partial R}{\partial t} = -\left(i\Delta + \frac{1}{T_2}\right)R - i\frac{d}{\hbar}EZ \tag{9.1.14}$$

$$\frac{\partial Z}{\partial t} = \frac{id}{2\hbar}(ER^* - E^*R) \tag{9.1.15}$$

where, as earlier, Z is the population difference between the levels, $R/2$ is the slowly varying part of the off-diagonal element of the atomic density matrix, d is the transition dipole moment and $\Delta = \omega_0 - \omega$ is the difference between the atomic resonance frequency and the carrier frequency of the incoming wave. The field strength E in (9.1.14) and (9.1.15) is determined by the macroscopic polarization P according to (9.1.11). On the other hand P itself is related to R by

$$P(x, t) = N_0 d \int g(\Delta) R(x, t, \Delta) \, d\Delta = N_0 d \langle R \rangle \qquad (9.1.16)$$

where N_0 is the atomic density and $g(\Delta)$ is the inhomogeneous line shape of the atoms in the medium. The angular brackets abbreviate the averaging over this distribution.

9.2 The linear and stationary régime: Fresnel's formulae

We shall treat the problem here in the linear approximation, when the incoming field is so weak that only a small fraction of the atoms becomes excited and so Z remains close to -1 during the whole process. Furthermore, we shall investigate the system on a time scale which is much longer than the relaxation time T_2, i.e. in the stationary limit. It is clear that in the linear approximation the propagation and reflection of a pulse during any finite duration can be described by the superposition of such solutions.

In the linear approximation we set $Z = -1$, thus equation (9.1.15) does not play any rôle, whilst equation (9.1.14) takes the form

$$\frac{\partial R}{\partial t} = \left(-i\Delta - \frac{1}{T_2} \right) R + i \frac{d}{\hbar} E. \qquad (9.2.1)$$

The stationary limit means that the relaxation term $-R/T_2$ has already damped R to its constant value, which can be obtained by setting $\partial R/\partial t = 0$. This yields the polarization term in the form

$$P = N_0 d \langle R \rangle = \chi E \qquad (9.2.2)$$

with

$$\chi = \frac{N_0 d^2}{\hbar} \left\langle \frac{\Delta + i/T_2}{\Delta^2 + 1/(T_2)^2} \right\rangle. \qquad (9.2.3)$$

This constitutes the relation between P and E with the frequency-dependent susceptibility χ.

Substituting (9.2.2) back into (9.1.11), we obtain an integral equation for the stationary amplitude E

$$E(x) = E_{\text{in}}(x) \, e^{ikx} + 2\pi i k \chi \int_0^L E(x') \, e^{ik|x-x'|} \, dx'. \qquad (9.2.4)$$

We shall seek the solution of this integral equation in the form

$$E = E_f e^{i\beta_1 x} + E_b e^{-i\beta_2 x} \qquad 0 < x < L. \tag{9.2.5}$$

As comparison with (9.1.9) shows, E_f is the amplitude of the forward wave and E_b is that of the backward wave within the medium. Substituting (9.2.5) into the right-hand side of (9.2.4) it must reproduce itself on the left-hand side for all $0 < x < L$. This requirement yields

$$\beta_1 = \beta_2 = k\sqrt{1 + 4\pi\chi} = kn \tag{9.2.6}$$

$$E_b = E_f \frac{n-1}{n+1} e^{2inkL} \tag{9.2.7}$$

and

$$E_f = \frac{2(n+1)}{(n+1)^2 - (n-1)^2 e^{2inkL}} E_{in} \tag{9.2.8}$$

where the usual notation

$$n = \sqrt{1 + 4\pi\chi} \tag{9.2.9}$$

has been used for the complex refractive index n. This means that within the medium both waves are propagating with phase velocity $v = c/n$. Note that the non-resonant background-susceptibility, resulting from other transitions, has been neglected here.

It is not difficult, either, to obtain the relations between the amplitudes of the incoming, transmitted and reflected waves. We substitute (9.2.5) back into the integral equation (9.2.4) and perform the integrations for $x < 0$. We obtain

$$E = E_{in} e^{ikx} + E_r e^{-ikx} \qquad x < 0 \tag{9.2.10}$$

with

$$E_r = \frac{(n^2 - 1)(e^{2inkL} - 1)}{(n+1)^2 - (n-1)^2 e^{2inkL}} E_{in}. \tag{9.2.11}$$

For $x > L$ equation (9.2.4) gives

$$E = E_{tr} e^{ikx} \qquad x > L \tag{9.2.12}$$

where

$$E_{tr} = \frac{4n e^{inkL}}{(n+1)^2 - (n-1)^2 e^{2inkL}} E_{in} e^{-ikL}. \tag{9.2.13}$$

Equations (9.2.11) and (9.2.13) are the expressions for the reflected and transmitted waves from a plane-parallel layer. To obtain the reflection and transmission coefficients corresponding to a semi-infinite medium, we have to take the limit $L = \infty$. In this case the imaginary part of n in the exponentials ensures the attenuation, and the terms containing e^{inkL} can be set to zero. Then (9.2.11) and (9.2.8) yield the Fresnel formulae for a single boundary in the case of normal incidence

$$E_r = \frac{1-n}{1+n} E_{\text{in}} \qquad E_{\text{tr}} = \frac{2}{1+n} E_{\text{in}}. \qquad (9.2.14)$$

These are well known from classical optics. Here they have been derived, however, without macroscopic boundary conditions: χ and n emerged from the microscopic level of the theory. Equation (9.2.4) and its consequences show, directly, that it is the superposition of the scattered and incident waves that gives rise to the change of the field amplitude as well as the velocity of propagation within the medium.

9.3   Cooperative transient properties of a resonant thin layer

In this section we consider the boundary value problem for an optically thin medium, i.e. when $L \ll \lambda$, which allows a significant simplification. In the case of a thin sample, in equation (9.1.11) we may put $\exp(ik|x - x'|) \approx 1$, and instead of integrating we can take the spatial average of the polarization. With these approximations, from equations (9.1.11), (9.1.16), (9.1.14) and (9.1.15) we obtain

$$\frac{\partial R}{\partial t} = -\left(i\Delta + \frac{1}{T_2}\right) R + \left(-i\frac{d}{\hbar} E_{\text{in}} + \frac{1}{T_R}\langle R \rangle\right) Z \qquad (9.3.1)$$

$$\frac{\partial Z}{\partial t} = \frac{id}{2\hbar}(E_{\text{in}} R^* - E_{\text{in}}^* R) - \frac{1}{2T_R}\left(\langle R^* \rangle R + \langle R \rangle R^*\right) \qquad (9.3.2)$$

where

$$T_R = \frac{\hbar c}{2\pi N_0 L \omega d^2} \qquad (9.3.3)$$

is the super-radiation time of the thin sample. We can see that in the present case, $L \ll \lambda$, T_R depends essentially on the surface atomic density, $N_0 L$.

When studying super-radiation, the terms containing E_{in} in (9.3.1) and (9.3.2) are not considered, whilst in the treatment of the coherent interaction of external radiation with the resonant medium, the terms containing $1/T_R$ are neglected. To obtain the reflected wave, however, both must be taken into account.

In order to obtain the reflected and transmitted amplitudes we have first to solve the ordinary differential equations (9.3.1) and (9.3.2). Then after reducing equations (9.1.12) and (9.1.13) to the thin medium limit, namely, putting $\exp(ik|x - x'|) \approx 1$ and taking the spatial average of the polarization, we can determine the reflected and transmitted waves in the following form

$$E_r = i\frac{\hbar}{dT_R}\langle R \rangle \tag{9.3.4}$$

$$E_{tr} = E_{in} + i\frac{\hbar}{dT_R}\langle R \rangle. \tag{9.3.5}$$

Equations (9.3.4) and (9.3.5) express that the atoms of the medium radiate in both directions equally. The reflected wave is identical with this secondary field emitted in the backward direction, whilst the transmitted wave is the superposition of the secondary field in the forward direction and the incident field.

9.3.1 Linear régime of transient reflection

We shall assume here that the excitation is weak, so that Z remains close to -1 during the whole process. In this case (9.3.2) can be omitted and (9.3.1) reduces to a single complex equation

$$\frac{\partial R}{\partial t} = -\left(i\Delta + \frac{1}{T_2}\right)R + i\frac{d}{\hbar}E_{in} - \frac{1}{T_R}\langle R \rangle \tag{9.3.6}$$

where Δ is a parameter.

Taking the Laplace transform of both sides of this linear equation yields

$$s\tilde{R} = -i\Delta\tilde{R} - \frac{1}{T_2}\tilde{R} - \frac{1}{T_R}\langle\tilde{R}\rangle + i\frac{d}{\hbar}\tilde{E}_{in} \tag{9.3.7}$$

where

$$\tilde{R} = \int_0^\infty e^{-st}R(t, \Delta)\,dt \qquad \tilde{E}_{in} = \int_0^\infty e^{-st}E_{in}(t)\,dt. \tag{9.3.8}$$

Expressing $\tilde{R}$ from (9.3.7) and then taking the average over the inhomogeneous line shape, we have

$$\langle\tilde{R}\rangle = \left\langle\frac{-1}{s + i\Delta + 1/T_2}\right\rangle\left(\frac{\langle\tilde{R}\rangle}{T_R} - i\frac{d}{\hbar}\tilde{E}_{in}\right). \tag{9.3.9}$$

For the sake of simplicity we assume the inhomogeneous broadening to have a Lorentzian line shape, the centre of which is at the carrier frequency of the incoming field

$$g(\Delta) = \frac{T_2^*}{\pi}\frac{1}{1 + \Delta^2 T_2^{*2}}. \tag{9.3.10}$$

In this case the operation $\langle\ \rangle$, i.e. the integration in (9.1.16), can be readily performed, and we obtain

$$\left\langle\frac{1}{s+i\Delta+1/T_2}\right\rangle = \int_{-\infty}^{\infty}\frac{g(\Delta)}{s+i\Delta+1/T_2}\,d\Delta = \frac{1}{s+1/\tau_2} \qquad (9.3.11)$$

where the notation

$$\frac{1}{\tau_2} = \frac{1}{T_2} + \frac{1}{T_2^*} \qquad (9.3.12)$$

has been introduced. τ_2 is the time constant characterizing the total effect of both the homogeneous and the inhomogeneous line broadenings.

After taking the inverse Laplace transform, we obtain

$$\langle R\rangle = i\frac{d}{\hbar}\int_0^t E_{\text{in}}(t')\exp\left[-(t-t')/\tau\right]dt' \qquad (9.3.13)$$

where

$$\frac{1}{\tau} = \frac{1}{\tau_2} + \frac{1}{T_R}. \qquad (9.3.14)$$

We see that in the linear régime both T_R and τ_2 equally contribute to the dynamics of the polarization $\langle R\rangle$. This, however, is not valid for the reflected and the transmitted waves, as the latter are determined by equations (9.3.4) and (9.3.5) containing only T_R.

If a step pulse of amplitude E_0 is switched on at $t=0$, then for the time dependence of the reflected wave we obtain, from (9.3.4) and (9.3.13)

$$E_r = E_0\frac{\tau}{T_R}(e^{-t/\tau} - 1). \qquad (9.3.15)$$

The stationary limit of the reflection coefficient is achieved during the time τ, and the value of this coefficient for the *intensity* is

$$\mathcal{R}_\infty = \frac{E_r^2(\infty)}{E_0^2} = \left(1+\frac{T_R}{\tau_2}\right)^{-2}. \qquad (9.3.16)$$

It also follows from equation (9.3.13) that after the incoming wave has been switched off, the reflected amplitude disappears exponentially with a time constant τ. To compare equation (9.3.16) with the appropriate Fresnel formula, we note that in the stationary case the index of refraction can be connected with τ_2/T_R. From equations (9.2.3), (9.2.9) and (9.3.3) we have

$$(n^2-1)kL = 2i\frac{\tau_2}{T_R}. \qquad (9.3.17)$$

Using (9.3.16), the expansion of the Fresnel coefficients (9.2.11) and (9.2.13) for a thin layer ($nkL \ll 1$) is in agreement with the results above.

We can also calculate the ratio η_r of the total energies of the reflected and incoming waves: W_r/W_{in}. Supposing that the latter is a step pulse of duration T, we obtain the following result

$$\eta_r = \frac{W_r}{W_{in}} = \mathcal{R}_\infty \left[1 + \frac{e^{-T/\tau} - 1}{T/\tau} \right]. \tag{9.3.18}$$

A similar calculation for the transmitted energy gives the result

$$\eta_{tr} = \frac{W_{tr}}{W_{in}} = 1 - \eta_r - 2\frac{T_R}{\tau_2}\eta_r. \tag{9.3.19}$$

The last term in this expression is obviously the ratio of the energy absorbed in the medium.

In order to analyse the behaviour of the reflection for the transient régime, we shall distinguish two main cases.

(i) First, let us suppose that $\tau_2 \ll T_R$, i.e. $\tau = \tau_2$, which is the usual situation. In this case

$$\eta_r = \mathcal{R}_\infty \left[1 + \frac{e^{-T/\tau_2} - 1}{T/\tau_2} \right] = \mathcal{R}_\infty f\left(\frac{T}{\tau_2}\right) \tag{9.3.20}$$

where f is the function $f(x) = 1 + (e^{-x} - 1)/x$. Now, the stationary reflection coefficient $\mathcal{R}_\infty = (\tau_2/T_R)^2$ is a small quantity. Reducing the duration of the pulse, this causes the reflection coefficient to be still less, because the medium has no time to respond to the excitation. Figure 9.1 shows $\eta_r/\mathcal{R}_\infty$ as a function of $\log_{10}(T/\tau_2)$. It can be seen that reflection and transmission depend relatively strongly upon T/τ_2. So, knowing the value of τ_2 for a material, this enables us to measure the duration of short pulses, and vice versa.

(ii) Now assume that $\tau_2 \gg T_R$, $(\tau = T_R)$, which is one of the conditions for the realization of super-radiance in the sample. The coefficient η_r in this case is

$$\eta_r = 1 + \frac{e^{-T/T_R} - 1}{T/T_R} = f\left(\frac{T}{T_R}\right). \tag{9.3.21}$$

It is determined by the same function f of T/T_R, so the dependence of the reflection on pulse duration can be deduced from figure 9.1 in this case as well. The transition from absolutely strong to weak reflection occurs at $T \approx T_R$. In the limits $\tau_2 > T > T_R$ and $T > \tau_2 > T_R$, the reflection coefficient is close to unity, whilst when $\tau_2 > T_R > T$ it goes to zero. The time dependence of the reflection coefficient on both $\log_{10} T/T_R$ and $\log_{10} T/\tau_2$ is shown in figure 9.2.

9.3.2 Cooperative nonlinear reflection and transmission

Let us turn now to the more general case, where the nonlinearities in equations (9.3.1) and (9.3.2) are to be accounted for. As we have noted in the introduction,

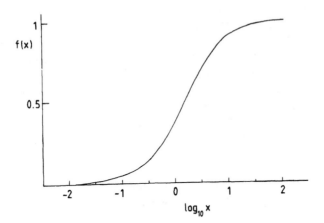

Figure 9.1. Graph of the function $f(x) = 1 + (e^{-x} - 1)/x$ versus $\log_{10} x$. For an optically thin layer, the ratio of the reflected and incident energies of a step pulse of duration T is $(\tau_2/T_R)^2 f(T/\tau_2)$, if $\tau_2 \ll T_R$, and $f(T/T_R)$ if $\tau_2 \gg T_R$.

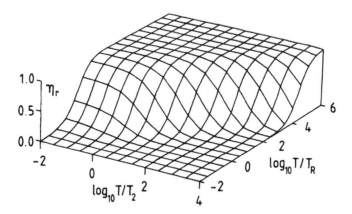

Figure 9.2. Ratio of the reflected and incident energies of a step pulse of duration T for an optically thin layer as a function of both $\log_{10}(T/\tau_2)$ and $\log_{10}(T/T_R)$.

the model considered in this chapter neglects the local field effect caused by the dipole–dipole interaction. As we shall see in Chapter 10, this effect produces a dynamic shift of the resonant frequency which depends upon the value of the inversion. This requires a modification of the equations if Z is assumed to change significantly. Therefore the results obtained in this chapter are generally valid only in the case of weak nonlinearity, when Z does not change considerably. We shall see, that under certain conditions, even for large-area pulses, the nonlinear equations (9.3.1) and (9.3.2), yield only a small change in the inversion, so that the present approximation is still valid.

We shall investigate first the case of exact resonance without polarization damping: $T_2 = \infty$. This allows us to obtain analytical results which can be used better to understand the effects of relaxation terms. This problem has been investigated by Rupasov and Yudson [RY82, RY87] and by two of the authors [BT86b, BT88]; see also [BN95]. Within the approximation introduced above, equations (9.3.1) and (9.3.2) can be written in the form

$$\frac{\partial R}{\partial t} = \left(-i\frac{d}{\hbar}E_{in} + \frac{1}{T_R}R\right)Z \tag{9.3.22}$$

$$\frac{\partial Z}{\partial t} = \frac{id}{2\hbar}(E_{in}R^* - E_{in}^*R) - \frac{1}{T_R}|R|^2. \tag{9.3.23}$$

Using these equations and (9.3.4) and (9.3.5), the law of conservation of energy can be deduced

$$\frac{c}{4\pi}|E_{in}|^2 = \frac{c}{4\pi}(|E_{tr}|^2 + |E_r|^2) + N_0 L\hbar\omega\frac{\partial Z}{\partial t}. \tag{9.3.24}$$

The incoming power flux is equal to the reflected and transmitted fluxes plus the rate of energy change of the atomic system. We shall solve equations (9.3.22) and (9.3.23) assuming that initially all atoms are in their ground state, $Z = -1$, and that the polarization is absent, $R = 0$. It can be easily seen that if E_{in} is real, R will remain purely imaginary during the evolution of the system, and we can set $R = -iV$, where V is real. As the length of the Bloch vector $|R|^2 + Z^2$ is a conserved quantity, it is straightforward to introduce the Bloch angle, with $V = -\sin\theta$ and $Z = -\cos\theta$. Now (9.3.22) and (9.3.23) can be recast into the single equation

$$\frac{d\theta}{dt} = \frac{d}{\hbar}E_{in} - \frac{1}{T_R}\sin\theta. \tag{9.3.25}$$

According to (9.3.4) and (9.3.5), the reflected and transmitted waves can now be obtained in the following form

$$E_r = \frac{\hbar}{dT_R}V = -\frac{\hbar}{dT_R}\sin\theta \tag{9.3.26}$$

$$E_{tr} = E_{in} - \frac{\hbar}{dT_R}V = \frac{\hbar}{d}\frac{d\theta}{dt}. \tag{9.3.27}$$

By omitting the term $(1/T_R) \sin \theta$ in (9.3.25) we would have the usual equation which describes the optical Rabi oscillations. On the other hand, in the most simplified model of super-radiation we have equation (9.3.25) with $E_{in} = 0$. The term $(1/T_R) \sin \theta$ takes into account the secondary field of the material, and in the absence of excitation this gives rise to super-radiation of the thin layer. As we can see, the same term is responsible for the reflected wave, and, in this sense, we may identify the coherent cooperative reflection with super-radiation.

We note that, in accordance with equation (9.3.25), the area of the incident pulse defined as

$$A = \frac{d}{\hbar} \int_{-\infty}^{\infty} E_{in} \, dt \qquad (9.3.28)$$

is not equal to the final value of θ, owing to the damping super-radiance term. $\theta(\infty) = 0 \pmod{2\pi}$, independently of A, as can be seen from (9.3.25), taking into account that $E_{in}(\infty) = 0$.

Equation (9.3.25) can be solved analytically for incoming pulses of certain specific forms. One of them is a step pulse E_{in} of constant amplitude [RY82]. However, the transmitted and reflected waves in this case exhibit infinite discontinuities when E_{in} drops to zero. Instead of giving the analytical form of the solution, we only note that if $(d/\hbar)E_{in} < 1/T_R$ then according to equation (9.3.25) θ has a stationary value, where $\sin \theta = (d/\hbar)E_{in}T_R$. If, now, the duration of the incident pulse is longer than T_R, the system then approximates this stationary state, and after $t > T_R$, $d\theta/dt$ will be zero, and consequently $E_r \approx -E_{in}$. The secondary field is found to be in the opposite phase compared with the incident wave, and that is why they cancel each other in the forward direction, and, accordingly, most of the incoming wave is reflected.

In actual experiments E_{in} is a continuous function of time. Therefore we shall now consider a more realistic and analytically tractable exciting pulse of the form

$$E_{in} = E_0 \operatorname{sech} \frac{t}{T} \qquad \text{where} \qquad \frac{dE_0}{\hbar} = \frac{2}{T}. \qquad (9.3.29)$$

Pulses with such parameters (E_0 can be varied) are just the '2π hyperbolic secant pulses' introduced in section 5.3. The solution of (9.3.25) corresponding to (9.3.29) has been obtained in [BT86b, BT88]

$$\tan \tfrac{1}{2}\theta = \frac{2}{((T/T_R) - 1)\exp(t/T_R) + ((T/T_R) + 1)\exp(-t/T_R)} \qquad (9.3.30)$$

and hence according to (9.3.26) and (9.3.27)

$$\frac{d}{\hbar} E_{tr}(t) = \frac{1}{T} \left(\operatorname{sech} \frac{t}{T} - \operatorname{sign}(T - T_R) \operatorname{sech} \frac{t - t_0}{T} \right) \qquad (9.3.31)$$

and

$$\frac{d}{\hbar} E_r(t) = \frac{1}{T} \left(-\operatorname{sech} \frac{t}{T} - \operatorname{sign}(T - T_R) \operatorname{sech} \frac{t - t_0}{T} \right) \qquad (9.3.32)$$

where

$$t_0 = T \ln \left| \frac{1 + T/T_R}{1 - T/T_R} \right|. \qquad (9.3.33)$$

These analytic results can be obtained also by the inverse scattering method [RY87, BN95].

In the limiting case $T \ll T_R$

$$E_{tr} \approx E_{in} \qquad \frac{d}{\hbar} E_r \approx \frac{2}{T_R} \operatorname{sech} \frac{t}{T} \tanh \frac{t}{T} \qquad \frac{E_{r_{max}}}{E_0} \approx \frac{T}{2T_R}. \qquad (9.3.34)$$

In this case $Z = -\cos\theta$ approaches close to $+1$, and therefore the problem cannot be regarded as being weakly nonlinear. Local field effects treated in the next chapter will therefore modify the results if $T \ll T_R$.

In the other limiting case, when $T \gg T_R$ (but $T < T_2$), the equations (9.3.31) and (9.3.32) above simplify to

$$\frac{d}{\hbar} E_{tr} \approx -\frac{2}{T_R} \operatorname{sech} \frac{t}{T} \tanh \frac{t}{T} \qquad E_r \approx -E_{in} \qquad \frac{E_{tr_{max}}}{E_0} \approx \frac{T_R}{2T}. \qquad (9.3.35)$$

The transmitted wave has two maxima in opposite phase, and their separation is $2T_R \ln(1 + \sqrt{2})$.

In this latter instance Z remains close to -1, hence the result is still valid when local field effects are included. The pulse shapes calculated from (9.3.31) and (9.3.32) for $T = 4T_R$ are shown in figure 9.3(a).

As is seen from equations (9.3.30)–(9.3.32), the character of the solution changes when $T = T_R$. This particular value of the parameters corresponds to the exceptional case where a pulse of the form (9.3.29) brings the system exactly into the completely inverted state, $Z = 1$. As can be easily verified, this happens not only for 2π hyperbolic secant pulses with $T = T_R$ but also for all pulses of the form of (9.3.29) provided that

$$\frac{dE_0}{\hbar} = \frac{1}{T} + \frac{1}{T_R} \qquad \mathcal{A} = \left(\frac{T}{T_R} + 1 \right) \pi. \qquad (9.3.36)$$

This looks like a rather artificial result. It corresponds to the situation in which a pendulum, started with an appropriate velocity from its lowest position, arrives exactly in the upper unstable equilibrium state over infinitely long time. Though this is valid, of course, only in the framework of the present mathematical model, it yields physical conclusions as well. It shows that if $dE_0/\hbar < 1/T + 1/T_R$ or if $\mathcal{A} < (1 + T/T_R)\pi$, then the system cannot reach its inverted state. For instance, if $T = 4T_R$ and $\mathcal{A} = 2\pi$, then E_0 is smaller than is required to invert the medium. Had we neglected the collective response—the term $(1/T_R) \sin\theta$ in (9.3.25)—the first half of a 2π pulse would have already inverted the atoms, whilst in fact even $Z = -0.75$ was not achieved with these parameters (see

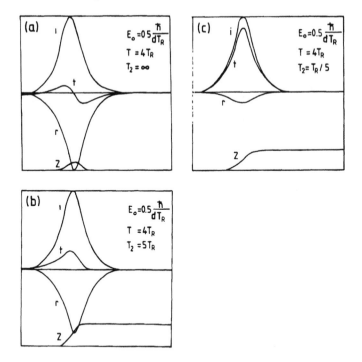

Figure 9.3. Time dependence of the transmitted (t) and reflected (r) amplitudes relative to the incident field strength (i) in the thin-medium limit. The time and amplitude scales are fixed by the incident pulse (i), having the form of $E_{in} = 0.5(\hbar/dT_R)\,\text{sech}\,(t/4T_R)$. Its area is $\mathcal{A} = 2\pi$. The relaxation time T_2 is ∞, $5T_R$ and $T_R/5$ for (a), (b) and (c) respectively. The vertical scale for the inversion (Z) runs from -1 to $+1$.

figure 9.3(a)). This also means that even for large-area pulses the behaviour of our system can be regarded as being weakly nonlinear.

Another—qualitatively different—situation where the inversion does not alter significantly is the case when both the peak amplitude and the duration are smaller than those for a 2π hyperbolic secant pulse. Figure 9.4(a) shows the result of the numerical solution of (9.3.25) with $dE_{in}/\hbar = (2/T_R)\,\text{sech}\,(t/T)$, for $T = 0.25T_R$ and $\mathcal{A} = 0.5\pi$. In the first part of the response of the medium, the effects of the self-field—and therefore the reflection—are relatively weak, and emission proceeds mainly in the forward direction. After the excitation is over— that is, in the time interval between T and T_R—the process can be regarded as collective radiation from a state which is not fully inverted. Therefore the tails of the forward and backward waves will be the same.

Let us now turn to the solution of equations (9.3.1) and (9.3.2) when the relaxation of the polarization is taken into account, $T_2 < \infty$. For the sake of simplicity we consider here only the case where $\Delta = 0$ for all the atoms. Quantitative results can be obtained here only numerically, but the form of

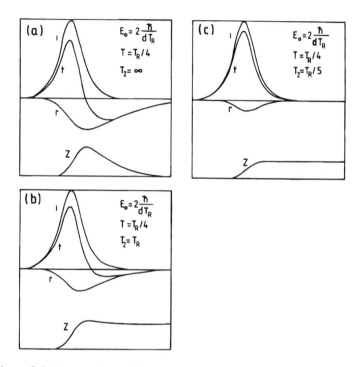

Figure 9.4. Same as figure 9.3 with $E_{\rm in} = 2(\hbar/dT_{\rm R})\,{\rm sech}\,(4t/T_{\rm R})$: $A = \pi/2$.

equation (9.2.13) shows that R, which gives rise to the reflected wave, is now damped, and, therefore, the reflection is reduced. This damping would be significant only if reflection were strong in the absence of relaxation, see figures 9.3(b) and 9.3(c). As $1/T_2$ grows, the reflected wave diminishes and the shape of the transmitted wave becomes closer to that of the exciting pulse. As is expected, for pulses of duration much less than $T_{\rm R}$ the relaxation with characteristic time $T_2 < T_{\rm R}$ will mainly affect the tails of the transmitted and reflected waves—that is, the super-radiant part—whereas the initial stimulated part will be changed only if $T > T_2$ (compare figure 9.4(a) with figures 9.4(b) and 9.4(c)). We also note that when the reflected wave is small in the undamped case the introduction of the term $1/T_2$ may then cause only a smaller effect: transmission is less sensitive to relaxation [BT88]. As is seen in the figures, if we introduce the damping term $-R/T_2$, then the atoms do not return to their ground state, $Z(\infty) > -1$, the field energy is transformed to the energy of the atoms. On a longer time scale Z goes, of course, back to -1, owing to a term in (9.1.15) which contains $-(Z+1)/T_1$. (This term has been neglected in the present model.) The atomic system dissipates the energy of the excitation.

9.4 Nonlinear transient reflection from the boundary of an extended medium

We now consider the more complicated problem of a medium of finite thickness. In this case we must solve equations (9.1.11), (9.1.14) and (9.1.15) numerically, and the reflected and transmitted waves can be determined from (9.1.12) and (9.1.13) respectively.

Here we have an extra parameter that can be chosen freely; it is the length of the sample L. The usual phenomenological treatment of reflection in linear optics disguises the physical origin of the reflected wave; nevertheless, it is well known that the latter arises as a back-scattering by the dipoles in the boundary layer of the medium. The same will be valid in the case of resonant interaction and weak nonlinearity. We expect that the depth of the material to be taken into account must be of the order of the wavelength. To show this explicitly, we substitute equation (9.1.11) into (9.1.14) and (9.1.15), and express the length of the medium in units of $1/k = \lambda/(2\pi)$. Introducing the notations $kx = y$, $kx' = y'$, we have

$$\frac{\partial R}{\partial t} = \left[-i\Delta - \frac{1}{T_2}\right] R + \left[-i\frac{d}{\hbar} E_{\text{in}}\, e^{iy} + \frac{1}{T_\lambda}\int_0^{kL} \langle R(y', t)\rangle\, e^{i|y-y'|} dy'\right] Z \quad (9.4.1)$$

$$\frac{\partial Z}{\partial t} = \left[i\frac{d}{2\hbar} E_{\text{in}}\, e^{iy} - \frac{1}{2T_\lambda}\int_0^{kL} \langle R(y', t)\rangle\, e^{i|y-y'|} dy'\right] R^* + \text{c.c.} \quad (9.4.2)$$

The solution of the coupled integro-differential equations (9.4.1) and (9.4.2) yields the reflected and transmitted waves as

$$E_{\text{r}} = \frac{i\hbar}{dT_\lambda}\int_0^{kL} \langle R(y', t)\rangle\, e^{iy'} dy' \quad (9.4.3)$$

$$E_{\text{tr}} = \left[E_{\text{in}} + \frac{i\hbar}{dT_\lambda}\int_0^{kL} \langle R(y', t)\rangle\, e^{-iy'} dy'\right] e^{ikL}. \quad (9.4.4)$$

Here T_λ is the super-radiation time of a layer of thickness $\lambda/(2\pi)$

$$T_\lambda = 2\pi\frac{L}{\lambda} T_R = \frac{\hbar}{2\pi d^2 N_0}. \quad (9.4.5)$$

The terms in (9.4.1) and (9.4.2) containing E_{in} represent the external driving field, whilst the integrals stand for the secondary field, originating from the dipoles of the material, and this gives rise to the reflected wave at $x = 0$. In the first approximation, if we omit these latter super-radiance terms, the first term, i.e. the incident wave, creates a polarization proportional to $R_s(x, t)\exp[i(kx - \omega t)]$, where $R_s(x, t)$ is slowly varying in time as well as in space. If, as the next approximation, we substitute R_s into the second, integral term, then we see that at $x = 0$ the integrand contains the expression $\exp(2ikx')$, and integration gives a net result which is approximately zero. Physically this means that the back-scattered waves from the different parts of the bulk of the medium suffer destructive interference. This is why we could apply the SVEAS if we were not interested in the reflected wave. However, our aim is to determine reflection, and we must not use this approximation. The consideration above may also suggest that in order to obtain just the reflected wave, it is enough to integrate up to $L \sim \lambda$, because the bulk of the medium, $L > \lambda$, will not contribute to the reflection. Our results below justify this assumption. We note that because R varies slowly with time, the retardation of its time argument can be neglected if $L \sim \lambda$.

The reasoning above, that the length scale to be used must be comparable with the wavelength, suggests also the relevant time parameter for the investigated effect. The time constant determining the magnitude of that part of the derivative which is responsible for the reflected waves is the super-radiation time T_λ of a layer of thickness $\lambda/(2\pi)$.

We note that in the semi-classical picture the terms containing the integral in (9.4.1) and (9.4.2) give rise to super-radiation when the relaxation times $T_2 = T_2^* = \infty$. Equation (9.4.3) shows that the reflected wave consists purely of this term, even in the presence of the external excitation. This fact implies that it is reasonable to regard transient reflection from an extended medium as super-radiation which originates from the boundary layer of the medium. The numerical calculations show that the depth of this layer is not larger than $\lambda/2$. We expect, therefore, that the results for an ideally thin medium, obtained in the previous section, will essentially retain their validity for an extended medium too. The reflected intensity will not depend upon the size of the sample beyond $\lambda/2$. We note that for very strong stationary fields this conclusion may not be valid and a special 'self-induced reflection' may occur [55, 56, 62–64]. This latter effect can also lead to an intrinsic optical bistability [B83, 64, MC95]. However, for field strengths usually applied in resonance experiments such effects are negligible.

The calculations yield the transmitted wave as well. In contrast to reflection, transmission does depend upon L. To determine the transmitted wave, however, it is not recommended to choose a length scale comparable with the wavelength, unless one is interested in the transmission of a thin layer. Our considerations above, and equations (9.4.1) and (9.4.2), show that if $L \gg \lambda$ it is then enough to use the traditional theoretical treatment [AE75, SSL74] with the SVEAS. It must be noted, however, that the initial pulse of SVEAS theory, is assumed to be already within the medium. Therefore the identification of the initial pulse with the external excitation is allowed only if reflection is weak.

We shall first discuss the problem without relaxation, $T_2 = \infty$. The conditions of strong reflection for the extended medium are similar to those for the thin medium case. The amplitude of the exciting pulse must satisfy $(d/\hbar)E_{in} < 1/T_\lambda$, otherwise in equations (9.4.1) and (9.4.2) the first term will dominate, and the polarization wave will propagate in the forward direction, leading to a weak reflection. The other condition is that the duration of the incoming pulse T must be greater than, or at least comparable with T_λ.

Figure 9.5 shows the results of numerical calculations for $\pi/2$ hyperbolic secant pulses of amplitude $E_0 = 0.5(\hbar/dT_\lambda)$ and duration $T = T_\lambda$, for $L = 0.5\lambda$,

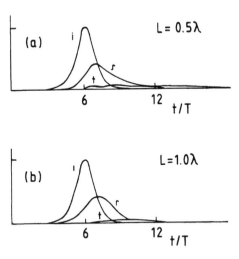

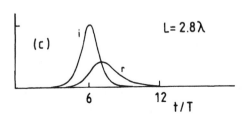

Figure 9.5. Time dependence of the transmitted (t) and reflected (r) intensities relative to the intensity of the incident wave (i). The latter is a $\pi/2$ hyperbolic secant pulse, $E_{\text{in}} = 0.5(\hbar/dT_\lambda)\,\text{sech}\,[(t - t_0)/T]$, $T = T_\lambda$, $t_0 = 6T$, $T_2 = \infty$. Increasing the length of the medium has no essential effect upon the reflection.

1.0λ and 2.8λ. The reflected waves are relatively strong and nearly equal in all cases; only the transmitted intensity is decreasing. This proves that the reflected wave originates in the boundary layer of the medium, from no deeper than $\lambda/2$.

For an excitation with a weak and short pulse the result will be very similar to those for the case of an infinitely thin layer. When the external pulse is shorter than the super-radiation time T_λ, but is not strong enough to bring the atoms back into their ground state, then, after the rapid coherent excitation, there remains an inversion and a polarization in the medium. This leads to super-radiation in both directions (see figure 9.6). The forward wave shows ringing, whilst the

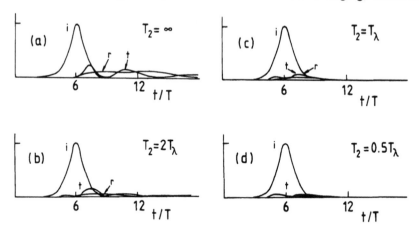

Figure 9.6. Same as figure 9.5 with $E_{in} = 2(\hbar/dT_\lambda)\,\mathrm{sech}\,[(t - t_0)/T]$, $T = T_\lambda/4$, $A = \pi/2$, $L = \lambda$ for different values of T_2. The secondary waves show a super-radiant character which becomes suppressed as T_2 decreases.

backward wave is a wide flat pulse, because it is the field of the boundary layer only.

When $(d/\hbar)E_{in} > T_\lambda^{-1}$ and $T < T_\lambda$ the reflected wave will then be relatively weak. Coherent propagation effects, e.g. self-induced transparency, are usually investigated under such conditions. To compare the results following from the present approach with those obtained by ignoring the rôle of the boundary, we consider the behaviour of a 2π hyperbolic secant pulse with duration less than T_λ (figure 9.6) [BT88]. The reflected wave diminishes if the amplitude of the excitation grows. The transmitted pulse is a single strong peak similar to the excitation. For layers of thickness of about λ we can already observe the delay of the transmitted pulse, a characteristic feature of coherent interaction. For wider pulses this delay increases. As the area of the propagating pulse within the medium is somewhat less than 2π, because of reflection, therefore according to the area theorem (5.3.12) its area must grow by up to 2π. This leads to the broadening of the transmitted pulse. The opposite effect, narrowing, has been obtained in the case of a 2.25π pulse, because then even the wave passing the immediate boundary has an area larger than 2π. Consequently, it has to decrease in order to reach this value. These large-amplitude pulses lead to a significant inversion of the medium, therefore the effects of the local field correction will modify these results [SBA88].

It is possible to investigate the influence of relaxation upon reflection and transmission. In equations (9.4.1) and (9.4.2) we can take into account the term R/T_2. We have seen already in section 9.2 that in the linear case, when $Z \simeq -1$, and for stationary excitation $T \gg T_2$, our equations yield the Fresnel formulae for a finite layer. From equations (9.2.3), (9.2.9) and (9.4.5) the resonant refractive index can be expressed as

$$n = \left[1 + 2i\frac{T_2}{T_\lambda}\right]^{1/2}. \tag{9.4.6}$$

This enables us to control the numerical results. As our calculations have shown [BT86a, BT88], in the case $T \gg T_2$ the transmitted and reflected pulses have the same form as the incident pulse. The reflection coefficient, defined as the reflected intensity divided by the incident intensity, could be calculated and showed the same dependence upon the length of the sample as the corresponding Fresnel formula (9.2.11). For $T = 8T_2$ and $T_2 = 0.25T_\lambda$ the difference between the analytical and numerical results was less than 10^{-3}.

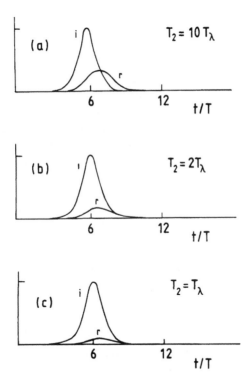

Figure 9.7. Effect of relaxation on the pulse shown in figure 9.3(b), $L = \lambda$.

Now let us turn to the non-stationary case. The inclusion of relaxation leads to the attenuation of the reflected wave and reduces the pulse delay, both being signs of the vanishing of coherent behaviour. In accordance with the results of the previous subsection, this effect upon the reflection is determined by the ratio of T_λ and T_2. Since the cooperative reflection can be regarded as super-radiation from the boundary layer, it will be reduced if we decrease T_2, and cannot be significant if $T_2 < T_\lambda$ (see figures 9.5(b) and 9.7).

Let us now summarize the conclusions based on the results presented so far. In the case of coherent interaction, similarly to the incoherent case, the reflected wave originates in the boundary layer of the medium, which is less than a wavelength deep. Therefore the character of the results for the ideally thin medium presented in section 9.3 is relevant for the extended medium, too.

To observe a strong cooperative reflection, the Rabi frequency of the incident field must be comparable with, or smaller than, the inverse of the super-radiation time T_λ of the medium of thickness $\lambda/(2\pi)$. Another condition is that the duration of the incident pulse must be shorter than the relaxation time of the macroscopic polarization, but longer than T_λ. Otherwise either the strong external field will force the atoms to radiate forward, or the macroscopic dipole moment of the system that creates the coherent reflection will be destroyed.

9.5 Internal reflection in an inverted medium: correlation of forward–backward super-radiance pulses

We shall consider here resonant reflection from the surfaces of the inverted medium of waves propagating inside the medium. We shall use this effect to explain the synchronization of the counter-propagating pulses in super-radiance which has been observed in $KCl:O_2^-$ [FSS84a] (see subsection 2.2.1). Recall that in this experiment a pair of super-radiant pulses is produced by the initial excitation. The pulses travel in opposite directions along the sample, both synchronous and identical in shape.

There are several proposals to explain why this synchronization of super-radiance appears in solids. It has been suggested that the reason might be backward scattering by density fluctuations of the active centres [BBKH89, JS92], or a spontaneously growing distributed feedback caused by a spatial pattern of characteristic length $\lambda/2$ [JS92, SSS89b, Se91b] and others. The idea that the reflection from the transversal surfaces of a sample may be the cause of the correlation of the counter-propagating super-radiance pulses was proposed by Haake *et al* [HKS92], who also investigated some other phenomenological models of coupling between forward–backward super-radiance waves. Earlier Lewenstein and Rząrzewski [LR82] and Schwan *et al* [SSS89b] obtained synchronization of counter-propagating pulses by considering the problem without SVEA in spatial coordinates. We proposed an explanation of this effect by resonant reflection from the transversal boundaries of the inverted medium [T94, MTS94]. Indeed, as we have shown in section 9.1, by dropping this approximation we can correctly describe reflections from the boundaries without additionally imposing phenomenological boundary conditions. The waves propagating in opposite directions mix with each other as a result of such a reflection, if the light transition time along the sample is much less than the super-radiance evolution time. Note that the boundaries of the resonant medium will be those of the active volume if we use a three- or four-level scheme for creating inversion by transversal pumping, as in the experiment described in [FSS84a].

It is reasonable to assume that the correlation has to appear while the depletion of inversion is negligible. In such a case the problem reduces to a linear one and can be treated as in section 9.2.

Let us substitute $Z = 1$ into (9.1.14) We then obtain (in the same manner as in section 9.2, see (9.2.9) and (9.2.3)) the complex refractive index and susceptibility

$$n = \sqrt{1 + 4\pi \chi} \qquad (9.5.1)$$

$$\chi = -\frac{N_0 d^2}{\hbar} \frac{\Delta + iT_2^{-1}}{\Delta^2 + (T_2)^{-2}} \qquad (9.5.2)$$

where as before $\Delta = \omega_0 - \omega$ is the de-tuning and T_2 is the homogeneous relaxation time. Here we neglect inhomogeneous broadening. We point out that in (9.5.2) the minus sign appears because we use the condition $Z = 1$ instead of $Z = -1$ as in (9.2.3), where the ground state of the resonant medium was considered. Since the complex value of χ now lies in the lower half of the complex plane, the real and imaginary parts of the refractive index, n, determined by equation (9.5.1) satisfy the following relation

$$\operatorname{Re} n \cdot \operatorname{Im} n < 0 \qquad (9.5.3)$$

corresponding to amplification. Note that for the case of $Z = -1$ the product $\operatorname{Re} n \cdot \operatorname{Im} n$ is positive, and this corresponds to absorption. From equations (9.5.1) and (9.5.2) it follows that the refractive index has a resonant character. Let us consider first the case where homogeneous broadening is relatively large and therefore the refractive index differs only slightly from unity. Then the square root in (9.5.1) is approximately

$$n \simeq 1 + 2\pi \chi = 1 - \frac{T_\lambda^{-1}(\Delta + iT_2^{-1})}{\Delta^2 + T_2^{-2}} \qquad (9.5.4)$$

where we have used (9.4.5) for T_λ

$$T_\lambda = T_R kL = \frac{\hbar}{2\pi d^2 N_0}. \qquad (9.5.5)$$

The reflection coefficient will then be in accordance with (9.2.14)

$$\frac{n - 1}{n + 1} = -\frac{1}{2} \frac{T_\lambda^{-1}(\Delta + iT_2^{-1})}{\Delta^2 + T_2^{-2}} \qquad (9.5.6)$$

with the maximum absolute value

$$\max \left| \frac{n - 1}{n + 1} \right| = \frac{1}{2} \frac{T_2}{T_\lambda}. \qquad (9.5.7)$$

This result is valid provided that $T_2 \ll T_\lambda$, which is the usual experimental situation. At the same time T_2 can be larger than T_R, as is needed for amplification.   Recall that $T_2/T_R = \alpha L/2$, where α is the amplification coefficient for the linear stationary régime (see (5.2.7)). Estimates given in [HKS92] show that the small value of the non-resonant reflection coefficient, about 10^{-3}–10^{-4}, is sufficient for pulse synchronization if N is of the order of 10^{10}–10^{15}. But in our case the reflection is resonant. Therefore in order to have identical delays and shapes of counter-propagating pulses it is necessary that all the actual spectral components of the pulse lie in the resonance width T_2^{-1}. The spectral width of the pulse is of the order of the reciprocal of the first lobe duration T_p. Hence the pulse must be long enough to obey the resonance condition

$$T_p > T_2. \tag{9.5.8}$$

This condition corresponds to the case of suppressed super-radiance that was observed in solids (see section 2.2). It is worth mentioning also that the maximum absolute value of the reflection coefficient (9.5.7) is larger in solids than in gases. Therefore counter-propagating pulse synchronization is more likely to appear in solids than in gases because of the larger value of the amplification coefficient α.

Let us consider now the forward–backward correlation for pure super-radiance, i.e. for the case $T_2 \gg T_\lambda$, Substituting $T_2^{-1} = 0$ into (9.5.2) we obtain with the help of (9.5.1)

$$n = \left(1 - \frac{T_\lambda}{\Delta}\right)^{1/2}. \tag{9.5.9}$$

We can see that for $0 < \Delta < T_\lambda$ the refractive index becomes purely imaginary and as a consequence the absolute value of the reflection coefficient is equal to unity. So the width of resonance will now be determined by T_λ^{-1} and not by T_2^{-1}. Since the duration T_p of the first lobe of pure super-radiance pulses is of the order $T_R \ln N$ [MGF76], the resonance condition then will be

$$T_R \ln N > T_\lambda \qquad \text{i.e.}\quad \ln N > kL. \tag{9.5.10}$$

This condition is hardly fulfilled for relatively extended systems, where kL is of the order of 10^3–10^4, as in the experiment discussed above.

The conclusions made in this section can be verified by direct numerical solution of the Maxwell–Bloch equations without the SVEA in the spatial coordinate.   Indeed, in section 9.1 we showed that by dropping this approximation we could correctly describe reflections from the boundaries without additionally imposing phenomenological boundary conditions.

The results of such numerical solutions are demonstrated in figure 9.8 [MTS94]. The initial polarization was chosen stochastically in each small interval Δx of the length of the sample with variance $(N\Delta x/L)^{-1/2}$. We can see the results of solutions for a typical stochastic realization of initial conditions

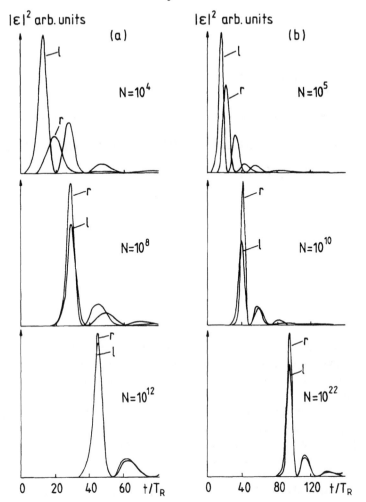

Figure 9.8. The establishment of correlations between SF pulses at the left (l) and right (r) ends of the sample calculated for a typical realization of initial polarization and different values of the number N of excited centres. (a) $kL = 30$; (b) 50.

with different values of N. The synchronization of the left and right pulses with increasing N is readily seen (figure 9.8), which is in qualitative agreement with condition (9.5.10).

Chapter 10

Resonant boundary value problem with local field effects

As is known, the field acting on the atoms in a medium is different from the averaged macroscopic field. The problem of the effective local field caused by dipole–dipole interaction has been discussed several times since Lorentz [49]. For a review see, for example, the work of van Krankendonk and Sipe [42]. The importance of the local field in time-dependent resonant phenomena and in the theory of super-radiation was recognized some time ago by Friedberg *et al* [FHM73]. In their work the reader can find several other references concerning the question of resonant frequency shifts (see also Chapter 6). Emelyanov and Zokhdi [EZ80], Hopf, Bowden and co-workers [7, 31–33] and Friedberg *et al* [21], discussed whether local field effects may give rise to an intrinsic optical bistability. In these works the stationary régime was the centre of interest, when the duration of the external excitation, T, was much longer than the atomic relaxation times T_2, T_2^*.

In the present chapter we will link these problems, and will investigate those optical properties of thin layers that are connected with both the local field correction and atomic coherent effects. We focus our attention on the problem of reflection and transmission of ultra-short pulses, $T \ll T_2$, T_2^* and, in contrast with the previous chapter, we take into account the effect of the local field correction upon these processes. Further on we will show that for a sufficiently thin layer the functional relation between the microscopic and macroscopic fields may lead to a new mechanism of nonlinear transparency of the medium. This nonlinearity generates an effect which is the analogue of the dispersive optical bistability in the stationary case, but it takes place on a much shorter time scale.

In the presentation of the problem we shall follow our paper [BMTZ91].

10.1 The modified Maxwell–Bloch equations with the local field correction

Let us consider the same boundary value problem as in the previous chapter. A linearly polarized plane wave, with amplitude slowly varying during one optical period, falls perpendicularly onto the surface of a plane-parallel, optically thin, resonant layer consisting of two-level atoms. The equations presented in section 9.1 remain valid, with the exception that the actual field strength $\mathcal{E}_{\text{eff}}$ acting on the atoms in the medium is different from the macroscopic field $\mathcal{E}$ obeying the inhomogeneous wave equation (9.1.2). As is known, they differ by a local field correction owed to the dipole–dipole interaction

$$\mathcal{E}_{\text{eff}} = \mathcal{E} + \frac{4\pi}{3}\mathcal{P}. \tag{10.1.1}$$

We will show in section 10.4, that this form of the correction is valid also in the case where the medium is an optically thin layer.

The slowly varying amplitude E of $\mathcal{E}$ can be expressed as in equation (9.1.11)

$$E(x,t) = E_{\text{in}}\left(t - \frac{x}{c}\right)e^{ikx} \tag{10.1.2}$$
$$+ \frac{2\pi i \omega}{c}\int_0^L P\left(x', t - \frac{|x - x'|}{c}\right)e^{ik|x-x'|}\,dx'$$

where E_{in}, as before, is the amplitude of the incoming wave, and the amplitude of the effective field strength can be written as

$$E_{\text{eff}} = E + \frac{4\pi}{3}P. \tag{10.1.3}$$

Our aim is to determine the amplitudes of the macroscopic reflected and transmitted waves. These macroscopic field strengths have the same expression as in Chapter 9

$$E_r(t) = E(0,t) - E_{\text{in}}(t) \qquad E_{\text{tr}}(t) = E(L,t). \tag{10.1.4}$$

The introduction of the local field, E_{eff}, influences the transmission and the reflection only implicitly, through the dynamics of the polarization determined by the optical Bloch equations

$$\frac{\partial R}{\partial t} = -\left(i\Delta + \frac{1}{T_2}\right)R - i\frac{d}{\hbar}E_{\text{eff}}Z \tag{10.1.5}$$

$$\frac{\partial Z}{\partial t} = \frac{id}{2\hbar}(E_{\text{eff}}R^* - E_{\text{eff}}^*R) - \frac{1}{T_1}(1 + Z) \tag{10.1.6}$$

and by the relation (9.1.16)

$$P(x,t) = N_0 d \int g(\Delta) R(x,t,\Delta)\,d\Delta. \tag{10.1.7}$$

Note that the driving terms in equations (10.1.5) and (10.1.6) contain E_{eff} instead of E. Otherwise we use the same notation as in the previous chapter. In the following we shall consider the system of equations (10.1.5), (10.1.6) and (10.1.2), which completely determine our problem.

Even without solving the equations above, it is possible to draw some conclusions about the physical consequences of the local field correction. In the case of negligible inhomogeneous broadening we have $P = dN_0 R$. Inserting $E_{\text{eff}} = E + (4\pi/3)dN_0 R$ into (10.1.5) and (10.1.6) we obtain

$$\frac{\partial R}{\partial t} = -\left(i\Delta' + \frac{1}{T_2}\right) R - i\frac{d}{\hbar}EZ \tag{10.1.8}$$

$$\frac{\partial Z}{\partial t} = \frac{id}{2\hbar}(ER^* - E^*R) - \frac{1}{T_1}(1 + Z) \tag{10.1.9}$$

where the following notations have been introduced

$$\Delta' = \omega_0 - \omega + \Delta_L Z = \Delta + \Delta_L Z \tag{10.1.10}$$

and

$$\Delta_L = \frac{4\pi}{3\hbar}d^2 N_0. \tag{10.1.11}$$

We see that initially, when the system starts from its ground state, the local field correction leads to a renormalization of the resonant frequency by an amount of $-4\pi d^2 N_0/(3\hbar)$. At later times, it will produce a dynamical shift in Δ', depending linearly upon the population difference Z. The amount of this frequency change is $8\pi d^2 N_0/(3\hbar)$ as Z ranges between -1 and $+1$. We note that $4\pi d^2 N_0/3$ is of the same order of magnitude as the interaction energy of two equal dipoles of strength d separated by a distance $(4\pi N_0/3)^{-1/3}$.

In the case of a narrow inhomogeneous line considered above, the correction $4\pi P/3$ does not directly affect the dynamics of the population, because it does not enter equation (10.1.9) for Z. But if we take into account the inhomogeneous broadening, the correction will appear explicitly in that equation, and will induce a redistribution of the inversion amongst the different frequencies. In what follows we shall consider only the case of negligible inhomogeneous line broadening.

As the local field correction causes a dynamical shift of the atomic resonant frequency, its effect upon the interaction will essentially depend upon the relative positions of the carrier frequency of the incoming wave, ω, and the renormalized resonant atomic frequency, $\omega_0' = \omega_0 - \Delta_L$. If $\omega \leq \omega_0'$ then during the excitation the system will be driven out of resonance even more, and will become more transparent. In the opposite case, if $\omega > \omega_0'$, the excitation will improve the resonance condition and, accordingly, the reflection will be enhanced. We will show below that this process of pulling into resonance has a threshold character.

10.2 Local field correction and the linear resonant refractive index

Let the incoming field be so weak that only a small fraction of the atoms becomes excited. Setting $Z = -1$, and taking the stationary limit $\partial R/\partial t = 0$ in (10.1.8), we obtain the usual result

$$P = \chi E \tag{10.2.1}$$

with the susceptibility

$$\chi = \frac{id^2 N_0}{\hbar} \frac{1}{i(\Delta - \Delta_L) + 1/T_2} = \frac{3i}{4\pi} \frac{\Delta_L}{i(\Delta - \Delta_L) + 1/T_2}. \tag{10.2.2}$$

(It is not difficult to incorporate into this result the inhomogeneous broadening: see equations (9.3.10)–(9.3.12).) It can be seen that the local field correction leads to a constant shift, Δ_L, of the line centre. If we change the frequency of the incoming wave to this value, we recover the resonant case as treated in the previous chapter, which justifies the treatment given there.

It is not difficult to see that the reflection and transmission in the linear and stationary régime are again given by the Fresnel formulae with the refractive index

$$n = \sqrt{1 + 4\pi\chi} = \left(1 + \frac{2}{kLT_R} \frac{1}{\Delta_0' - i/T_2}\right)^{1/2} \tag{10.2.3}$$

where

$$\Delta_0' = \Delta - \Delta_L \tag{10.2.4}$$

and T_R is the super-radiation time of the medium

$$T_R^{-1} = \frac{2\pi d^2 N_0 kL}{\hbar}. \tag{10.2.5}$$

Note the following connection between T_R and Δ_L

$$T_R^{-1} = \tfrac{3}{2}\Delta_L kL. \tag{10.2.6}$$

In the case of exact resonance, $\Delta_0' = 0$, we have

$$n = \left(1 + i\frac{2T_2}{kLT_R}\right)^{1/2} = n_1 + in_2. \tag{10.2.7}$$

If $T_2/kLT_R \gg 1$, we obtain $n_1 \approx n_2 \approx (T_2/kLT_R)^{1/2}$, i.e. both the real and imaginary parts of the refraction index are large. If in this case the thickness of the sample is large enough, so that $n_2kL \gg 1$, then according to the appropriate Fresnel formula, the absolute values of the reflected and incident waves are nearly equal, $E_r \approx E_{in}$, the incoming wave is almost totally reflected, and the field penetrates into the medium only to a depth of $\lambda/(2\pi n_1)$.

If the incident pulse is ultra-short, $T \ll T_2$, then the considerations above need to be refined. The spectral width of such a pulse, T^{-1}, is much larger than the width of the absorption line T_2^{-1}. In this case the index of refraction must be considered as a function of the frequency components of the pulse. So one has to determine the reflection and transmission coefficients for each spectral component, and then for the whole pulse. The other possibility is to remain in the time domain and solve the coupled equations (10.1.2), (10.1.8) and (10.1.9). Going over to the nonlinear case below, we shall follow this latter route.

10.3 Nonlinear transmission and mirror-less bistability

Turning to the investigation of strong pulses already causing a significant inversion, we restrict ourselves to the case of a thin layer, for which the spatial dependences of the field and the polarization can be neglected. For a thin layer the system of equations (10.1.2), (10.1.8) and (10.1.9) takes an essentially simpler form. In this case the integral equation for the macroscopic field strength reduces to a simple algebraic equation

$$E = E_{in} + 2\pi i k L P = E_{in} + 2\pi i k L N_0 d \langle R \rangle. \qquad (10.3.1)$$

When deriving (10.3.1) from (10.1.2) we have approximated the exponential in the integrand by unity, and we have neglected the dependence of all the amplitudes on the spatial variable x. At first sight the condition $L < \lambda/2\pi$ (and not $L < \lambda/2\pi |n|$) seems to be sufficient to validate this approximation. In order to to understand the situation, let us estimate the omitted terms. It is clear that they are of the order of $2\pi d N_0 (kL)^2 \langle R \rangle$, and they must be small compared with the terms kept in the equation. Therefore the condition that the layer can be considered to be optically thin has the form

$$kL \left| \frac{\langle R \rangle}{d E_{in} T_R/\hbar - \langle R \rangle} \right| \ll 1. \qquad (10.3.2)$$

It is not difficult to prove that in the linear case, when $\langle R \rangle = (\chi/dN_0)E$ (for $\chi \gg 1$), condition (10.3.2) is equivalent to $|n|kL \ll 1$.

Having established this criterion, we can determine now the effective field strength in the limit of a thin layer

$$E_{eff} = E + \frac{4\pi}{3} P = E_{in} + 2\pi i k L P + \frac{4\pi}{3} P. \qquad (10.3.3)$$

In section 10.4 we will give a complete systematic derivation of equation (10.3.3).

For the thin medium the transmitted and the reflected waves are given by

$$E_{tr} = E_{in} + 2\pi i k L P \qquad (10.3.4)$$

$$E_r = 2\pi i k L P. \qquad (10.3.5)$$

To simplify notation, instead of the field amplitudes we shall use the following quantities: $\epsilon_{in} = dE_{in}/\hbar$, $\epsilon = dE/\hbar$, $\epsilon_{eff} = dE_{eff}/\hbar$. They have the dimension of frequency (the Rabi frequency for real and constant E [AE75]). With these notations our system of equations describing the interaction of the light pulse with the thin layer, takes the following form

$$\frac{\partial R}{\partial t} = -\left(i\Delta + \frac{1}{T_2}\right)R - i\epsilon_{eff}Z \qquad (10.3.6)$$

$$\frac{\partial Z}{\partial t} = \tfrac{1}{2}i(\epsilon_{eff}R^* - \epsilon_{eff}^*R) - \frac{1}{T_1}(1 + Z) \qquad (10.3.7)$$

$$\epsilon_{eff} = \epsilon_{in} + \left(\Delta_L + \frac{i}{T_R}\right)\langle R\rangle \qquad (10.3.8)$$

and the reflected and transmitted amplitudes are proportional to the quantities

$$\epsilon_r = \frac{i}{T_R}\langle R\rangle \qquad (10.3.9)$$

$$\epsilon_{tr} = \epsilon_{in} + \frac{i}{T_R}\langle R\rangle. \qquad (10.3.10)$$

Here again the angular brackets denote the average over the inhomogeneous line.

If $kL \ll 1$, then from equation (10.2.6) it follows that $\Delta_L \gg T_R^{-1}$. Therefore for a thin layer the local field contribution, $\Delta_L\langle R\rangle$, to the effective field, ϵ_{eff}, is dominant over the radiation field $i\langle R\rangle/T_R$, as far as the absolute values are concerned. Nevertheless, both terms are important, because of the $\pi/2$ phase shift between them. The local correction generates a shift in the resonant frequency, whilst the radiation field, as it will be shown below, induces a collective relaxation of the polarization and of the inversion with a time constant T_R.

Let us consider an ultra-short light pulse, the duration of which is less than the relaxation times, but is longer than the super-radiation time, $T_R < T < T_2, T_2^*, T_1$. In the linear case this relationship between the time constants has been the condition of strong reflection. What happens if we increase the intensity of the incoming pulse, so that nonlinearity is supposed to play a rôle? For an ultra-short pulse, i.e. when the relaxation terms can be neglected, the system of equations (10.3.6) and (10.3.7) can be written in the form

$$\frac{\partial R}{\partial t} = \left(-i\Delta' + \frac{1}{T_R}Z\right)R - i\epsilon_{in}Z \qquad \Delta' = \Delta + \Delta_L Z \qquad (10.3.11)$$

$$\frac{\partial Z}{\partial t} = \tfrac{1}{2}i\epsilon_{in}(R^* - R) - \frac{1}{T_R}|R|^2 \qquad (10.3.12)$$

where the amplitude ϵ_{in} was taken to be real and $\langle R\rangle = R$. As follows from equation (10.3.11), if the phase memory of the atomic system is conserved ($T_1, T_2, T^* \to \infty$), then the time scale of the radiation relaxation is T_R/Z. This quantity, as well as the shift of the resonant frequency $\Delta_L Z$, depend upon the inversion Z. Therefore the width of the resonance will be determined by the quantity $|Z|/T_R$.

10.3.1 Nonlinear transparency by resonant excitation

Let the incoming field be de-tuned from the transition frequency, ω_0, so that $\omega = \omega_0 - 4\pi d^2 N_0/3\hbar$, i.e. $\Delta'_0 = \Delta - \Delta_L = 0$. If the amplitude of the exciting wave is not too large, $\epsilon_{\text{in}} < T_R^{-1}$, then, as we have seen in Chapter 9, the pulse is strongly reflected from the layer. If we increase the incoming amplitude, and hence the degree of the excitation, then the atomic system will be driven out of resonance, and, instead of reflecting, the layer will be transparent. To estimate the de-tuning Δ' at which the transparency becomes significant, let us use the linear approximation, equation (10.2.3). We also suppose that $T_2/T_R \gg 1$. Taking the thin-layer limit ($nkL \ll 1$) of the Fresnel formulae (9.2.11) and (9.2.9), it is easy to see that the transmission will be comparable with or larger than the reflection if $|n^2 - 1|kL/2 \leq 1$. According to equation (10.2.3) this is equivalent to the relation $\Delta'T_R \geq 1$, and thus the transparency of the layer will be significant when the frequency shift induced by the field is comparable with or larger than the width of the resonance

$$\Delta' \geq \frac{1}{T_R}. \qquad (10.3.13)$$

As

$$\Delta' = (1 + Z)\Delta_L = 2\rho_{ee}\Delta_L \qquad (10.3.14)$$

where ρ_{ee} is the population of the upper state, using (10.2.6) we can obtain an estimate for the appropriate population of the upper state

$$\rho_{ee} \simeq \frac{1}{\Delta_L T_R} \simeq kL \ll 1 \qquad (10.3.15)$$

because $kL \ll 1$ in the present case.

As is known, for extended systems transparency is connected either with saturating fields equalizing the ground- and excited-state populations ($\rho_{ee} \simeq \frac{1}{2}$) or with the effect of self-induced transparency (see section 5.3). In both cases the variation of the population ρ_{ee} is large, it is of the order of unity. In contrast to this, in our case transparency already appears at a negligible inversion $\rho_{ee} \simeq kL \ll 1$.

In order to obtain a more detailed picture of the response of the resonant nonlinear thin layer, we have calculated numerically the time dependences of the transmission coefficient, $T = |\epsilon_{\text{tr}}/\epsilon_{\text{in}}|^2$, and of the inversion, Z. The results are shown in figure 10.1. This figure demonstrates the build-up of a stationary T and Z, after switching on the external field with different constant amplitudes.

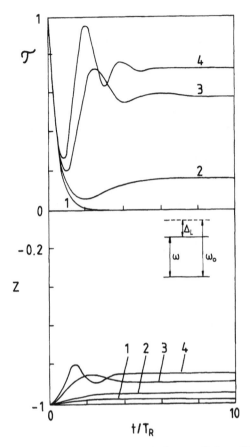

Figure 10.1. Build-up of the stationary transmission coefficient, $\mathcal{T} = |\epsilon_{tr}/\epsilon_{in}|^2$, and of the inversion Z of a thin layer, $kL = 0.066\,667$, $\Delta_L T_R = 2/(3kL) = 10$; after switching on a resonant ($\Delta_0' = 0$) step pulse of constant amplitude, $\epsilon_{in} = \epsilon_0/T_R$. Curve (1) $\epsilon_0 = 0.1$; curve (2) $\epsilon_0 = 0.3$; curve (3) $\epsilon_0 = 0.7$; curve (4) $\epsilon_0 = 1.0$.

We must emphasize that stationarity is understood here on a time scale longer than T_R, but shorter than T_2. As can be seen, the numerical results justify the conclusion about the growth of the transparency caused by increasing the incident amplitude. A weak incoming wave will be totally reflected after a time T_R, needed for the build-up of the polarization (curve (1) in figure 10.1). On increasing the amplitude, the layer becomes more and more transparent (curves (2–4)). This transparency arises at a negligible excitation of the atoms, the inversion remains close to its initial value -1. The calculations also show that the stationary transmission and inversion are achieved not monotonically, but exhibiting damped optical nutations. The expression for the stationary value for the reflected and transmitted waves, as well as for the transmission coefficient

T, can be found from the system of equations (10.3.11) and (10.3.12). Setting the derivatives equal to zero, we obtain from the first equation

$$R_{st} = \frac{iZ_{st}}{Z_{st}/T_R - i\Delta'}\epsilon_{in}.$$

(10.3.16)

The second equation is satisfied identically. The system, however, has a constant of motion, $|R|^2 + Z^2 = 1$, and this allows us to obtain an expression for Z

$$(1 - Z_{st}^2)\left[(\Delta')^2 + \frac{Z_{st}^2}{T_R^2}\right] - \epsilon_{in}^2 Z_{st}^2 = 0.$$

(10.3.17)

According to (10.3.9) and (10.3.10) the amplitudes of the reflected and transmitted fields can be written in the form

$$\epsilon_r = \frac{i}{T_R}R_{st} = \frac{iZ_{st}/T_R}{Z_{st}/T_R - i\Delta'}\epsilon_{in}$$

(10.3.18)

$$\epsilon_{tr} = \epsilon_{in} + \frac{i}{T_R}R_{st} = \frac{i\Delta'}{Z_{st}/T_R - i\Delta'}\epsilon_{in}$$

(10.3.19)

respectively. This yields the following transmission coefficient

$$T = \frac{(\Delta')^2}{(\Delta')^2 + (Z_{st}/T_R)^2}.$$

(10.3.20)

Equation (10.3.20) shows that the transmission will be comparable with the reflection, $T \simeq \frac{1}{2}$, when

$$(\Delta')^2 = Z_{st}^2/T_R^2$$

(10.3.21)

i.e. when the frequency shift Δ' induced by the field becomes equal to the width of the resonant frequency, $|Z|/T_R$. Taking into account that now $\Delta = \Delta_L$, and hence $\Delta' = (1 + Z_{st})\Delta_L$, from the equation $(1 + Z_{st})\Delta_L = -Z_{st}/T_R$ we find that

$$Z_{st} = -1 + \frac{1}{\Delta_L T_R} \qquad \rho_{ee} = \frac{1}{2\Delta_L T_R}$$

(10.3.22)

where the relation $1/(\Delta_L T_R) = \frac{3}{2}kL \ll 1$ has been exploited. Equation (10.3.22) shows that the onset of the transparency already begins at small values of the inversion, which is in accordance with our previous qualitative arguments.

It is more convenient to express the transmission coefficient by the population of the excited state, ρ_{ee}. Using the smallness of this quantity, we obtain from equation (10.3.20)

$$T = \frac{\rho_{ee}^2}{\rho_{ee}^2 + (3kL/4)^2}.$$

(10.3.23)

The equation for ρ_{ee} follows from (10.3.17), and in the limit $\rho_{ee} \ll 1$ it has the form

$$\rho_{ee}[\rho_{ee}^2 + (\tfrac{3}{4}kL)^2] = (\tfrac{3}{8}kL)^2(\epsilon_{in}T_R)^2. \qquad (10.3.24)$$

With the help of this result we can estimate the amplitude of the field inducing the transparency. Setting the limit between low and high transmission at $T = \tfrac{1}{2}$, i.e. at $\rho_{ee} = 3kL/4$, then for the threshold for high transparency we obtain the condition

$$T \geq \tfrac{1}{2} \qquad \text{if} \quad \epsilon_{in}T_R \geq (6kL)^{1/2}. \qquad (10.3.25)$$

10.3.2 Bistability of reflection and transmission by non-resonant excitation

We have considered above a resonant excitation: the frequency of the external field was equal to the renormalized transition frequency, $\omega = \omega_0 - \Delta_L$. Now let us turn to the non-resonant case $\Delta_0' = \Delta - \Delta_L \neq 0$. It is clear that if the frequency of the external field, ω, falls into the range below the renormalized transition, then the incoming wave drives the atoms further out of resonance, and therefore weakens the reflection and amplifies the transmission. Essentially new effects arise in the opposite case, when the frequency of the exciting pulse is above the renormalized resonance frequency, $\omega > \omega_0 - \Delta_L$ (see figure 10.2). In

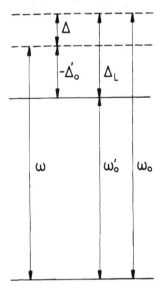

Figure 10.2. The scheme of non-resonant excitation for observing optical bistability in the transmission of the resonant layer.

this case the dynamical phase shift can compensate the initial de-tuning, and this leads to a stronger reflection. As will be shown below, in this case it is possible to obtain a sudden transition between states of low and high transmissions, depending either upon the initial de-tuning or upon the external field amplitude.

Suppose that the the incoming wave has a constant amplitude, ϵ_{in}, and is tuned above resonance, $\Delta_0' < 0$. Let us consider the quasi-stationary régime of the transmission

$$\frac{\partial R}{\partial t} = 0 \qquad \frac{\partial Z}{\partial t} = 0. \qquad (10.3.26)$$

Then in equations (10.3.16)–(10.3.20) we have

$$\Delta' = \Delta_0' + (1 + Z_{st})\Delta_L \qquad \Delta_0' = \Delta - \Delta_L < 0. \qquad (10.3.27)$$

We will be interested only in that range of the parameters for which the inversion remains close to its initial value, $Z = -1$. We recast equation (10.3.17) by expressing the inversion through the population of the upper state, $\rho_{ee} = (1 + Z)/2$, and assume that $\rho_{ee} \ll 1$. Taking into account also that $\Delta' = \Delta_0' + 2\rho_{ee}\Delta_L$, we obtain, for the stationary value of the upper level, ρ_{ee}, the following equation

$$4\rho_{ee}\left[(\Delta_0' + 2\rho_{ee}\Delta_L)^2 + \left(\frac{1}{T_R}\right)^2\right] = \epsilon_{in}^2. \qquad (10.3.28)$$

This equation has either a single real root or three real roots. In the first case the function $\rho_{ee}(\epsilon_{in})$ is single-valued. In the second case, however, to each ϵ_{in} there correspond three different populations of the upper level. This latter case indicates the possibility of bistable behaviour of the system.

The condition for obtaining three real roots can be found by determining the zeros of the derivative

$$\frac{d\epsilon_{in}^2}{d\rho_{ee}} = 0. \qquad (10.3.29)$$

Introducing the notation

$$z = \Delta_0' + 2\rho_{ee}\Delta_L \qquad (10.3.30)$$

we obtain

$$3z^2 - 2\Delta_0'z + \left(\frac{1}{T_R}\right)^2 = 0. \qquad (10.3.31)$$

Equation (10.3.31) will have three different real roots if

$$\Delta_0' < -\frac{\sqrt{3}}{T_R}. \qquad (10.3.32)$$

In this case the plot of the function $\rho_{ee}(\epsilon_{in})$ shows a wrinkle. The boundaries of the interval where this function is three-valued can be determined by substituting the roots of equation (10.3.31) into (10.3.28). The smallness of the roots of equation (10.3.28) is ensured by demanding the inequality $|\Delta_0'| \ll \Delta_L$. According to (10.3.22) this means that the quantity $|\Delta_0'|T_R$ can be of the order of unity.

In the following it will be more convenient to use the dimensionless quantity

$$\epsilon_0 = \epsilon_{in}T_R. \qquad (10.3.33)$$

Figure 10.3 shows how Z_{st} depends on ϵ_0 and Δ_0', illustrating the above analytical considerations. The figures were drawn using the exact equation (10.3.17), which is of fourth order, in contrast to the approximate equation (10.3.28), which is only of third order. Therefore in a certain range of the variables ϵ_0^2 and Δ_0' the function $Z_{st}(\epsilon_0, \Delta_0')$ is four-valued. The unstable values have been plotted with dotted lines.

The dependence of the transmission coefficient T upon the dimensionless intensity, ϵ_0^2, is shown on the upper half of figure 10.3(a). It can be seen that, beginning from some value of Δ_0' (in the figure at $\Delta_0' = -2/T_R$), there is a hysteresis in the transmission.

The S-shaped form of the function Z_{st} means that there is a multi-valued (bistable) response of the system. As an illustration, in figure 10.4 we show the build-up of the stationary transmission coefficient, T, and inversion, Z_{st}, for a thin layer ($kL = 0.06667$, $\Delta_L T_R = 10$) after switching on a non-resonant field ($\Delta_0'T_R = -2.2$) of constant amplitude $\epsilon_{in} = \epsilon_0/T_R$. As can be seen from figure 10.4, when ϵ_0 becomes larger than a certain value the character of the response suddenly changes from an almost total transmission to a strong reflection. For the parameters we have chosen, this threshold is at $\epsilon_0 = 0.725$. The corresponding jump can be observed also in the behaviour of the inversion Z. We can calculate numerically the dynamical frequency shift during the process. In the range above the threshold, in the case shown in the figure it is only when $\epsilon_0 > 0.725$ that the initial de-tuning is almost perfectly compensated by the dynamical phase shift. For the stationary value of $(\Delta' - \Delta)T_R$ we have obtained 2.3. This means that in the stationary régime the atoms are in resonance with the external field. This causes a strong reflection, while the departure of the inversion from its initial value remains relatively small. The bistability in the transmission of the resonant layer considered here bears a purely dispersive character, because we did not take into account the relaxation processes, and the terms $1/T_1$, $1/T_2$ were set equal to zero. This approach is therefore different from that of references [7, 31, 32], where bistability was studied on the time scale of the phase relaxation T_2.

Everything that has been stated above is valid for light pulses which are longer than the time of stationary response T_R but shorter than the relaxation times T_1 and T_2. Let us recall that there is another restriction that has been imposed: the incident wave can be reflected only if $\epsilon_{in} < T_R^{-1}$. This is related

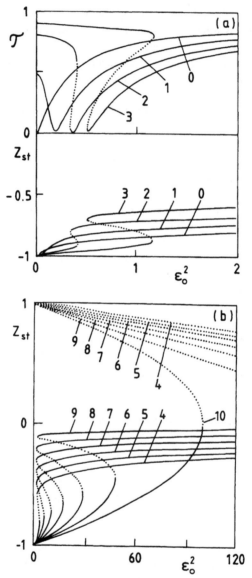

Figure 10.3. The dependence of the inversion Z_{st} and that of the transmission coefficient T upon the dimensionless intensity of excitation, $\epsilon_0^2 = (\epsilon_{in} T_R)^2$. The numbers denote the corresponding values of the de-tuning, $-\Delta_0' T_R$.

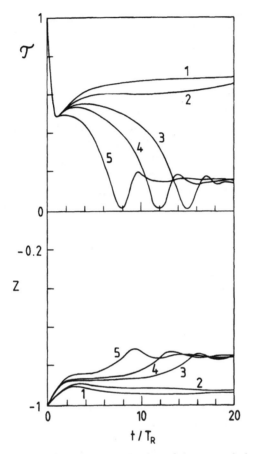

Figure 10.4. The sudden change in the kinetics of the transmission coefficient of the thin layer ($kL = 0.066\,667$) in the case of a non-resonant excitation ($\Delta_0' T_R = -2.2$) of constant amplitude $\epsilon_{in} = \epsilon_0 / T_R$. Curve (1) $\epsilon_0 = 0.72$; curve (2) $\epsilon_0 = 0.725$; curve (3) $\epsilon_0 = 0.7275$; curve (4) $\epsilon_0 = 0.73$; curve (5) $\epsilon_0 = 0.74$.

to the fact that the induced polarization of the layer cannot generate a secondary field with an amplitude larger than T_R^{-1}. Therefore the contrast in bistability will be high only if $\epsilon_{in} < T_R^{-1}$.

10.4 The local field correction for a thin layer: microscopic derivation

In the previous sections of this chapter we have been using the following form for the effective field strength

$$\mathcal{E}_{eff} = \mathcal{E} + \frac{4\pi}{3} P \qquad (10.4.1)$$

where $\mathcal{E}$ is the solution of the macroscopic Maxwell equations in the medium. $\mathcal{E}_{\text{eff}}$ differs from $\mathcal{E}$ by the local field correction. In the present section we are going to prove that this classical form [49] is valid also for an optically thin layer.

The total field acting upon the atoms consists of the superposition of the external field and the field originating from the dipoles of the medium itself. Here we shall calculate the latter, the effective electric field strength at the place of the ith atom generated by all the other dipoles. We assume that each dipole along the layer points in the same direction, characterized by the unit vector l, and will neglect inhomogeneous broadening. We consider a plane disk-shaped layer of thickness $L \ll \lambda$ and radius R_0.

The amplitude of the electric field originating from the jth dipole, that oscillates with frequency ω, $d_j = d_j \, e^{-i\omega t}$ can be written in the form (see equation (6.2.3) or [9, 34])

$$E_j^d = ik^3 \left[(d_j - (nd_j)n) \, F_1(kr) + n(d_j n) F_2(kr) \right] \qquad (10.4.2)$$

where n is the unit vector pointing from the site of the jth dipole towards the point of observation which is at a distance r, and $k = \omega/c$. The functions F_1 and F_2 are defined in the following way

$$F_1(\xi) = e^{i\xi} \left[\frac{1}{\xi^2} - \frac{i}{\xi} + \frac{i}{\xi^3} \right] \qquad (10.4.3)$$

$$F_2(\xi) = -e^{i\xi} \left[\frac{2i}{\xi^3} + \frac{2}{\xi^2} \right]. \qquad (10.4.4)$$

We shall calculate that component of the resulting field strength at r_i, which points in the direction of the dipoles

$$E^d = \frac{d_i \, E^d}{d_i} = l_i \sum_{j \neq i} E_j^d (r_i). \qquad (10.4.5)$$

It is slowly varying in time and is the sum of the fields originating from all other dipoles of the medium.

We choose the origin of coordinates to be in the centre of the disk and we calculate the field acting upon an atom lying on the symmetry axis (the x direction). We cut out a spherical volume of a small, but macroscopic, radius δ around the location of the ith atom. The field of the dipoles generated by the atoms within the sphere is zero [34] at the site r_i. The field of the dipoles outside the sphere can be calculated in the continuous approximation.

Then

$$E^d = ik^3 N_0 d R B \qquad (10.4.6)$$

with

$$B = \int_{V-V_\delta} B(r_{ij}) \, dV_j \qquad (10.4.7)$$

where the function $B(r_{ij})$ is given by the expression

$$B(r_{ij}) = \left(\left[(l_i l_j) - (l_j n_{ij})(l_j n_{ij}) \right] F_1(kr_{ij}) + (l_i n_{ij}) F_2(kr_{ij}) \right). \qquad (10.4.8)$$

The integration in (10.4.7) is to be performed over the volume of the cylindrical medium of length L and radius R_0, with the spherical hole of volume V_δ in it. Here l_i is the unit vector pointing in the direction of d_i (in our case $l_i = l_j = l$), $r_{ij} = r_i - r_j$, $n_{ij} = r_{ij}/r_{ij}$.

In cylindrical coordinates (ρ, φ, x)

$$l_i n_{ij} = -\frac{\rho_j}{r_{ij}} \cos \varphi_j \qquad r_{ij}^2 = \rho_j^2 + (x_j - x_i)^2. \qquad (10.4.9)$$

The ρ_i coordinate does not enter the expression (10.4.9), because the ith atom is assumed to be on the axis of the layer, $\rho_i = 0$. After integrating in (10.4.7) over φ_j and using (10.4.9) we obtain

$$\begin{aligned}
B = \pi &\left[\left(\int_{-L/2}^{x_i - \delta} dx_j + \int_{x_i + \delta}^{L/2} dx_j \right) \int_0^{R_0} \rho_j \, d\rho_j \right. \\
&+ \int_{x_i - \delta}^{x_i + \delta} dx_j \int_{[\delta^2 + (x_i - x_j)^2]^{1/2}}^{R_0} \rho_j \, d\rho_j \bigg] \\
&\times \left[F_1(kr_{ij}) + F_2(kr_{ij}) \right. \\
&+ \frac{(x_i - x_j)^2}{r_{ij}^2} [F_1(kr_{ij}) - F_2(kr_{ij})] \bigg].
\end{aligned} \qquad (10.4.10)$$

Substituting $\rho_j^2 = r_{ij}^2 - (x_j - x_i)^2$ and using the relations

$$\xi[F_1(\xi) + F_2(\xi)] = -\frac{d}{d\xi} \left[e^{i\xi} \left(1 - \frac{i}{\xi} \right) \right] \qquad (10.4.11)$$

$$\frac{1}{\xi}[F_1(\xi) - F_2(\xi)] = -\frac{d}{d\xi} \left[e^{i\xi} \left(\frac{i}{\xi^3} + \frac{1}{\xi^2} \right) \right] \qquad (10.4.12)$$

we can perform the integration over ρ_j in (10.4.10). We obtain

$$\begin{aligned}
B = \frac{\pi}{k^2} &\left(\int_{-L/2}^{x_i - \delta} dx_j + \int_{x_i + \delta}^{L/2} dx_j \right) \\
&\times \left\{ \Phi(k[R_0^2 + (x_i - x_j)^2]^{1/2}, k(x_j - x_i)) \right. \\
&- \Phi(k|x_j - x_i|, k(x_j - x_i)) \big\} \\
&+ \frac{\pi}{k^2} \int_{x_i - \delta}^{x_i + \delta} dx_j \left\{ \Phi(k[R_0^2 + (x_i - x_j)^2]^{1/2}, k(x_j - x_i)) \right. \\
&- \Phi(k, \delta(x_j - x_i)) \big\}
\end{aligned} \qquad (10.4.13)$$

where

$$\Phi(\xi, \eta) = -\left[1 - \frac{i}{\xi} + \frac{\eta^2}{\xi^2}\left(1 + \frac{i}{\xi}\right)\right] e^{i\xi}. \tag{10.4.14}$$

The integrations of the functions $\Phi(k|x_j - x_i|, k(x_j - x_i))$ and $\Phi(k\delta, k(x_j - x_i))$ also can be performed in explicit form, so that for B we obtain

$$B = \frac{\pi}{k^2}\left[\frac{8i}{3k}(1 - k\delta) e^{ik\delta} - \frac{4i}{k} e^{ikL/2} \cos kx_i \right.$$
$$\left. + \int_{-L/2}^{L/2} dx_j \, \Phi(k[R_0^2 + (x_i - x_j)^2]^{1/2}, k(x_j - x_i))\right]. \tag{10.4.15}$$

So far we have not used any approximation in the calculations, therefore the expression for B is exact in the framework of the present model. Let us turn our attention to the fact that B remains finite even in the limit $\delta \to 0$.

$$B = \frac{\pi}{k^2}\left[\frac{8i}{3k} - \frac{4i}{k} e^{ikL/2} \cos kx_i \right.$$
$$\left. + \int_{-L/2}^{L/2} dx_j \, \Phi(k[R_0^2 + (x_i - x_j)^2]^{1/2}, k(x_j - x_i))\right]. \tag{10.4.16}$$

From this equation in the limit of a thin layer ($kL \ll 1, k|x_i| \ll 1$) it follows that

$$B = -\frac{4\pi i}{3k^3} + \frac{2L\pi}{k^2}$$
$$+ \frac{\pi}{k^2}\int_{-L/2}^{L/2} dx_j \Phi(k[R_0^2 + (x_i - x_j)^2]^{1/2}, k(x_j - x_i)). \tag{10.4.17}$$

The third term, containing the integral, can be shown to be negligible in the limit $R_0 \to \infty$. Therefore, according to (10.4.6) the amplitude of the effective field strength in the optically thin layer is

$$E^d = \frac{4\pi}{3} N_0 dR + 2\pi ik L N_0 dR. \tag{10.4.18}$$

It is the first term which embodies the difference between the effective and the macroscopic field, and it is in full accordance with the standard result for the local field correction $4\pi P/3$. The second term in (10.4.18) is the reaction field $2\pi ikLP$, which can also be obtained from the solution of the macroscopic Maxwell equations in the limit of a thin medium, as in equation (10.3.1).

10.5   Concluding remarks

In describing the nonlinear optical properties of optically thin resonant layers, it is of principal importance to take into consideration the dipole–dipole interaction of the atoms. This effect can be effectively taken into account in the optical Bloch equations by replacing the macroscopic mean field, $\mathcal{E}$, with the effective local field, $\mathcal{E}_{\text{eff}} = \mathcal{E} + 4\pi \mathcal{P}/3$. This correction may lead to several interesting physical effects. We have specifically discussed those that may be important in some future applications, namely:

(i)   the nonlinear transparency of a thin resonant layer remaining near its ground state;

(ii) the bistability of the transmission with an intensity-dependent threshold.

These effects manifest themselves when certain relations between the parameters of the atomic system and the incident field are fulfilled. The duration of the excitation, T, must be larger than the super-radiation time of the layer, T_R, and at the same time it is assumed to be shorter than the relaxation times of the atomic system. In addition, the Rabi frequency corresponding to the amplitude of the incident pulse must be less than T_R^{-1}. Thus the conditions for observing the predicted effects are $T_R < T < T_2^*, T_2$ and $\epsilon_{\text{in}} < T_R^{-1}$.

We expect that it will be possible to observe the effects analysed in the present chapter, particularly in the excitonic lines of aromatic compounds [14] (such as naphtalene and anthracene), or in materials containing unoccupied d or f orbitals [68], such as Cr_2O_3 or MnO_2. The lowest electronic excitations in these materials are excitons of small radius. The presence of the excitonic lines shows that the dipole–dipole interaction is larger than the homogeneous and inhomogeneous broadenings. This gives reason to hope that the required inequalities are satisfied in such systems.

Chapter 11

New sources and applications of super-radiance

In previous chapters we have given the theory of several aspects of super-radiance, including those which find application in systems studied intensively at present in several laboratories. The source of super-radiance that we had primarily in mind was that from atoms with discrete energy spectra. However, super-radiance is also possible in other systems. It acquires essentially new features where the radiation source has a continuous energy spectrum. Two such sources will be discussed in this last chapter: the free-electron laser and the electron–hole plasma in semiconductors. Atomic nuclei that may super-radiate in the gamma-ray domain may at least in principle be another non-conventional source of super-radiance. This will be discussed in section 11.2.

From section 11.4 onwards we return to consider some fundamental questions connected with cooperative emission. First we describe sub-radiance (SBR), a phenomenon closely related to super-radiance. In the next section we discuss an interesting new experiment of DeVoe and Brewer [DVB96], who have directly observed for the first time the effect of two-atom collective emission, just as predicted by Dicke in 1954 when the whole subject of super-radiance was launched. We have also included in this final chapter a short overview on squeezing and non-classical light in super-radiance.

11.1 High-gain super-radiant régimes in FEL

Recently Bonifacio and co-workers [BCCSPP90, BCSPP92] have discussed the emergence of high-gain super-radiant régimes in a free-electron laser (FEL). The discussion followed their earlier work [BC85, BN88, BNP89] on super-radiance in high-gain FELs. The possibility of such régimes is a very important example of cooperative radiation–matter interaction and we shall outline the results of Bonifacio and co-workers in some detail.

The free-electron laser was first suggested by Madey [51]. It is an experimental set-up in which a beam of relativistic electrons of velocity v, $\beta = v/c \sim 1$, passes through a device (wiggler) that produces a spatially periodic (with a period of $\lambda_w = L_w/N_w$) transverse magnetostatic field B (see figure 11.1). Because of the transverse oscillations of the electron in such a field, *spontaneous* synchrotron radiation is emitted by the electron into a narrow cone around the beam axis. The wavelength λ_s of this radiation is determined by the velocity of the electron, as well as by the wiggler parameters. To a good approximation, it is given by a simple relation

$$\lambda_s = \lambda_w/2\gamma^2 \tag{11.1.1}$$

where γ is the electron energy expressed in units of its rest mass mc^2.

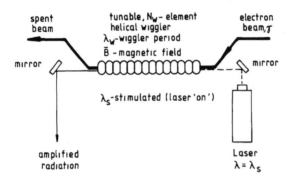

Figure 11.1. Schematic diagram of an FEL amplifier. A laser beam of wavelength λ co-propagates with a beam of electrons through the wiggler of length L_w. The energy of the electrons, γ, is tuned to satisfy the resonance condition, $\lambda_s = \lambda$. The amplified (stimulated) emission is registered at the wiggler exit. Resonance tuning is also possible by varying the magnetic field B of the wiggler.

If an external resonant field with $\lambda \approx \lambda_s$ co-propagates with the electron beam, the interaction of each electron with this field induces emission and decelerates the electron, or else the absorption process leads to their acceleration. The outcome will depend upon the relation between the phases of the oscillations of the electron and the co-propagating field. In a high-gain régime the interaction of N electrons with the self-radiation field produces periodic self-bunching of the electrons on the scale of the radiation wavelength λ. This sets up a correlation between the oscillations of different electrons. As a result they produce coherent cooperative emission (synchrotron radiation) with an enhanced intensity that is generally proportional to N^α, where $1 \leq \alpha \leq 2$ (the case $\alpha = 1$ corresponds to the spontaneous emission of uncorrelated electrons).

Bonifacio *et al* numerically solved the nonlinear equations which describe a one-dimensional model of the FEL (the total system of the relativistic electron beam and the co-propagating field, in the presence of a wiggler). In this way, they were able to analyse in detail the evolution of the coherent synchrotron emission, in particular the emergence of super-radiance régimes.   This analysis was carried out for conditions corresponding to different experimental arrangements used for the realization of the FEL.

One possible realization of a super-radiant régime is when the electron beam is pre-bunched by an external source, e.g. by a strong laser field. In this case, the radiation fields emitted by individual electrons sum up coherently thus resulting in the N^2 scaling of the stimulated synchrotron radiation.

This régime may arise, for example, in the final stage of a transverse optical klystron. However, the super-radiant régime may also evolve from an electron beam which is initially (i.e. at the wiggler entrance) un-bunched and becomes bunched during the passage through the wiggler. In the latter case the propagation effects (slippage) due to the difference between the beam velocity v and the velocity c of the propagating radiation should be taken into account.

The slippage parameter S is defined as follows

$$S = L_s/L_b \qquad (11.1.2)$$

where $L_s = N_w\lambda$, N_w is the number of periods in the wiggler and L_b is the length of the electron pulse. S is a measure of the advance (slippage) of the radiation front ahead of the pulse, at the end of the wiggler. The second, super-radiant parameter K relates the cooperation length of the system, L_c [BC85] to the dimension of the pulse

$$K = L_c/L_b = S/G \qquad (11.1.3)$$

where G is the exponential gain for the length of the wiggler. Then $\bar{L}_b = 1/K$ is the scaled pulse length as expressed in units of L_c.

Using the same scaling by L_c, we introduce the coordinates $\bar{z}$ for the position in the wiggler and $\bar{z}_1$ for the position in the frame moving with the velocity of the electron beam. We assume that the point $\bar{z} = 0$ corresponds to the wiggler entrance and that the pulse is in the interval $0 \le \bar{z}_1 \le 1/K$. This interval is usually referred to as the interaction region. Outside it, the electromagnetic radiation propagates in a vacuum. For an electron pulse moving along the wiggler axis with scaled velocity $\bar{v}$, we shall assume that it extends over the region $0 \le \bar{z} - \bar{v}t \le 1/K$.

Bonifacio *et al* considered solutions of the high-gain ($G \gg 1$) FEL equations, with a small initial fluctuation of the field.   Below we outline their results obtained in the approximation of the Compton régime, that is for sufficiently low electron densities in the beam and for sufficiently fast electrons, for three different assumptions regarding the length of the electron pulse.

(i) *An infinitely long uniform density beam.* Here the propagation effects are not taken into account ($S = 0$). The corresponding solution is a steady-state radiation that depends on the position in the wiggler, $\bar{z}$, but not on the position in the beam, $\bar{z}_1$ (all points in the beam are equivalent).

The stimulated cooperative radiation grows exponentially, starting from the wiggler entrance, with the saturated peak intensity scaling as $N^{4/3}$.

(ii) *Short beam pulses.* The slippage is taken into account ($L_b \ll L_c$; $S \gg K^{3/2} > 1$) [BC85, BN88]. The development of the radiation pulse intensity I and the change in the average momentum $\langle p \rangle$ of the beam are depicted in figure 11.2. They are shown there as a function of $\bar{z}_1$, at three different positions $\bar{z}$ in the wiggler. Therefore, the lower windows show the development of the radiation pulse in time. As the electron pulse moves through the wiggler at a velocity $v < c$, the radiation escapes from it. It shows an exponential growth, up to the first peak, with the peak intensity scaling as N^2. However, its intensity does not build up to the saturation level of the steady-state régime (i). This is the case of 'weak' super-radiance [BC85].

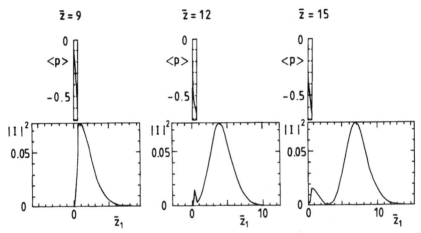

Figure 11.2. The short-pulse régime of an FEL (weak super-radiance) [BCCSPP90]. Electron (upper window) and radiation (lower window) pulses as functions of $\bar{z}_1$ at three positions in the wiggler: $\bar{z} = 9, 12, 15$. The parameters are $K = 1.66$, $G = 15$.

(iii) *Long beam pulses* The slippage is taken into account ($L_b \gg L_c$; $K \ll S \approx 1$). [BC85, BN88]. A typical solution is shown in figure 11.3, in the same format as in figure 11.2 for the short-pulse case. For the radiation intensity I there are three distinctly different regions of $\bar{z}_1$: (a) the slippage (instability) region, $0 < \bar{z}_1 < \bar{z}$; (b) the steady-state region $\bar{z} < \bar{z}_1 < 1/K$; and (c) the no-interaction region $1/K < \bar{z}_1 < 1/K + \bar{z}$ which represents the radiation escaping from the pulse. In the slippage region (a), the radiation pulse evolves as a short super-radiant pulse similar to (ii). The head edge of this pulse moves at velocity c slipping over the electron pulse. If the pulse is sufficiently long the radiation does not escape it immediately as in the case of weak super-radiance. It is continuously fed by the spontaneous radiation emitted by the electrons that enter the slippage region. In this

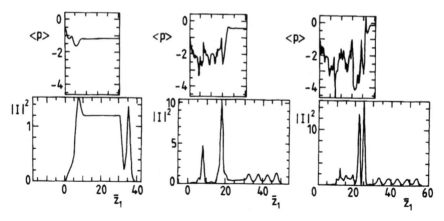

Figure 11.3. The long-pulse régime of an FEL (strong super-radiance) [BCCSPP90]. Electron and radiation pulses as functions of $\bar{z}_1$, at three positions in the wiggler: $\bar{z} = 12$, 24, 30. The parameters are $K = 0.03$, $G = 30$.

way, the initially small super-radiant pulse builds up, without saturation, to produce typical spikes with peak intensity scaling as N^2. As can be seen in figure 11.3, this is the case of 'strong' super-radiance [BCCSPP90]. The peak intensity of the spikes is well (by a factor of ten) above the saturation level in the next, steady-state region (*b*); the latter is unaffected by the instability régime of the first stage. In the third, no-interaction region (*c*), radiation propagates in vacuum, ahead of the electronic pulse and keeping track of the oscillatory behaviour of the steady-state region. For the electron pulse profiles (upper windows of figure 11.3), there are two different regions: the slippage region (*a*) which shows spikes in $\langle p \rangle$, corresponding to energy transfer between the beam and the radiation pulse; and the steady-state region (*b*) where the electronic pulse is uniform. The computer simulation showed [BCCSPP90] that, if de-tuning is introduced, it may considerably enhance the energy output of the super-radiant FEL régime.

In induction-driven modern linear accelerators (linacs) the 'slippage region' $\bar{\lambda} N_w$, where $\bar{\lambda}$ is the scaled wavelength of the radiation, is typically a small fraction of the beam length $\bar{L}_b$. Hence the radiation field is practically uniform over the pulse length. This is the case of the steady-state régime (i) considered above. Recently, the steady-state régime has been realized in experiments on the induction linac of the Lawrence Livermore laboratory [OAFPS85, OACFP86].

The opposite case of super-radiance régimes (ii) and (iii) may occur in FELs driven by an RF accelerator. In such a device the 'slippage region' can be much longer than the dimension of the electron pulse, $\bar{\lambda} N \gg \bar{L}_b$, owing to the long wavelength of the radiation. This is the condition which is essential for the departure from the steady-state régime.

Although the super-radiance régimes in high-gain FELs have not yet been demonstrated experimentally, it is expected that such régimes can be realized in experiments currently planned on modern linear accelerators [BCCSPP90, BCSPP92].

The super-radiance effects in synchrotron radiation have also been discussed in [ZKK87, GS91]. The quantum approach to the problem of FELs is outlined in [P88].

11.2 Gamma-ray super-radiance

Recoilless nuclear transitions in solids (the Mössbauer effect) were suggested as a source of intensive coherent radiation in the sub-nanometre range more than three decades ago [BNT63]. However, the realization of gamma-ray lasers is a challenge that requires resolving many physical and technical problems. Various features of a possible design have been discussed since then [BSG81, BS95].

An interesting physical point was made by Baldwin, Feld and co-workers [BF86, BFHHT86]. As good mirrors are unavailable at gamma-ray wavelengths, a possible source of intensive coherent radiation might be super-radiance. These authors discussed the minimal requirements that had to be imposed on a hypothetical experimental set-up which would be capable of producing observable high-intensity emission on a nuclear transition.

The gamma-ray device is expected to work on a recoilless Mössbauer transition from an excited level pumped from a pure sample of a long-lived isomer (storage state). Baldwin and Feld pointed out that the rate of the stimulated Mössbauer transition in heavy nuclei such as ^{119}Sn or ^{133}Ba would have to compete with the rates of two other processes de-populating the upper level: spontaneous emission and internal conversion (see figure 11.4). However, in Mössbauer transitions, the internal conversion rate is known usually to exceed that of spontaneous emission. Consequently, ordinary laser amplification, with an emission line of nearly natural width, would require too long a build-up time, compared with de-population due to internal conversion. Therefore, such a gamma-ray laser could not be realized. The difficulty could be overcome if it were possible to produce a super-radiant gamma-ray pulse with a sufficiently short cooperative time T_R. Therefore the arrangement has to be a single-pass system with very high gain. This can be accomplished by using a long crystal (Mössbauer host) capable of facilitating photon channelling (Borrmann effect). Doping of the crystal with active nuclei has to be sufficiently high to ensure observable output but not too high to disturb the lattice and deteriorate the Mössbauer and Borrmann effects.

Taking these conditions, together with the super-radiance régime of a bright propagating pulse, imposes a considerable restriction on parameters of the super-radiance transition, doping concentration and geometry of the crystal. By considering super-radiance in the usual semiclassical approximation described

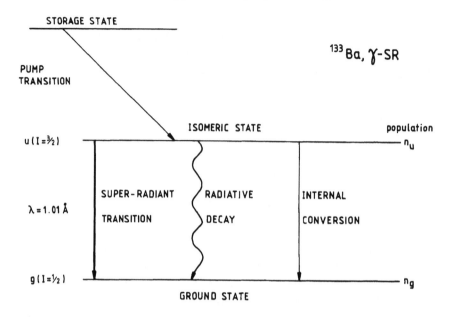

Figure 11.4. An energy-level diagram showing the Mössbauer transition considered by Baldwin and Feld [BF86]. The storage state can lie either above or below the upper level u. The energies of these two levels should be comparable, in order to minimize the energy requirement of the transfer step.

in Chapter 5, Baldwin and Feld carried out the analysis of the kinetics of super-radiance gamma-ray emission and established the critical (lowest) number of active nuclei and the dimensions of the crystal needed to contain them. Figure 11.5 shows the case of the ^{133}Ba dopant placed into a cubic-symmetry boron nitride (Borazon) host. The relevant M1 Mössbauer transition with a wavelength of 1.01×10^{-16} cm is from the $I = \frac{3}{2}$ upper level to the $I = \frac{1}{2}$ ground level of ^{133}Ba. According to the estimates of Baldwin and Feld, fewer than 10^{13} active nuclei would be sufficient for producing super-radiance.

The discussion in [BF86, BFHHT86], however, omitted other requirements that had to be imposed in order to achieve gamma-ray amplification based on nuclear transitions in solids. These requirements, along with several principal concepts of gamma-ray lasers, have been critically discussed in full in a recent review by Baldwin and Solem [BS95]. One of the outstanding problems is realization of the fast pumping into a laser level from the storage state.

One of the proposals reducing the pumping requirements (harmful to the crystal host) is based on the idea of 'lasing without inversion' by creating a coherent superposition of a close pair of nuclear states that both couple to an upper state. For such a level scheme, amplification (lasing) without inversion (see section 5.10 and the papers of Scully [Sy94] and Kocharovskaya and Mandel

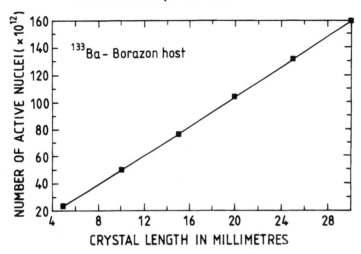

Figure 11.5. The number of active upper-level ^{133}Ba nuclei required for the super-radiance threshold, as a function of length, for unit Fresnel number.

in [75]) can be achieved in the optical region. This method, if it can be applied to nuclear transitions, may possibly offer alternative ways of designing a coherent source of gamma-rays, thus circumventing the population inversion problem emphasized earlier [BF86].

11.3 Recombination super-radiance in semiconductors

In this section we are going to discuss another system that has a continuous energy spectrum and which is a possible source of super-radiance. It is the electron–hole plasma in semiconductors.

It is known that super-radiance may be treated in a sense as a *limiting* régime of the operation of a pulsed laser, which imposes a principal restriction on the minimum duration of the generated coherent pulses. As for the semiconductor lasers, they still do not exhibit super-radiant generation. The attempts [BV89] to achieve collective spontaneous emission by interband recombination of *free* carriers apparently only add new questions to the issue. The pulses obtained had a duration of several picoseconds and intensity, i.e. energy flux density, of the order of 100 to 500 MW cm^{-2}. In the non-super-radiant régimes of special semiconductor lasers [22], pulses of approximately the same power and duration were obtained [BMN76, KIN91, LOD91].

A rigorous treatment of the super-radiance problem implies that at the initial moment $t = 0$ the coherent (resonant) external field is absent, and the non-equilibrium (inverted) state of active oscillators, i.e. of degenerate electron–hole (e–h) pairs in a semiconductor, is created by incoherent pumping. In such a system, the following three kinds of relaxation times are essential. (All quantitative estimates below are given for GaAs, with the band gap $E_g \simeq 1.4$ eV and effective masses $m_h \simeq 10$, $m_e \simeq 0.7 m_0$; m_0 is the mass of a free electron.)

(i)   The homogeneous (dynamical) time T_2 of spontaneous radiative recombination of a single e–h pair, $T_2 \simeq A_{e.h}^{-1} \sim 250$ ps, where $A_{e.h}$ is the Einstein coefficient.

(ii) The kinetic time τ_2 (homogeneous and inhomogeneous) of intraband relaxation caused by electron–phonon or electron–electron scattering of carriers within conduction and valence bands $\tau_2 \sim 0.05$–0.5 ps [BV89, E83, LO81] and [22, 17]. Since $\tau_2 \ll T_2$, the kinetic time τ_2 is reputed to be the effective time of 'transverse' relaxation of polarization.

(iii) The inhomogeneous (static) time T_2^*, defined by the bandwidth of eigenfrequencies of active oscillators, both initially inverted and self-consistently excited in the super-radiance process. In a semiconductor $T_2^* = \hbar/\Delta \sim 0.01$–$1$ ps, where Δ is determined by the Fermi energy or, more generally, by the energy bandwidth occupied by degenerate electrons.

To achieve super-radiant generation the rate of collective spontaneous recombination, which is characterized by the growth rate ω'' of self-consistent oscillations of the electromagnetic field and polarization proportional to $\exp(-i\omega' t + \omega'' t + i\mathbf{k}\mathbf{r})$, should exceed the above relaxation rates: $\omega'' \gg T_2^{-1}$, τ_2^{-1}, $(T_2^*)^{-1}$. However, as is shown in [BKK91, BKK92], this requirement normally cannot be met. The value of ω'' grows too slowly with the density of active oscillators, and when the strong kinetic relaxation is overcome, $\omega'' > \tau_2^{-1}$, we usually have $\omega'' \ll (T_2^*)^{-1}$. As a result, strong inhomogeneous broadening suppresses the collective effects in spontaneous emission of free electrons and holes in semiconductors. To achieve the super-radiant régime we have to take some special measures, e.g. suppressing the intraband relaxation by cooling to helium temperatures, or the enhancement of the recombination rate by creating the spectral singularities of e–h excitations that increase the spectral density of states. In this context it is interesting to analyse 'impure' super-radiance at the band-to-impurity-level transition in doped semiconductors [BV89, LO81], recombination super-radiance of resonant plasmons or photons at indirect band-to-band transitions in strongly non-equilibrium semiconductors with participation of phonons [LO81, KK91, XRW90], and excitonic super-radiance, especially at the resonance on specific active centres or in quantum structures with reduced dimension [LO81, MYK91, PSM91].

In [BKK91, BKK92] the simplest possibility was considered, namely, the imposition of a homogeneous magnetic field on an idealized intrinsic isotropic semiconductor. It was shown that the singularities, arising in the density of e–h states at Landau levels, strongly enhance the growth rate of collective recombination, and make recombination super-radiance feasible even for an unchanged dipole transition moment. Both linear and nonlinear stages of collective spontaneous recombination in the uni-directional (without resonator) and single-mode (in a low Q resonator) régimes were studied, and the limiting parameters of the ultra-short pulses of the recombination super-radiance were found. According to estimates for GaAs, in magnetic fields of the order of 10^5–10^6 G coherent optical pulses of duration 0.1–1 ps and intensity 100 MW cm^{-2} can be generated in samples of typical sizes 3 μm^2 × 30 μm and electron–hole density 5 × 10^{17} cm^{-3}.

The question of whether the established parameters impose fundamental limitations is equivalent, from the practical point of view, to the following one: is it possible to achieve generation of more powerful and shorter pulses in semiconductors? The answer turns out to be ambiguous, even within the employed model of an idealized semiconductor without taking into account any additional spectral peculiarities of e–h states. The definite 'no' seems to be correct only in the super-radiant régime, that is, when the generation of coherent pulses is fully independent of kinetic relaxation of electrons and holes. At the same time, for a given value of the super-radiance growth rate, ω'', we may turn to the usual laser (rate) generation régime by increasing the kinetic relaxation up to the value τ_2^{-1} with the use of phonons, plasmons, impurities or defects. Then the employment of the well known methods (mode locking, Q- and ω-modulated resonators, etc) makes it possible to obtain more powerful and shorter pulses of duration up to $\tau_2 < 0.1$ ps. Two experiments are worth mentioning here. According to [GLP89] a nonlinear saturated absorber in a hetero-laser with passive mode locking can produce very short pulses of duration of the order of 0.2 ps and peak power of the order of 20 W. In the more sophisticated scheme of a semiconductor laser ([DLF90]), supplied by quantum wells, external resonator, active mode locking and pulse compression, pulses of duration of the order of 0.5 ps and peak power of the order of 70 W were obtained. Nevertheless, in practice the duration of pulses generated in modern hetero-lasers with high-Q resonators usually exceeds 1 ps [BV89, BMN76, KIN91, LOD91] and [22]. That is why the real opportunity to shorten the pulse duration entirely as a result of the coherent process of super-radiance, i.e. collective spontaneous recombination of e–h pairs, without the use of special resonators and complicated schemes of laser generation with subsequent pulse compression, appears attractive.

11.4 Sub-radiance

There is a coherent effect that in some sense is the opposite of super-radiance, which is usually referred to as *sub-radiance* (SBR). Sub-radiance occurs when there exist metastable states of a collection of identical radiators (atoms) that are in the condition of having a long phase memory, i.e. the same condition as for the realization of super-radiance. The concept of SBR was also proposed by Dicke in his classical paper [D54]. A comprehensive discussion of SBR can be found in a series of papers by Crubellier and co-workers [CLPP85, CLP86, C87, CP87], who also observed the SBR phenomenon for the first time [PCPCL85]. If the super-radiance effect is caused by the constructive interference of atomic states, SBR demonstrates the result of destructive interference. Radiative transitions from SBR states are not allowed.

The simplest example of an SBR state is the antisymmetric state of a system of two identical two-level atoms (see section 1.1). In a collection of N identical atoms the SBR states are the eigenstates of the square of the total pseudo-spin, $\widehat{R}^2$, with eigenvalue $J(J+1)$, where $J < \frac{1}{2}N$, and of the projection of the total pseudo-spin, $\widehat{R}_3$, with minimal eigenvalue $-J$. Indeed, since $\widehat{R}^2$ is a constant of motion, radiative transitions are possible only between states that have the same J. Consequently $|J, -J\rangle$ is an excited state with energy $(\frac{1}{2}N - J)\hbar\omega_0$, which cannot spontaneously decay in the cooperative régime (here we have used the notation of Chapter 1). Certainly, for times longer than the de-phasing time interval, the coherence of the SBR state is destroyed, and the usual spontaneous decay proceeds.

The questions of most importance are:

(i)   how to populate such states;
(ii) how to release the stored energy through a cooperative radiation process.

For full inversion, only the state $|\frac{1}{2}N, \frac{1}{2}N\rangle$ is populated, and then the super-radiance cascade of transitions to the ground state follows. However, for incomplete inversion where some intermediate levels of energy $\hbar\omega_0 M$ $(0 < M < \frac{1}{2}N)$ are initially populated, it is possible to obtain an SBR state as a result of the consequent super-radiance transition. Indeed, as has been shown by Dicke [D54], the population of the level $M\hbar\omega_0$, with a full statistical mixture of states, leads mainly to populating the eigenstates of $\widehat{R}^2$ with quantum number J near M, because of their maximum statistical weight. The cascading transition is then permissible only to the level $-\hbar\omega_0 M$ and the corresponding state is just the SBR state.

In the next section we discuss a recent experiment [DVB96], where the antisymmetric state of two two-level atoms that form a stable system (a 'crystal') in a trap, was initially populated. By varying the inter-atomic separation, the collective decay rate could be changed from a value *higher* than that for the single-atom constant, γ, to a value *lower* than γ. This corresponded to the transition from a super-radiant to a sub-radiant régime of emission.

Sub-radiant states can also be populated when there is total inversion, provided that the atomic levels are degenerate [CLPP85, CLP86]. This is explained by the possibility of partially antisymmetric states in the upper level, whilst in the non-degenerate case the upper state $|\frac{1}{2}N, \frac{1}{2}N\rangle$ is purely symmetrical. The group theoretical analysis of this problem can be found in [C87, CP87, ARS88].

In the recent work of Keitel *et al* [KSS92] the principal possibility of mutual switching between SBR and super-radiance is considered for a system of N three-level atoms in the 'V' configuration. Each atom is characterized by three states: $|g\rangle$, the ground state; $|e\rangle$, the excited state with a transition frequency in the microwave domain; $|u\rangle$, the second excited state of the system in the optical region, with the transition $|u\rangle \longrightarrow |e\rangle$ forbidden. If the system is in a symmetric super-radiant state, the microwave pulse converting state '*g*' into state '*e*' is able to decrease the decay rate by a factor of the order of N, owing to the destruction of interference between the '*u*' and '*e*' states. Conversely, the partially antisymmetric state can be converted in the same manner into a super-radiant state.

The theory of this effect is worked out in [KSS92] for the ideal Dicke system of three-level atoms. If it is possible to produce such a system, it will present a device in which the release of very high energy is triggered by a low-energy microwave pulse.

11.5 Experimental observation of super-radiance and SBR of two trapped ions

The fast development of experimental techniques for trapping and detection of single ions has recently allowed DeVoe and Brewer [DVB96] to carry out the first direct observation of a microscopically resolved interaction between two laser-cooled ions in a microtrap. In the experiment, the authors directly observed the collective spontaneous emission of two trapped ions as a function of the ion–ion separation, r. The few-atom technique made it possible to avoid averaging over statistical distribution of atom–atom separations, as well as other unknown variables that are needed in experiments with macroscopic samples. Therefore, masking effects that had limited earlier experiments (see Chapter 2) were eliminated in [DVB96].

The principle of the DeVoe–Brewer experiment is to measure the spontaneous emission rate of the two-ion crystal formed in an RF Paul trap, as a function of the ion–ion separation, r. The measured rates were compared with the corresponding rates for a single trapped ion. The set-up of the experiment is shown in figure 11.6 where two $^{138}\text{Ba}^+$ ions forming a stable 'crystallic' system are cooled by using two frequency-stabilized carrier-wave (CW) dye lasers. The ionic crystal was viewed, through a sapphire window, by a microscope objective with a compensation for the spherical aberration of the window. A typical diffraction-limited image of the crystal recorded in the experiment is shown in figure 11.7.

The results of this experiment are clearly described by the QED theory of Chapter 3 within the two two-level atom model. This model produces four collective states of the system: the ground and doubly excited states as well as two intermediate, symmetric (triplet) and antisymmetric (singlet), states. The elementary theory of section 1.1 yields a vanishing decay rate via the collective antisymmetric singlet state. However, if the correct r dependence of the electromagnetic interaction between the two atoms is taken into account (see section 3.5), then both the symmetric and the antisymmetric channels have a non-vanishing decay rate (compare figures 1.1 and 3.1). According to equation (3.5.22) the effective decay rates via the symmetric $|s\rangle$ and antisymmetric $|a\rangle$

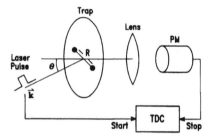

Figure 11.6. The set-up of the De Voe–Brewer experiment. A two-ion crystal is condensed in the radial plane of an 80 μm radius planar ion trap (shown schematically by a ring). The laser beam makes an angle θ with the trap axis and excites the crystal at $t = 0$. The start of the exciting pulse and the arrival of the spontaneous photons are recorded on a time-to-digital converter (TDC), which is fitted to an exponential decay [DVB96].

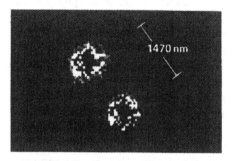

Figure 11.7. Diffraction-limited image of a two-ion ^{138}Ba$^+$ crystal with separation $r = 1470$ nm. The crystal is inclined at an angle of 40° to the (horizontal) plane containing k and the trap axis. Then the antisymmetric state ($\Phi = \pi$) is populated when the laser pulse is directed at $\theta = 16°$ to the trap axis. The set-up enables a no-free-parameter fit [DVB96].

channels are given by

$$\Gamma_\pm = \gamma \pm \tilde{\gamma} \approx \gamma(1 \pm 3/2 \sin(kr)/kr) \qquad (11.5.4)$$

see figure 3.1. Here $\tilde{\gamma}$ is the same as γ_{ij} in equation (3.4.9). In the experiments of DeVoe and Brewer the ion–ion separation was $r \approx 1500$ nm, and spontaneous emission was monitored on the $6^2P_{1/2} \rightarrow 6^2S_{1/2}$ transition at $\lambda = 2\pi/k = 493$ nm. The maximal value of $\tilde{\gamma}$ used in the equation above assumes $\sin \theta = 1$ in (3.4.9). This value could be achieved by choosing the polarization ϵ_l of the exciting laser so that $\epsilon_l \perp r$.

The multiexponential decay predicted by (3.5.22), with both fast and slow components, can be clearly identified only if $kr < 1$. With the parameters of the experiment we have $kr > 10$, $\tilde{\gamma} \ll \gamma$, and there is no appreciable contribution

of the fast decaying component. Therefore, in order to be able to observe the
r dependence of the decay rate and to compare it with the theory, DeVoe and
Brewer have chosen the following approach. Instead of placing the system in
the uppermost symmetrical state, they populated the intermediate antisymmetric
collective state. The exciting pulse with wavevector k induces dipole moments
in the pair of ions with the phase difference $\Phi = kr$. As shown in [DVB96],
for weak excitation the population of the uppermost state can be neglected and,
by choosing $\Phi = 0$ or $\Phi = \pi$, either the intermediate $|s\rangle$ or the $|a\rangle$ states can be
populated. In the experiment the r dependence of Γ_- has been measured. The
required value $\Phi \approx \pi$ corresponding to the antisymmetric state was achieved
by adjusting the angle θ between the direction of the laser beam and the trap
axis, at a fixed r.

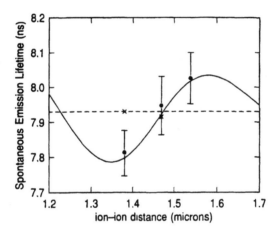

Figure 11.8. Comparison of theory (solid curve) with experimental values of lifetimes
at $r = 1380$, 1470 and 1540 nm. The ion–ion separation is independently known by
measuring the secular oscillation frequency of one ion. The lifetimes are calibrated by
comparison with 7.930 ± 0.03 nm measured for a single ion in the same trap (dashed
horizontal line). (i) Full circles with error bars: $\epsilon_l \perp r$. (ii) Crosses (error bars omitted):
$\epsilon_l \| r$. Note the strong dependence on the polarization. The point for which the lifetime
is smaller than the single atom value is a signature of super-radiance, while the point
above that value corresponds to sub-radiance [DVB96].

The resulting signal is a single exponential decay with the constant

$$\Gamma_-(r) = \gamma(1 - \alpha \sin(kr)/kr) \qquad (11.5.5)$$

where $\alpha = \frac{3}{2}$ for an ideal non-degenerate two-level system. The authors
of [DVB96] have taken into account level degeneracy and the so-called
micromotion of the ions (which is a consequence of the trapping mechanism).
These effects lead to a modified factor $\alpha = 0.33$.

The decay rate was measured by a transient technique where the ion crystal was excited by a short laser pulse at $t = 0$ and the time of arrival of spontaneous photons was recorded on a time-to-digital converter (TDC).

Fitting the data to a one-exponential formula $A \exp(-\Gamma t) + B$ yields the corresponding decay constant. The comparison of the experimental and theoretical results for $\Gamma_-(r)$ is seen in figure 11.8. We note that in order to get statistically reliable estimates of decay rates the ion crystal in the trap had to be kept under constant observation for several hours.

The dependence of the emission rate on the laser polarization could be studied as well. As shown by equation (3.4.9), if $\epsilon_l \| r$, which means that induced dipole moments are parallel to r, then $\sin \theta = 0$ and super-radiant effects are reduced by a factor of $kr > 10$. This has been confirmed by changing the polarization of the exciting wave at a fixed distance of the ions (see figure 11.8).

The apparatus is designed to be scaled to lower ion–ion separations, into the region 100 nm $< r < 1000$ nm where super-radiance is expected to dominate spontaneous emission. As the authors pointed out, the use of ion traps permits precise control of ion–ion distances so that the r dependence of the emission rate can be observed directly. This opens up opportunities in the near future for new types of microscopically resolved, detailed and precise tests of QED and other atomic interactions.

11.6   Super-radiance and non-classical light

It is interesting to point out the relation between super-radiance and such non-classical optical phenomena as sub-Poissonian statistics of photons and squeezing.

Sub-Poissonian (or photon-number squeezed) light has fewer fluctuations in the photon flux than a coherent light wave. A light field mode with a minimum uncertainty product but with an unequal variance of fluctuations in the sine and cosine field quadratures is said to be quadrature squeezed. The properties of such non-classical light fields are discussed, for example, in the review article [69] and in the topical issues [72–74]. Non-classical light is attracting significant interest both in theory and experiment since its applications are promising for quantum noise reduction in optical measurements and in communication systems.

The sub-Poissonian statistics of photons in super-radiance have been predicted in [BSH71a, BSH71b] for a pencil-shaped polyatomic system coupled to a single mode of radiation with effective damping (this system is in essence equivalent to the Dicke model). As was shown in section 1.3, for a special initial condition (binomial distribution of the atomic population near the totally inverted state) the dispersion of the number of emitted photons is less than the corresponding Poissonian value during a small fraction of the pulse duration. If the initial state is an eigenstate of $\widehat{R}_3$ with $M = 0$, then the reduction in photon noise increases [L85].

Many current theoretical investigations are related to the non-classical statistical properties of the collective emission and absorption by a system of identical two-level atoms inside a high-Q single-mode resonant cavity. This system combines the Dicke model discussed above with the well-known Jaynes–Cummings model of the interaction of one atom with a single mode of the field. A variety of squeezing effects have been established in the cavity field in the process of the radiative evolution, starting from the various collective states, $|J, M\rangle$, of the atomic subsystem (eigenstates of $\widehat{R}^2$ and $\widehat{R}_3$ with eigenvalues $J(J + 1)$ and M, see section 1.2) and both classical and non-classical states of the cavity oscillator (see e.g. [LT91, S95a]). Sub-Poissonian statistics and squeezing are present in the weak collective fluorescence field from N identical two-level atoms, driven by a strong classical light wave. Knight and Trang Quang [KTQ90] have shown that in the limit of purely collective evolution (i.e. when atomic relaxation can be neglected) the light field in the sidebands of the fluorescence triplet becomes significantly squeezed.

Non-classical effects have also been studied in a system of many atoms with three (or more) energy levels involved in the process of collective radiation. For example, Haake *et al* [HKFGR93] have presented a novel laser scheme, in which perfect squeezing of the laser field could be attained as a result of the presence of collective spontaneous emission on an adjacent optical transition in a system of many identical three-level atoms.

The simplest model of the super-radiant laser (SRL) consists of N identical atoms with the ground, intermediate and upper levels 0, 1 and 2. A classical monochromatic wave resonant with the transition $0 \leftrightarrow 2$ provides a pump. As in the conventional laser, the quantized mode of laser radiation with the annihilation and creation operators $\widehat{a}$ and $\widehat{a}^{\dagger}$ is interacting with the polarization of the atomic ensemble on the transition $2 \leftrightarrow 1$. Lasing polarization is described by the collective operators $\widehat{R}_-^{(12)}$ and $\widehat{R}_+^{(12)}$, which are defined in analogy to the case of N two-level atoms. The collective polarizations of the transitions $0 \leftrightarrow 2$ and $0 \leftrightarrow 1$ are defined in a similar way. The Hamiltonian

$$\widehat{H} = i\hbar g \left(\widehat{a}\widehat{R}_+^{(12)} - \widehat{a}^{\dagger}\widehat{R}_-^{(12)} \right) + i\hbar\Omega \left(\widehat{R}_+^{(20)} - \widehat{R}_-^{(20)} \right) \qquad (11.6.6)$$

is responsible for the interaction of the atomic ensemble with the lasing mode and the pump wave of Rabi frequency Ω. An essential feature of the model is the cooperative (super-radiant) nature of the atomic relaxation from the intermediate state 1 to the ground state 0. This relaxation process can be completely determined by the collective polarization operators $\widehat{R}_-^{(01)}$ and $\widehat{R}_+^{(01)}$ in the channel $0 \leftrightarrow 1$ in analogy to (3.4.21). The observed laser field described by $\widehat{a}_{\text{out}}$ is emitted from the cavity through the outcoupling mirror. This implies a finite damping of the cavity-field oscillator.

If certain conditions on the coupling and damping constants of the SRL are met, the output light field is perfectly squeezed: its low-frequency amplitude noise is suppressed far below the vacuum level. The Mandel parameter

$$Q = \frac{\langle (\hat{n} - \langle \hat{n} \rangle)^2 \rangle}{\langle \hat{n} \rangle} - 1 \tag{11.6.7}$$

[53] (see also section 1.3) for this field is negative, $Q \to -\frac{1}{2}$. This means that the fluctuations in the intra-cavity laser field are sub-Poissonian ($Q = 0$ for the coherent state and conventional laser far above threshold). The dependence of the linewidth of the SRL on the number of atoms N is $\Delta\nu \sim 1/N^2$, which is different from $\Delta\nu \sim 1/N$ for a conventional laser.

It is interesting to compare the super-radiant laser with a similar but non-collectively radiating one [61]. The low-frequency squeezing in the ordinary laser is far from the perfect noise reduction in an SRL. The scaling of radiation linewidth with N is also different in a non-collective laser.

The effects on super-radiance caused by a light field in the state of a squeezed vacuum have also attracted significant interest in the last few years. A squeezed vacuum is a field for which the expectation value of both quadratures of the electric field operator is zero, but the variance in one of the quadratures is smaller than that of the vacuum field. The average number of photons in a squeezed vacuum is not zero, and the distribution of phase is not uniform. In observations, such a field looks like a wide-band noise field with specific non-classical amplitude–phase correlations.

The spontaneous emission of a single atom in the presence of a squeezed vacuum was first considered by Gardiner [23]. He has shown that if the field is in a state consisting of a band of squeezed modes that are nearly resonant with the atomic transition, then the temporal evolution of the atomic state becomes different from that for ordinary spontaneous emission. The component of the polarization that is in phase with the low-noise quadrature component of the field decays more slowly than the other one. The idea that initiation of super-radiance by a squeezed vacuum instead of a true vacuum will modify the dynamics of the cooperative process has been put forward by Palma and Knight [PK89] and by Palma *et al* [PVLOKL90]. They have shown that in the presence of a squeezed vacuum the delay time statistics of super-radiance depend on the phase of the emitted light and on the amount of initial squeezing.

The cooperative radiative evolution of atoms in a broad-band squeezed vacuum has been considered by Agarwal and Puri [AP90]. These authors have determined the stationary population of the collective atomic (Dicke) states and the intensity of the emitted radiation.

Super-radiance references and further reading

1927

[D27]   Dirac P A M 1927 Emission and absorption of radiation *Proc. R. Soc.* A **114** 243–65

1930

[WW30]   Weisskopf V and Wigner E 1930 Berechnung der natürlichen Linienbreite auf Grund der Diracschen Lichttheorie *Z. Phys.* **63** 54–73; 1930 Über die natürliche Linienbreite in der Strahlung des Harmonischen Oscillators *Z. Phys.* **65** 18–29

1954

[BP54]   Bloembergen N and Pound R V 1954 Radiation damping in magnetic resonance experiments *Phys. Rev.* **95** 8–12

[D54]   Dicke R H 1954 Coherence in spontaneous radiation processes *Phys. Rev.* **93** 99–110

1958

[CWC58]   Chester P F, Wagner P E and Castle J G Jr 1958 Two-level solid state maser *Phys. Rev.* **110** 281–2

[F58]   Fain V M 1958 Quantum phenomena in the radiofrequency range *Usp. Fiz. Nauk.* **64** 273–313

[FGBGT58]   Feher B, Gordon J, Buehler E, Gere E and Thurmond C 1958 Spontaneous emission of radiation from an electron spin system *Phys. Rev.* **109** 221–2

1959

[F59]   Fain V M 1959 Contribution to the theory of coherent spontaneous radiation *Zh. Eksp. Teor. Fiz.* **36** 798–802 (in Russian)

1960

[Y60]   Yariv A J 1960 Spontaneous emission from an inverted spin system *J. Appl. Phys.* **31** 740–1

1963

[BNT63]   Baldwin G C, Neissel J P and Tonks L 1963 *Proc. IEEE* **51** 1247

1964

[D64]   Dicke R H 1964 The coherence brightened laser *Quantum Electronics III* vol 1, ed N Bloembergen and P Grivet (New York: Columbia University Press) pp 35–54

1965

[AB65]   Arecchi F T and Bonifacio R 1965 Theory of optical maser amplifiers *IEEE J. Quant. Electron.* **QE-1** 169–78

1966

[F66]   Fleck J A 1966 Quantum theory of laser radiation. I. Many–atom effects. *Phys. Rev.* **149** 309–21

1967

[MH67]   McCall S L and Hahn E L 1967 Self-induced transparency by pulsed coherent light *Phys. Rev. Lett.* **18** 908–11

1968

[ES68]   Ernst V and Stehle P 1968 Emission of radiation from a system of many excited atoms *Phys. Rev.* **176** 1456–80

[NK68]   Nagibarov V P and Kopvilem Y Kh 1968 Superadiation of a boson avalanche *Zh. Eksp. Teor. Fiz. Pis.* **54** 312–7

1969

[A69]   Agarwal G S 1969 Master equations in phase-space formulation of quantum optics *Phys. Rev.* **178** 2025–35

[BC69]   Burnham D C and Chiao R Y 1969 Coherent resonance fluorescence excited by short light pulses *Phys. Rev.* **188** 660–75

[ER69]   Eberly J H and Rehler N E 1969 Dynamics of super-radiant emission *Phys. Lett.* **29A** 142–3

[E69]   Ernst V 1969 Coherent emission of a photon by many atoms *Z. Phys.* **218** 111–28

[IL69]   Icsevgi A and Lamb W E Jr 1969 Propagation of light pulses in a laser amplifier *Phys. Rev.* **185** 517–45

[L69]   Lamb G L 1969 Π pulse propagation in a lossless amplifier *Phys. Lett.* **29A** 507–8

[MH69]   McCall S L and Hahn E L 1969 Self-induced transparency *Phys. Rev.* **183** 457–85

1970

[A70]   Agarwal G S 1970 Master equation approach to spontaneous emission *Phys. Rev.* A **2** 2038–47; 1971 *Phys. Rev.* A **3** 1783–93; 1971 *Phys. Rev.* A **4** 1791–1801

[AC70]   Arecchi F T and Courtens E 1970 Cooperative phenomena in resonant electromagnetic propagation *Phys. Rev.* A **2** 1730–7

[AK70]   Arecchi F and Kim D 1970 Line shifts in cooperative spontaneous emission *Opt. Commun.* **2** 324–8

[AKS70] Arecchi F T, Kim D and Smith I 1970 Role of quantum interference in super-radiant decay *Lett. Nuovo Cimento* **3** 598–9

[BP70]   Bonifacio R and Preparata G 1970 Coherent spontaneous emission *Phys. Rev.* A **2** 336–47

[C70]   Crisp M D 1970 Propagation of small area pulses of coherent light through a resonant medium *Phys. Rev.* A **1** 1604–11

[DK70a] Davidson R and Kozak J J 1970 On the relaxation to quantum-statistical equlibrium of the Wigner–Weisskopf atom in a one dimensional radiation field. I. A study of spontaneous emission *J. Math. Phys.* **11** 189–202

[DK70b] Davidson R and Kozak J J 1970 On the relaxation to quantum-statistical equlibrium of the Wigner–Weisskopf atom in a one dimensional radiation field. II. Finite systems *J. Math. Phys.* **11** 1420–36

[ER70]   Eberly J H and Rehler N E 1970 Super-radiant intensity fluctuations *Phys. Rev.* A **2** 1607–10

[H70]   Haken H 1970 Laser theory *Handbuch der Physik* **XXV/2c**, ed L Genzel (Berlin: Springer)

[L70a]   Lehmberg R H 1970 Radiation from an *N*-atom system. I General formalism *Phys. Rev.* A **2** 883–8

[L70b]   Lehmberg R H 1970 Radiation from an *N*-atom system. II Spontaneous emission from a pair of atoms *Phys. Rev.* A **2** 889–96

1971

[A71]   Abrosimov G V 1971 Spatial and temporal coherence of the radiation of pulsed Ne and Tl vapour lasers *Opt. Spektrosk.* **31** 106–10 (Engl. Transl. 1971 *Sov. Phys.–Opt. Spectrosc.* **31** 54–6)

[BSH71a]   Bonifacio R, Schwendimann P and Haake F 1971 Quantum statistical theory of super-radiance. I *Phys. Rev.* A **4** 302–13

[BSH71b]   Bonifacio R, Schwendimann P and Haake F 1971 Quantum statistical theory of super-radiance. II *Phys. Rev.* A **4** 854–64

[CS71]   Chang C S and Stehle P 1971 Resonant interaction between two neutral atoms *Phys. Rev.* A **4** 630–40

[Do71]   Degiorgio V 1971 Statistical properties of super-radiant pulses *Opt. Commun.* **2** 362–4

[E71]   Eberly J H 1971 Inhomogeneous broadening in super-radiant fluorescence *Nuovo Cimento* **1** 182–4

[PS71]   Pike E R and Swain S 1971 A general approach to nonequlibrium quantum statistics *J. Phys. A: Math. Gen.* **4** 555–63

[RE71]   Rehler N E and Eberly J H 1971 Super-radiance *Phys. Rev.* A **3** 1735–51

[RRW71]   Ricardo R and Wills C 1971 Description of super-radiant damping by multi-mode Maxwell–Bloch equation *Phys. Lett.* **37A** 302–5

[V71]   Varfolomeev A A 1971 Coherent effects in spontaneous emission by non-identical atoms *Zh. Eksp. Teor. Fiz* **59** 1702–10 (Engl. Transl. 1971 *Sov. Phys.–JETP* **32** 926)

1972

[ACGT72]   Arecchi F T Courtens E Gilmore R and Thomas H Atomic coherent states in quantum optics *Phys. Rev.* A **6** 2211–37

[D72]   Duncan G 1972 Application of Green's functions to radiative cooperative effects in multi-atom systems *Phys. Rev.* A **6** 947–58

[Ey72]   Eberly J H 1972 Super-radiance revisited *Am. J. Phys.* **40** 1374–83

[F72]   Fain V M 1972 *Photons and Non-Linear Media* (Moscow: Soviet Radio)

[FHM72]   Friedberg R, Hartmann S R and Manassah J T 1972 Limited super-radiant damping of small samples *Phys. Lett.* **40A** 365–6

[RT72]   Ressayre E and Tallet A 1972 Conditions for the appearance of super-radiance in the semi-classical model *Lett. Nuovo Cimento* **5** 1105–8

[SST72]   Smirnov D F, Sokolov I V and Trifonov E D 1972 Collective effects in spontaneous emission by two atoms *Zh. Eksp. Teor. Fiz.* **63** 2015–112 (Engl. Transl. 1973 *Sov. Phys.–JETP* **36** 1111–4)

[SELM72]   Stroud C R Jr, Eberly J H, Lama W L and Mandel L 1972 Super-radiant effects in systems of two-level systems *Phys. Rev.* A **5** 1094–104

[V72]   Varfolomeev A A 1972 Collective radiative processes involving non-identical atoms *Zh. Eksp. Teor. Fiz* **61** 111–8 (Engl. Transl. *Sov. Phys.–JETP* **35** 59–62)

1973

[FHM73]   Friedberg R, Hartmann S R and Manassah J T 1973 Frequency shifts in emission and absorbtion by resonant systems of two-level atoms *Phys. Rep.* C **7** 101–79

[M73]   Morawitz H 1973 Super-radiant level shift and its possible detection in a transient optical experiment *Phys. Rev.* A **7** 1148–60

[PW73]   Picard R H and Willis C R 1973 Coupled super-radiance master equations *Phys. Rev.* A **8** 1536–61

[RT73]   Ressayre E and Tallet A 1973 Effects of inhomogeneous broadening on cooperative spontaneous emission of radiation *Phys. Rev. Lett.* **30** 1239–41

[SB73]   Sanders R and Bullough R K 1973 Perturbation theory of super-radiance. super-radiant emission *J. Phys. A: Math. Gen.* **6** 1348–59

[Sm73]   Stenholm S 1973 Quantum theory of electromagnetic fields *Phys. Rep.* C **6** 1–122

[Sn73]   Schwendimann P 1973 Coherent spontaneous emission from a large system *Z. Phys.* **265** 267–84

[SHMF73]   Skribanowitz N, Herman I P, MacGillivray J C and Feld M S 1973 Observation of Dicke super-radiance in optically pumped HF gas *Phys. Rev. Lett.* **30** 309–12

[ST73]   Sokolov I V and Trifonov E D 1973 Collective spontaneous emission by polyatomic systems *Zh. Eksp. Teor. Fiz.* **65** 74–81 (Engl. Transl. 1974 *Sov. Phys.–JETP* **40** 37–40)

[Z73]   Zardecki A 1973 Brownian motion model for super-radiance fluctuation *Phys. Lett.* **44A** 363–4

1974

[AKNS74]   Ablowitz M J, Kaup D J, Newell A C and Segur H 1974 Coherent pulse propagation, a dispersive irreversible phenomenon *J. Math. Phys.* **15** 1852–8

[A74]   Agarwal G S 1974 *Quantum Optics (Springer Tracts in Modern Physics 70)* (Berlin: Springer)

[BB74]   Banfi G and Bonifacio R 1974 Super-fluorescence and cooperative frequency shift *Phys. Rev. Lett.* **33** 1259–63

[FFH74]   Flusberg A, Friedberg R and Hartmann S R 1974 Comment on the calculation of super-radiance *Phys. Rev.* A **10** 1904–5

[FH74a]   Friedberg R and Hartmann S R 1974 Temporal evolution of super-radiance in a small sphere *Phys. Rev.* A **10** 1728

[FH74b]   Friedberg R and Hartmann S R 1974 super-radiant stability in specially shaped small samples *Opt. Commun.* **10** 298–301

[JJGS74]   Jaegle P, Jamelat G, Garillon A and Surean A 1974 Super-radiant line in the soft x-ray range *Phys. Rev. Lett.* **33** 1070–5

[JM74a]   Jodoin R and Mandel L 1974 Super-radiance in an inhomogeneously broadened atomic system *Phys. Rev.* A **9** 873–84

[JM74b]   Jodoin R and Mandel L 1974 Super-radiance and optical free induction *Phys. Rev.* A **10** 1898–903

[LL74]   Lee Y C and Lee P S 1974 Coherent radiation from thin films *Phys. Rev.* B **10** 344–8

[M74]   De Martini F 1974 Dicke super-radiance from atomic system with dipole coupling *Lett. Nuovo Cimento* **10** 275–85

[MP74]   De Martini F and Preparata G 1974 Dicke super-radiance and long range dipole–dipole coupling *Phys. Lett.* **48A** 43–4

[NCB74a]   Narducci L, Coulter C and Bowden C 1974 Comments on some recent solutions of the super-radiant master equations *Phys. Rev.* A **9** 999–1003

[NCB74b]   Narducci L, Coulter C and Bowden C 1974 Exact diffusion equation for a model for super-radiant emission *Phys. Rev.* A **9** 829–45

[RT74]   Ressayre E and Tallet A 1974 Symmetrized master equation approach to spontaneous emission *Nuovo Cimento* B **21** 325–53

[SSL74]   Sargent M, Scully M O and Lamb W E 1974 *Laser Physics* (Reading, MA: Addison-Wesley)

[Ss74]   Schuurmans M 1974 Radiative decay of a pair of atoms *Phys. Lett.* **47A** 493–4

[Sv74]   Sokolov I V 1974 Angular correlations in collective spontaneous emission *Vest. Leningr. Univ.* No 4 21

[ST74]   Sokolov I V and Trifonov E D 1974 Collective spontaneous emission by polyatomic systems *Zh. Eksp. Teor. Fiz.* **65** 74–81 (Engl. Transl. *Sov. Phys.–JETP* **38** 37–40)

[ZR74]   Zakowicz W and Rzążewski K 1974 Collective radiation by harmonic oscillators *J. Phys. A: Math. Gen.* **7** 869–80

[Z74]   Zardecki A 1974 Super-radiant intensity fluctuations in neoclassical radiation theory *Can. J. Phys.* **52** 2469–78

1975

[AE75]   Allen L and Eberly J H 1975 *Optical Resonance and Two-Level Atoms* (New York: Wiley)

[BB75]   Banfi G and Bonifacio R 1975 Super-fluorescence and cooperative frequency shift *Phys. Rev.* A **12** 2068–82

[BaL75]   Benza V and Lugiato L A 1975 Adiabatic formulae for super-fluorescence *Nuovo Cimento* B **30** 80–6

[BHMS75]   Bonifacio R, Hopf F A, Meystre P and Scully M O 1975 Steady-state pulses and super-radiance in short-wavelength, swept-gain amplifiers *Phys. Rev.* A **12** 2568–73

[BL75a]   Bonifacio R and Lugiato L A 1975 Cooperative radiation processes in two-level systems: Super-fluorescence I *Phys. Rev.* A **11** 1507–21

[BL75b]   Bonifacio R and Lugiato L A 1975 Cooperative radiation processes in two-level systems: Super-fluorescence II *Phys. Rev.* A **12** 587–98

[Ga75]   Gambardella P I 1975 Non-linear Dicke super-radiance models *Physica* B **80** 550–60

[Gt75]   Grabett H 1975 Generalized Fokker–Planck equation in super-radiance theory *Z. Phys.* B **21** 99–103

[GL75]   Gronchi M and Lugiato L A 1975 *c*-number equation for oscillatory super-fluorescence *Lett. Nuovo Cimento* **14** 315–24

[HM75]   Hopf F A and Meystre P 1975 Quantum theory of a swept gain laser amplifier *Phys. Rev.* A **12** 2534–48

[II75]   Ikeda A and Ito H 1975 A quasi-probability approach to cooperative spontaneous emission *Prog. Theor. Phys.* **54** 654–68

[K75]   Korolev F A, Oditson A I, Turkin N G and Yakunin V P 1975 Spectral structure of pulse super-luminescence lines of gases *Kvant. Elektron.* **5** 413–7 (Engl. Transl. 1975 *Sov. J. Quantum Electron.* **5** 237–39)

[Lb75]   Lamb G L 1975 Amplification of coherent optical pulses *Phys. Rev. A* **11** 2052–9

[Le75]   Lee C T 1975 Diagrammatic technique for calculating matrix elements of collective operators in super-radiance *Phys. Rev. A* **12** 575–86

[M75]   Mallory W R 1975 Cooperative interaction of atoms with a radiation field *Phys. Rev. A* **11** 2036–42

[MK75]   Milonni P W and Knight P L 1975 Retarded interaction of two non-identical atoms *Phys. Rev. A* **11** 1090–2

[NS75]   Nakamura K and Sugano S 1975 Cooperative radiation from highly excited magnetic insulator *J. Phys. C: Solid State Phys.* **8** 4071–82

[Sn75]   Schwendimann P 1975 Some comments on the use of phased angular momentum states in the theory of coherent spontaneous emission from large system *Z. Phys. B* **21** 301–2

[ST75]   Sokolov I V and Trifonov E D 1975 Angular correlation of photons in super-radiance *Zh. Eksp. Teor. Fiz.* **67** 481–6 (Engl. Transl. 1975 *Sov. Phys.–JETP* **40** 238)

[Su75]   Su J 1975 Pulse propagation and super-radiance *Nuovo Cimento* B **25** 59–77

[T75a]   Thompson B V 1975 Low lying energy states of the Dicke Hamiltonian *Phys. Lett.* **54A** 271–2

[T75b]   Thompson B V 1975 Comment on the optical Dicke model *J. Phys. A: Math. Gen.* **8** 115–6

1976

[AT76]   Agarwal G S and Trivedi S S 1976 Super-radiance from three-level systems in atomic coherent state representation *Opt. Commun.* **18** 417–20

[AW76]   Aarstma T J and Wiersma D A 1976 Photon echo spectroscopy of organic mixed crystals *Phys. Rev. Lett.* **36** 1360–3

[B76]   Babiker M 1976 Super-radiance near conducting and plasma surfaces *J. Phys. A: Math. Gen.* **9** 799–813

[BMN76]   Basov N G, Molchanov A G, Nasibov A S, Obidin A Z, Pechenov A N and Popov Yu M 1976 Solid state streamer lasers *Zh. Eksp. Teor. Fiz.* **70** 1751–61 (in Russian)

[BM76]   Bonifacio R and Morawitz H 1976 Cooperative emission of an excited molecular monolayer into surface plasmons *Phys. Rev.* **36** 1559–62

[EFMSS76] Elias L R, Fairbank W M, Madey J M J, Schwettman H A and Smith T I 1976 Observation of stimulated emission of radiation by relativistic electrons in a spatially periodic transverse magnetic field *Phys. Rev. Lett.* **36** 717–20

[EK76]   Emel'yanov V I and Klimontovich Yu L 1976 Time evolution and fine structure of Dicke super-radiation and super-luminosity in the system of two-level atoms *Opt. Spektrosk.* **41** 913–9 (Engl. Transl. 1976 *Sov. Phys.–Opt. Spectrosc.* **41** 541–5)

[FMH76]   Flusberg A, Mossberg T and Hartmann S R 1976 Observation of Dicke super-radiance at 1.30 μm in atomic Tl vapor *Phys. Lett.* **58A** 373–4

[FC76]   Friedberg R and Coffey B 1976 Single-mode super-fluorescence theory compared with experiment *Phys. Rev.* A **13** 1645–7

[FH76]   Friedberg R and Hartmann S 1976 Super-radiant life time: its definitions and relation to absorbtion length *Phys. Rev.* A **13** 495–6

[GH76]   Glauber R I and Haake F 1976 Super-radiant pulses and directed angular momentum states *Phys. Rev.* A **13** 357–66

[GL76]   Gronchi M and Lugiato L A 1976 Phase-space description of oscillatory super-fluorescence *Phys. Rev.* A **13** 830–52

[GFPH76]   Gross M, Fabré C, Pillet P and Haroche S 1976 Observation of near-infrared Dicke super-radiance on cascading transitions in atomic sodium *Phys. Rev. Lett.* **36** 1035–8

[HMM76]   Hopf F, Meystre P and McLaughlin D W 1976 Quantum theory of a swept gain amplifier *Phys. Rev.* A **13** 777–83

[Le76a]   Lee C T 1976 Transition from incoherence to coherence in the spontaneous emission of extended systems *Phys. Rev.* A **13** 1657–9

[Le76b]   Lee C T 1976 Transition point of super-radiance *Phys. Rev.* A **14** 1926–8

[Ln76]   Longren K E 1976 On a solution to the single mode model of super-radiance *Phys. Rev.* A **58** 285–6

[MGF76]   MacGillivray J C and Feld M S 1976 Theory of super-radiance in an extended, optically thick medium *Phys. Rev.* A **14** 1169–89

[M76]   Milonni P W 1976 Semi-classical and quantum electrodynamical approaches in non-relativistic radiation theory *Phys. Rep.* C **25** 1–81

[NS76]   Nakamura K and Sugano S 1976 Super-fluorescence from magnetic insulator as a model crystal *J. Lumin.* **12/13** 195–200

[RT76]   Ressayre E and Tallet A 1976 Basic properties for cooperative emission of radiation *Phys. Rev. Lett.* **37** 424–7

[SWR76]   Stone I P, Witzan A and Ross I 1976 Super-radiance end energy transfer within a system of atoms *Physica* A **84** 1–47

[Su76]   Su J Y 1976 Pulse propagation and super-radiance *Nuovo Cimento* B **33** 635–64

[Si76]   Suzuki M 1976 Fluctuations in super-radiance *Physica* A **84** 48–67

[TNS76]   Takena S, Nagashina M and Sugimoto J 1976 Coherent states and photon dynamics in the Dicke model Hamiltonian *J. Phys. Soc. Japan* **41** 921–8

1977

[BPM77]   Bösiger P, Brun E and Meier D 1977 Solid state nuclear spin-flip maser pumped by dynamic nuclear polarization *Phys. Rev. Lett.* **38** 602–5

[C77]   Crubellier A 1977 Level-degeneracy effects in super-radiance theory *Phys. Rev.* A **15** 2430–8

[G77]   Gibbs H M 1977 Super-fluorescence experiments *Coherence in Spectroscopy and Modern Physics* ed F T Arecchi, R Bonifacio and M O Scully (New York: Plenum)

[GVH77]   Gibbs H M, Vrehen Q H F and Hikspoors H M J 1977 Single-pulse super-fluorescence in cesium *Phys. Rev. Lett.* **39** 547–50

[L77]   Lee C T 1977 Exact solution of the super-radiance master equation I. II *Phys. Rev. A* **15** 2019–31; 1977 *Phys. Rev. A* **16** 301–12

[LS77]   Lee H and Stehle P 1977 Independent atom picture for a super-radiant system *J. Phys. B: At. Mol. Phys.* **10** 357–9

[O77]   Orszag M 1977 Spontaneous end stimulated emission from atoms prepared in the super-radiant states *J. Phys. A: Math. Gen.* **10** 159–62

[RC77]   Rautian S G and Chernobrod B M 1977 Cooperative effect in Raman scattering *Zh. Eksp. Teor. Fiz.* **72** 1342–8 (Engl. Transl. 1977 *Sov. Phys.–JETP* **49** 705)

[RT77]   Ressayre E and Tallet A 1977 Quantum theory for super-radiance *Phys. Rev. A* **15** 2410–23

[SI77]   Steudel H 1977 Solitons in stimulated Raman scattering *Ann. Phys., Lpz.* **34** 188–202

[TM77]   Trifonov E D and Malikov R F 1977 Intensity fluctuation in super-radiance *Proc. Herzen Pedagog. Inst.* pp 5–6 (in Russian)

[TZ77]   Trifonov E D and Zaïtsev A I 1977 Semi-classical theory of cooperative radiation of a poly-atomic system *Zh. Eksp. Teor. Fiz.* **72** 1407–13 (Engl. Transl. 1977 *Sov. Phys.–JETP* **45** 739)

[VHG77]   Vrehen Q H F, Hikspoors H M J and Gibbs H M 1977 Quantum beats in super-fluorescence in atomic caesium *Phys. Rev. Lett.* **38** 763–7

1978

[BBM78]   Bösiger P, Brun E and Meier D 1978 Ruby NMR laser: A phenomenon of spontaneous self-organization of a nuclear spin system *Phys. Rev. A* **18** 671–84

[CF78]   Coffey B and Friedberg R 1978 Effect of short range Coulomb interaction on cooperative spontaneous emission *Phys. Rev. A* **17** 1033–48

[CLP78]   Crubellier A, Liberman S and Pillet P 1978 Doppler-free super-radiance experiments with Rb atoms: polarization characteristics *Phys. Rev. Lett.* **41** 1237–40

[GH78]   Glauber R and Haake F 1978 The initiation of super-fluorescence *Phys. Lett.* **68A** 29–32

[GL78]   Gronchi M and Lugiato L 1978 Quantum statistics of oscilatory super-fluorescense *Phys. Rev. A* **18** 689–96

[GRH78]   Gross M, Raimond J M and Haroche S 1978 Doppler beats in super-radiance *Phys. Rev. Lett.* **40** 1711–4

[GS78]   Grubellier A and Schweighofer M 1978 Level degeneracy effects in super-radiance theory *Phys. Rev. A* **18** 1797–815

[RT78]   Ressayre E and Tallet A 1978 Markovian model for oscillatory super-fluorescence *Phys. Rev. A* **18** 2196–203

[SPV78]   Schuurmans M F H, Polder D and Vrehen Q H F 1978 Super-fluorescence: QM derivation of Maxwell–Bloch description with fluctuating field source *J. Opt. Soc. Am.* **68** 699–700

1979

[BFN79]   Bonifacio R, Farina J D and Narducci L M 1979 Transverse effects in super-fluorescence *Opt. Commun.* **31** 377–82

[CST79]   Cahuzac Ph, Sontag H and Toschek P E 1979 Visible super-fluorescence from atomic europium *Opt. Commun.* **31** 37–41

[ES79a]   Emelyanov V I and Seminogov V N 1979 Super-radiance under Raman scattering *Zh. Eksp. Teor. Fiz.* **76** 34–45 (Engl. Transl. 1979 *Sov. Phys.–JETP* **49** 17)

[ES79b]   Emelyanov V I and Seminogov V N 1979 Effect of pump depletion on super-radiance process under Raman scattering *Kvant. Elektron.* **6** 635–8 (Engl. Transl. 1979 *Sov. J. Quantum Electron.* **9** 383)

[GHFB79]   Gounand F, Hugon M, Fournier P R and Berlande J 1979 Super-radiant cascading effects in rubidium Rydberg levels *J. Phys. B: At. Mol. Phys.* **12** 547–53

[GGFHR79]   Gross M, Goy P, Fabré C, Haroche S and Raimond J M 1979 Maser oscillation and microwave super-radiance in small systems of Rydberg atoms *Phys. Rev. Lett.* **43** 343–6

[He79]   Haake F 1979 Theory of super-fluorescence *Laser Spectroscopy IV* ed H Walther and K W Rothe (Berlin: Springer) p 451

[HKSHG79]   Haake F, King H, Schröder G, Haus J and Glauber R 1979 Fluctuations in super-fluorescence *Phys. Rev. A* **20** 2047–63

[HKSHGH79]   Haake F, King H, Schröder G, Haus J, Glauber R and Hopf F 1979 Macroscopic quantum fluctuations in super-fluorescence *Phys. Rev. Lett.* **42** 1740–3

[HB79]   Hermann J A and Bullough R K 1979 Oscillating super-fluorescence: the shape and delay of the leading pulse *Opt. Commun.* **31** 219–22

[Hf79]   Hopf F A 1979 Phase-wave fluctuation in super-fluorescence *Phys. Rev. A* **20** 2064–73

[M79]   Marek J 1979 Observation of super-radiance in Rb vapour *J. Phys. B: At. Mol. Phys.* **12** 229–34

[PRSFC79]   Pivtsov V S, Rautian S G, Safonov V P, Folin C G and Chernobrod B M 1979 I. Observation of cooperative effect in Raman scattering *Pis. Zh. Eksp. Teor. Fiz.* **30** 342–5

[PSV79]   Polder D, Schuurmans M F H and Vrehen Q H F 1979 Super-fluorescence: quantum mechanical derivation of Maxwell–Bloch description with fluctuating field source *Phys. Rev. A* **19** 1192–203

[RC79]   Rautian S G and Chernobrod B M 1979 Effect of phase self-modulation of light on cooperative Raman scattering *Kvant. Elektron.* **6** 2645–6 (Engl. Transl. 1979 *Sov. J. Quantum Electron.* **9** 1571)

[SP79]   Schuurmans M F H and Polder D 1979 Super-fluorescence and amplified spontaneous emission: a unified theory *Phys. Lett.* **72A** 306–8

[TZM79] Trifonov E D, Zaĭtsev A I and Malikov R F 1979 Super-radiance of an extended system *Zh. Eksp. Teor. Fiz.* **76** 65–77 (Engl. Transl. 1979 *Sov. Phys.–JETP* **49** 33)

[V79] Vrehen Q H F 1979 Experiments on the initiation and coherence properties of super-fluorescence *Laser Spectroscopy IV* ed H Walther and K W Rothe (Berlin: Springer) p 471

[VS79] Vrehen Q H F and Schuurmans M F H 1979 Direct measurement of the effective initial tipping angle in super-fluorescence *Phys. Rev. Lett.* **42** 224–7

1980

[AEI80] Andreev A V, Emel'yanov V I and Il'inskiĭ Y A 1980 Collective spontaneous emission (Dicke super-radiance) *Usp. Fiz. Nauk* **131** 653–94 (Engl. Transl. 1980 *Sov. Phys.–Usp.* **23** 493)

[CJSGH80] Carlson N W, Jackson D J, Schawlow A L, Gross M and Haroche S 1980 Super-radiance triggering spectroscopy *Opt. Commun.* **32** 350–4

[C80] Chernobrod B M 1980 Effect of the process of light propagation on cooperative Raman scattering *Opt. Spektrosk.* **49** 692–8 (Engl. Transl. 1980 *Sov. Phys.–Opt. Spectrosc.* **49** 378–81)

[CLP80] Crubellier A, Liberman S and Pillet P 1980 Super-radiance and sub-radiance in three-level systems *Opt. Commun.* **33** 143–8

[EZ80] Emelyanov V I and Zokhdi Z 1980 Bistability and hysteresis of static polarisation induced by laser illumination *Kvant. Elektron.* **7** 1510–5 (Engl. Transl. 1980 *Sov. J. Quantum Electron.* **10** 869–73)

[FM80] Feld M S and MacGillivray J C 1980 Super-radiance *Coherent nonlinear optics (Springer Topics in Current Physics 21)* ed M S Feld and V S Letokhov (Berlin: Springer) pp 7–57

[G80] Green J M 1980 Demonstration of laser action and super-fluorescence in CF_4 optically pumped by a continuously tunable CO_2 laser *J. Phys. D: Appl. Phys.* **13** 2217–22

[HHJSG80] Haake F, Haus J, King H, Schröder G and Glauber R 1980 Delay time statistics and inhomogeneous line broadening in super-fluorescence *Phys. Rev. Lett.* **45** 558

[MMT80] Malikov R F, Malyshev V A and Trifonov E D 1980 Semi-classical theory of cooperative emission by an extended system *Theory of Cooperative, Coherent Effects in Radiation* ed E D Trifonov (Leningrad: Herzen Pedagogical Institute) pp 3–32 (in Russian)

[MR80] Marek J and Ryschka M 1980 Quantum beats in super-radiance in sodium vapours *J. Phys. B: At. Mol. Phys.* **13** L491–6

[NW80] Nakamura K and Wahimiya S 1980 Dynamical behaviour of super-radiance in a model magnetic insulator *J. Phys. C: Solid State Phys.* **13** 3483–91

[RC80] Rautian S G and Chernobrod B M 1980 Cooperative resonance light scattering under field induced atom level splitting *Zh. Eksp. Teor. Fiz.* **78** 1365–75 (Engl. Transl. 1980 *Sov. Phys.–JETP* **51** 687)

| [SP80] | Schuurmans M F H and Polder D 1980 Theory of super-fluorescence *J. Opt. Soc. Am.* **70** 609–10 |

[SP80] Schuurmans M F H and Polder D 1980 Theory of super-fluorescence *J. Opt. Soc. Am.* **70** 609–10

[Ss80] Schuurmans M F H 1980 Super-fluorescence and amplified spontaneous emission in an inhomgeneously broadened system *Opt. Commun.* **34** 185–9

[Sv80] Shamrov N I 1980 Transformation of a semiinfinite step pulse by a resonant absorbing medium *Cooperative Coherent Effects in Radiation* ed E D Trifonov (Leningrad: Herzen Pedagogical Institute) pp 76–84 (in Russian)

[Sl80] Steudel H 1980 The initial process of super-fluorecence in microscopic description *Ann. Phys., Lpz.* **37** 57–66

[TTS80a] Trifonov E D, Troshin A S and Shamrov N I 1980 Cooperative Raman scattering *Opt. Spektrosk.* **48** 1036–9 (Engl. Transl. 1980 *Sov. Phys.–Opt. Spectrosc.* **48** 1980 567–9)

[TTS80b] Trifonov E D, Troshin A S and Shamrov N I 1980 Theory of cooperative Raman scattering *Cooperative Coherent Effects in Radiation* ed E D Trifonov (Leningrad: Herzen Pedagogical Institute) pp 43–75 (in Russian)

[VW80] Vrehen Q H F and Der Weduwe J J 1980 Experiments on quantum fluctuations in super-fluorescence *J. Opt. Soc. Am.* **70** 610–1

[Z80] Zakharov V E 1980 On propagation of an amplified pulse in a two-level medium *Pis. Zh. Eksp. Teor. Fiz.* **32** 603–7 (Engl. Transl. 1980 *Sov. Phys.–JETP Lett.* **32** 359)

1981

[AAI81] Andreev A V, Arutyunyan R V and Il'inskiĭ Yu A 1981 Kinetics of super-fluorescent decay in multi-level systems *Opt. Spekrosk.* **50** 1050–6 (Engl. Transl. 1981 *Sov. Phys.–Opt. Spectrosc.* **50** 578–81)

[AS81] Adam G and Seke J 1981 Closed equations of motion for a system of *N* two-level atoms in the case of spontaneous emission *Phys. Rev. A* **23** 3118–27

[BSG81] Baldwin G C, Solem J C and Gol'danskii V I 1981 Approaches to the development of gamma ray lasers *Rev. Mod. Phys.* **53** 687–744

[CLDT81] Chung H K, Lee J B and De Temple T A 1981 Lethargic gain *Opt. Commun.* **39** 105–9

[CLPS81] Crubellier A, Liberman S, Pillet P and Schweighofer M G 1981 Experimental study of quantum fluctuations of polarisation in super-radiance *J. Phys. B: At. Mol. Phys.* **14** L177–82

[EZ81] Egorov V S and Zatserkovnyuk N M 1981 Investigation of the characteristics of neon super-radiation at a wavelength of 614.3 nm in a pulsed discharge in a capillary *Opt. Spektrosk.* **50** 858–64 (Engl. Transl. 1981 *Sov. Phys.–Opt. Spectrosc.* **50** 469–73)

[HHKSG81] Haake F, Haus J, King H, Schröder G and Glauber R 1981 Delay time statistics of super-fluorescent pulses *Phys. Rev. A* **23** 1322–33

[LV81] Leonardi C and Vaglica A 1981 Coherent trapping and beats in super-fluorescence *J. Phys. B: At. Mol. Phys.* **14** L307–11

[LO81]   Levanyuk A P and Osipov V V 1981 Edge luminescence of direct-gap
            semiconductors *Usp. Fiz. Nauk.* **133** 427–77 (Engl. Transl. 1981 *Sov.
            Phys.–Usp.* **24** 187–215)

[MGF81a]   MacGillivray J C and Feld M S 1981 Limits of super-radiance as a
            process of achieving short pulses of high energy *Phys. Rev.* A **23**
            1334–49

[MGF81b]   MacGillivray J C and Feld M S 1981 Super-radiance in atoms and
            molecules *Contemp. Phys.* **22** 299–310

[MMT81]   Malikov R F, Malyshev V A and Trifonov E D 1981 Shape of a super-
            radiation spectrum *Opt. Spektrosk.* **51** 406–10 (Engl. Transl. 1981
            Sov. Phys.–Opt. Spectrosc. **51** 225–7)

[MGMF81]   Mattar F P, Gibbs H M, McCall S L and Feld M S 1981 Transverse
            effects in super-fluorescence *Phys. Rev. Lett.* **46** 1123–6

[RT81]   Rosenberger A T and DeTemple T A 1981 Far-infrared super-radiance
            in methyl fluoride *Phys. Rev.* A **24** 868–82

[RM81]   Ryschka M and Marek J 1981 Observation of quantum beats in super-
            radiance on the $5\,^2D_{3/2}$–$6\,^2P_{1/2}$ transition in caesium vapours *Phys.
            Lett.* **86A** 98–100

[SHLR81]   Schröder G, Haake F, Lewenstein H and Rzążewski K 1981 Spectral
            properties of super-fluorescent pulses *Opt. Commun.* **39** 194–6

[SVP81]   Schuurmans M F H, Vrehen G F H, Polder D and Gibbs H M 1981
            Super-fluorescence *Adv. At. Mol. Phys.* **17** 167–228

[VW81]   Vrehen Q H F and der Weduwe J J 1981 Quantum fluctuations in
            super-fluorescence delay times *Phys. Rev.* A **24** 2857–60

1982

[AI82]   Andreev A V and Il'inskiĭ Yu A 1982 Super-radiation of extended
            systems *Izv. Akad. Nauk. Ser. Fiz. SSSR* **46** 985–9 (Engl. Transl.
            1982 Bull. Acad. Sci. USSR Phys. Ser. **46** 150–4

[BKS82]   Bogolyubov N N Jr, Fam Le Kien F and Shumovskii A S 1983
            Dynamics of a two-level system and estimate of the relaxation time
            Teor. Mat. Fiz. **53** 108–13 (Engl. Transl. 1982 *Theor. Math. Phys.*
            53 1014–8)

[DE82]   Drummond P D and Eberly J H 1982 Transverse coherence and scaling
            in four-dimensional simulations of super-fluorescence *Phys. Rev.* A
            25 3446–8

[FSS82]   Florian R, Schwan L O and Schmid D 1982 Super-radiance and high
            gain mirrorless laser activity of O_2^- centres in KCl *Solid State
            Commun.* **42** 55–7

[GH82]   Gross M and Haroche S 1982 Super-radiance: an essay on the theory
            of collective spontaneous emission *Phys. Rep.* **93** 301–96

[HR82]   Haake F and Reibold R 1982 Two-color super-fluorescence from three-
            level systems *Phys. Lett.* **92A** 29–31

[KKN82]   Karnyukhin A V, Kuz'min R N and Namiot V A 1982 Contribution
            to semi-classical theory of super-radiation in one-dimensional
            crystalline structures *Zh. Eksp. Teor. Fiz.* **82** 561–72 (Engl. Transl.
            1982 *Sov. Phys.–JETP* **55** 334)

[L82]   Lee C T 1982 Fluctuations in tipping angles and delay times of simple super-fluorescence *Appl. Phys. Lett.* **41** 821–3

[LPV82]   Leonardi C, Peng J S and Vaglica A 1982 Beats in Dicke super-radiant emission *J. Phys. B: At. Mol. Phys.* **15** 4017–28

[LR82]   Lewenstein M and Rząžewski K 1982 Coupling between left- and right-going waves in the initial stage of super-fluorescence *Phys. Rev. A* **26** 1510–5

[MMT82]   Malikov R F, Malyshev V A and Trifonov E D 1982 Effect of relaxation on cooperative-radiation dynamics system *Opt. Spektrosk.* **53** 652–9 (Engl. Transl. *Sov. Phys.–Opt. Spectrosc.* **53** 387–91)

[MB82]   Mattar F P and Bowden C M 1982 Light control by light with an example in coherent pump dynamics, propagation, transverse and diffraction effects in three-level super-fluorescence *Appl. Phys. B* **29** 149–51

[RGGFH82]   Raimond J M, Goy P, Gross M, Fabré C and Haroche S 1982 Statistics of millimetre wave photons emitted by a Rydberg atom maser: an experimental study of fluctuations in single-mode super-radiance *Phys. Rev.* **49** 1924–7

[RY82]   Rupasov V I and Yudson V I 1982 Boundary value problems in nonlinear optics of resonant media *Kvant. Elektron.* **9** 2179–86 (Engl. Transl. 1982 *Sov. J. Quantum Electron.* **12** 1415–9)

[VG82]   Vrehen Q H F and Gibbs H M 1982 Super-fluorescence experiments *Dissipative Systems in Quantum Optics (Springer Topics in Current Physics 21)* ed R Bonifacio (Berlin: Springer) pp 111–47

1983

[B83]   Benedict M G 1983 *Synergetics: Proc. Int. Conf. (Tallinn, 1982)* (Tallinn: Valgus) pp 234–8

[BL83]   Bonifacio R and Lugiato L A 1983 Super-fluorescence in a cavity *Opt. Commun.* **47** 79–83

[BS83]   Bowden C M and Sung C C 1983 Initiation of super-fluorescence in coherently pumped three-level systems *Phys. Rev. Lett.* **50** 156–9

[E83]   Yeliseyev P G 1983 *Introduction to the Physics of Injection Lasers* (Moscow: Nauka) (in Russian)

[GZM83]   Gabitov I R, Zakharov V E and Mikhailov A V 1983 Super-fluorescence pulse shape *Pis. Zh. Eksp. Teor. Fiz.* **37** 234–7 (Engl. Transl. 1983 *JETP Lett.* **37** 279–82)

[KKN83]   Karnyukhin A V, Kuz'min R N and Namiot V A 1983 Super-radiance in a two-dimensional model *Zh. Eksp. Teor. Fiz.* **84** 878–91 (Engl. Transl. 1983 *Sov. Phys.–JETP* **57** 509)

[LRS83]   Laptev V D, Reutova N M and Sokolov I V 1983 Influence of transverse inhomogeneity of the radation field and of the active medium on the dynamics of super-radiance from an extended system *Kvant. Elektron.* **10** 2060–5 (Engl. Transl. 1983 *Sov. J. Quantum Electron.* **13** 1372)

[MBKS83]   Mantsyzov B I, Bushuev V A, Kuz'min R N and Serebryakov S L 1983
             Super-radiance in extended media *Zh. Eksp. Teor. Fiz.* **85** 862–8
             (Engl. Transl. 1983 *Sov. Phys.–JETP* **58** 498)

[MB83]   Mattar F P and Bowden C M 1983 Coherent pump dynamics,
             propagation, transverse, and diffraction effects in three-level super-
             fluorescence and control of light by light *Phys. Rev. A* **27** 345–59

[MGGRFH83]   Moi L, Goy P, Gross M, Raimond J M, Fabré C and Haroche S 1983
             Rydberg-atom masers I and II. *Phys. Rev. A* **27** 2043–64, 2065–81

[MS83a]   Mostowski J and Sobolewska B 1983 Delay time statistics in super-
             fluorescence for large Fresnel numbers *Phys. Rev. A* **28** 2573–5

[MS83b]   Mostowski J and Sobolewska B 1983 Initiation of super-fluorescence
             from a sphere *Phys. Rev. A* **28** 2943–52

[NSS83]   Naboikin Yu V, Samartsev V V and Silaeva N B 1983 Super-radiance
             in impure molecular crystals *Izv. Akad. Nauk SSSR: Ser. Fiz.* **47**
             1328–32 (Engl. Transl. *Bull. Acad. Sci. USSR Phys. Ser.* **47** 74–8)

[R83]   Richter Th 1983 Cooperative spontaneous emission from an initially
             fully excited systems of three identical two-level atoms *Ann. Phys.,
             Lpz.* **40** 234–61

[TTS83]   Trifonov E D, Troshin A S and Shamrov N I 1983 Resonant coherent
             Raman scatering in an extended system *Opt. Spektrosk.* **54** 966–71
             (Engl. Transl. 1983 *Sov. Phys.–Opt. Spectrosc.* **54** 573–6)

[WGMCCMF83] Watson E A, Gibbs H M, Mattar F P, Cormier M, Claude Y,
             McCall S L and Feld M S 1983 Quantum fluctuations and transverse
             effects in super-fluorescence *Phys. Rev. A* **27** 1427–34

[ZMT83]   Zaïtsev A I, Malyshev V A and Trifonov E D 1983 Super-radiance
             of multi-atomic system with allowance for the Coulomb interaction
             Zh. Eksp. Teor. Fiz. **84** 475–86 (Engl. Transl. 1983 *Sov. Phys.–JETP*
             57 275)

[ZLNSSS83]   Zinoviev P V, Lopina S V, Naboikin Yu V, Silaeva M B, Samartsev
             V V and Sheibut Yu E 1983 Super-radiance in a diphenyl crystal
             containing pyrene *Zh. Eksp. Teor. Fiz.* **85** 1945–52 (Engl. Transl.
             1983 *Sov. Phys.–JETP* **58** 1129)

[Z83]   Zverev V V 1983 Calculating the statistical characteristics of the super-
             radiance from an ensemble of multilevel molecules *Opt. Spektrosk.*
             54 733–6 (Engl. Transl. 1983 *Sov. Phys.–Opt. Spectrosc.* **54** 432)

1984

[BG84]   Benedict M G and Gyémánt I 1984 On the interaction of an ultrashort
             light pulse with a thin resonant medium *Acta Phys. Chem. Szeged.*
             30 115–9

[BCM84]   Bonifacio R, Casagrande F and Milani M 1984 Superradiance and
             superfluorescence in Josephon junction arrays *Phys. Lett. A* **101**
             427–31

[CG84]   Cremer C and Gerber G 1984 Observation of super-fluorescence and
             stimulated emission in BiI after non-resonant two-photon pumping
             Appl. Phys. B **35** 7–10

[CLP84]   Crubellier A, Liberman S and Pillet P 1984 Super-radiance fluctuations in a $j = \frac{1}{2} \rightarrow j' = \frac{1}{2}$ atomic system *J. Phys. B: At. Mol. Phys.* **17** 2771–80

[FSS84a]   Florian R, Schwan L O and Schmid D 1984 Time-resolving experiments on Dicke super-fluorescence of O_2^- centres in KCl. Two-colour super-fluorescence *Phys. Rev.* A **29** 2709–15

[FSS84b]   Florian R, Schwan L O and Schmid D 1984 Two-color super-fluorescence of O_2^- centres in KCl *J. Lumin.* **31&32** 169–71

[GZM84]   Gabitov I P, Zakharov V E and Mikhailov A V 1984 Non-linear theory of super-fluorescence *Zh. Eksp. Teor. Fiz.* **86** 1204–16 (Engl. Transl. 1984 *Sov. Phys.–JETP* **59** 703)

[HR84]   Haake F and Reibold R 1984 Interplay of super-fluorescence and incoherent processes in multilevel systems *Phys. Rev.* A **29** 3208–17

[L84]   Lee C 1984 Q-representation of the atomic coherent states and the origin of fluctuations in super-fluorescence *Phys. Rev.* A **30** 3308–18

[MT84]   Malikov R F and Trifonov E D 1984 Induced super-radiance in activated crystals *Opt. Commun.* **52** 74–6

[MS84a]   Mostowski J and Sobolewska B 1984 Transverse effects in stimulated Raman scattering *Phys. Rev.* A **30** 610–2

[MS84b]   Mostowski J and Sobolewska B 1984 Three-dimensional theory of initiation of super-fluorescence *Phys. Rev.* A **30** 1392–400

[RY84]   Rupasov V I and Yudson V I 1984 Contribution to the exact theory of Dicke super-radiation *Zh. Eksp. Teor. Fiz.* **86** 819–25

[Sn84]   Schwendimann P 1984 Damping effects in two-colour super-fluorescence *Opt. Acta* **31** 107–14

[Sv84a]   Shamrov N I 1984 Non-resonant cooperative Raman effect in an extended system *Opt. Spektrosk.* **57** 43–9 (Engl. Transl. 1984 *Sov. Phys.–Opt. Spectrosc.* **57** 26–9)

[Sv84b]   Shamrov N I 1984 Effects of phase relaxation in non-resonant cooperative Raman scattering *Opt. Spektrosk.* **57** 627–33 (Engl. Transl. 1984 *Sov. Phys.–Opt. Spectrosc.* **57** 380–4)

[VKLMMT84]   Varnavsky O P, Kirkin A N, Leontovich, Malikov R F, Mozharovsky A M and Trifonov E D 1984 Coherent amplification of ultrashort pulses in activated ccrystals *Zh. Eksp. Teor. Fiz.* **86** 1227–39 (Engl. Transl. 1984 *Sov. Phys.–JETP* **59** 716–23)

[ZKK84]   Zheleznyakov V V, Kocharovsky V V and Kocharovsky Vl V 1984 Super-radiance and dissipative instability in an inverted two-level system *Zh. Eksp. Teor. Fiz.* **87** 1565–81 *Sov. Phys.–JETP* **60** 897–905

[ZLNS84]   Zinoviev P V, Lopina S V, Naboikin Yu V and Silaeva M B 1984 Experimental observation of super-radiance in pyrene-doped diphenyl crystals *Fiz. Nizk. Temp.* **10** 510–7 (Engl. Transl. 1984 *Sov. J. Low Temp. Phys.* **10** 266)

1985

[AEI85]   Arutyunyan R V, Enaki N A and Il'inskiĭ Yu A 1985 Splitting of a
               chain of equations for super-radiance of a two-level system *Opt.
               Spektrosk.* **58** 252–6 (Engl. Transl. 1985 *Sov. Phys.–Opt. Spectrosc.*
               58 151–4)

[AZM85]   Avetisyan Yu A, Zaĭtsev A I and Malyshev V A 1985 On the theory
               of super-radiance of many-atom systems. Allowance for resonant
               dipole–dipole interaction of atoms *Opt. Spektrosk.* **59** 967–74 (Engl.
               Transl. 1985 *Sov. Phys.–Opt. Spectrosc.* **59** 582–6)

[BG85]   Benedict M G and Gyémánt I 1985 On the interaction of an ultrashort
               light pulse with a thin resonant medium II *Acta Phys. Chem. Szeged.*
               31 695–8

[BT85]   Benedikt M G and Trifonov E D 1985 Effect of relaxation on laser
               lethargy and super-radiance *Opt. Spektrosk.* **59** 161–6 (Engl. Transl.
               1985 *Sov. Phys.–Opt. Spectrosc.* **59** 95–8)

[BC85]   Bonifacio R and Casagrande F 1985 The superradiant régime of a free
               electron laser *Nucl. Instrum. Meth. Phys. Res.* A **239** 36–42

[BBKS85]   Bogoliubov N N, Bashkirov E K, Fam Le Kien and Shumovskii A S
               1985 Super-radiance allowing for the pumping process *Physica* A
               133 413

[CLPP85]   Crubellier A, Liberman S, Pavolini D and Pillet P 1985 Super-
               radiance and sub-radiance: I. Interatomic interference and symmetry
               properties in three-level systems *J. Phys. B: At. Mol. Phys.* **18** 3811–
               33

[HTF85]   Heinzen D J, Thomas J E and Feld M S 1985 Coherent ringing in
               super-fluorescence *Phys. Rev. Lett.* **54** 677–80

[HR85]   Haroche S and Raimond J M 1985 Radiative properties of Rydberg
               states in resonant cavities *Adv. At. Mol. Phys.* **20** 347–411

[KK85]   Kocharovsky V V and Kocharovsky Vl V 1985 Super-radiance
               statistics for three-dimensional samples *Opt. Commun.* **53** 345–8

[KoKn85]   Kudenko Y A and Kuzmin E V 1985 Superradiance for the study of
               gamma-ray lasers in a system with a change of angular momentum
               Laser and Particle Beams **3** 109–18

[L85]   Lee C T 1985 Photon antibunching in the spontaneous emission from
               Dicke's super-radiant state *Opt. Commun.* **56** 136–40

[NAZMSSS85]   Naboikin Yu V, Andrianov S N, Zinoviev P V, Malyukin Yu V,
               Samartsev V V, Silaeva M B and Sheibut Yu E 1985 Relaxation
               processes in a pyrene-doped diphenyl crystal by the Dicke super-
               radiance method *Zh. Eksp. Teor. Fiz.* **89** 1146–54 (Engl. Transl. 1985
               Sov. Phys.–JETP **62** 660–4)

[OAFPS85]   Orzechowski T J, Anderson B R, Fawley W M, Prosnitz D,
               Scharlemann E T, Yarema S M, Hopkins D, Paul A C, Sessler A M
               and Wurtele J S 1985 Microwave radiation from a high-gain free-
               electron laser amplifier *Phys. Rev. Lett.* **54** 889–92

[PCPCL85]   Pavolini D, Crubellier P, Pillet P, Cabaret L and Liberman S 1985
               Experimental evidence for subradiance *Phys. Rev. Lett.* **54** 1917–20

[PY85]   Popov V N and Yarunin V S 1985 *Collective Effects in Quantum Statistics of Matter and Radiation* (St Petersburg: St Petersburg State University Press) (in Russian)

[PG85a]   Prasad S and Glauber R J 1985 Diffractive effects in pulse propagation through a resonant medium *Phys. Rev.* A **31** 1575–82

[PG85b]   Prasad S and Glauber R J 1985 Initiation of super-fluorescence in a large sphere *Phys. Rev.* A **31** 1583–97

[SAH85]   Seke J, Adam G and Hittmair O 1985 Time-dependent projection operator technique in super-radiance theory *Acta Phys. Austriaca* **56** 225–37

[Sl85]   Steudel H 1985 Super-fluorescence from small samples *Ann. Phys., Lpz.* **42** 54–8

[ZLM85]   Zinoviev P V, Lopina S V, Malyukin Yu V, Naboikin Yu V, Silaeva N B and Samartsev V V 1985 Delay of a super-radiance pulse in a pyrene-diphenyl crystal *Zh. Prikl. Spektrosk.* **43** 587–90 (in Russian)

[ZMNS85]   Zinoviev P V, Malyukin Yu V, Naboikin Yu V and Silaeva N B 1985 Temperature dependence of super-radiance in pyrene-doped diphenyl crystals in the temperature range from 1.5 K to 60 K *Fiz. Nizk. Temp.* **11** 210–2 (Engl. Transl. 1985 *Sov. J. Low Temp. Phys.* **11** 113)

1986

[A86a]   Avetisyan Yu A 1986 On taking into account diffraction effects in super-radiance *Cooperative Radiation and Photon Statistics* (ed E D Trifonov and A S Troshin) (Leningrad: State Pedagogical Institute Press) pp 44–62 (in Russian)

[A86b]   Avetisyan Yu A 1986 Mode structure of super-radiance *Cooperative Radiation and Photon Statistics* (ed E D Trifonov and A S Troshin) (Leningrad: State Pedagogical Institute Press) pp 62–73 (in Russian)

[ATS86]   Andreev A V, Tikhomirov O Yu and Shaĭymkulov M O 1986 The dynamics of super-radiation a planar crystalline layer *Izv. Akad. Nauk. SSSR Ser. Fiz.* **50** 1507–12 (Engl. Transl. 1986 *Bull. Acad. Sci. USSR Ser. Phys.* **50** 437–52)

[AZMNRSSS86]   Andrianov S N, Zinoviev P V, Malyukin Yu V, Naboikin Yu V, Rudenko E N, Samartsev V V, Silaeva M B and Sheibut Yu E 1986 Dicke optical super-radiance of solid solution of pyrene in diphenyl upon local heating of a sample *Fiz Nizk. Temp.* **12** 985–99 (Engl. Transl. 1986 *Sov. J. Low Temp. Phys.* **12** 558)

[AZMNSSS86]   Andrianov S N, Zinvoviev P V, Malyukin Yu V, Naboikin Yu V, Samartsev V V, Silaeva N B and Sheibut Yu E 1986 Effect of non-equilibrium phonons on optical Dicke super-radiance *Zh. Eksp. Teor. Fiz.* **91** 1990–2000 (Engl. Transl. 1986 *Sov. Phys.–JETP* **64** 1180–6)

[BF86]   Baldwin G C and Feld M S 1986 Kinetics of nuclear super-radiance *J. Appl. Phys.* **59** 3665–71

[BFHHT86] Baldwin G C, Feld M S, Hannon J P, Hatton J T and Trammell G T 1986 Mössbauer–Borrmann superradiance *J. Physique* **47** NC-6 299–308

[BT86a] Benedict M G and Trifonov E D 1986 Threshold condition for super-radiance *Opt. Spektrosk.* **61** 681–2 (Engl. Transl. 1986 *Sov. Phys.–Opt. Spectrosc.* **61** 425)

[BT86b] Benedict M G and Trifonov E D 1986 Cooperative effects in reflection of ultra-short pulses from boundary of a resonant medium *Cooperative Radiation and Photon Statistics* (ed E D Trifonov and A S Troshin) (Leningrad: Herzen Pedagogical Institute) pp 3–32 (in Russian)

[CLP86] Crubellier A, Liberman S and Pillet P 1986 Super-radiance theory and random polarisation *J. Phys. B: At. Mol. Phys.* **19** 2959–71

[CP86] Crubellier A and Pavolini D 1986 Super-radiance and sub-radiance. II. Atomic system with degenerate transitions *J. Phys. B: At. Mol. Phys.* **19** 2109–38

[ELRS86] Egorov V S, Laptev V D, Reutova N M and Sokolov I V 1986 Asymmetry of super-radiance under delayed excitation *Kvant. Elektron.* **13** 878–80 (Engl. Transl. 1986 *Sov. J. Quantum Electron.* **16** 474)

[LPV86] Leonardi C, Persico F and Vetri G 1986 Dicke model and theory of driven and spontaneous emission *Riv. Nuovo Cimento* **9** 1–85

[MS86] Mostowski J and Sobolewska B 1986 Wave guide effects in super-fluorescence and stimulated Raman scattering *Phys. Rev.* A **34** 3109–20

[OACFP86] Orzechowski T J, Anderson B R, ClarkJ C, Fawley W M, Paul A C, Proznitz D, Scharlemann E T, Yarema S M, Hopkins B D, Sessler A M and Wurtele J 1986 High-efficiency extraction of micro-wave radiation from a tapered-wiggler free-electron laser *Phys. Rev. Lett.* **57** 2172–5

[RSC86] Rautian S G, Safonov V P and Chernobrod B M 1986 Cooperative light scattering *Izv. Akad. Nauk SSSR: Ser. Fiz.* **50** 640–6 (in Russian) (Engl. Transl. 1986 *Bull. Acad. Sci. USSR Phys. Ser.* **50** 14–20)

[R86] Reibold R 1986 Statistics of composed pulses in super-fluorescence *Phys. Lett.* **115A** 325–8

[Se86a] Seke J 1986 The counter-rotating terms in the super-radiance *Nuovo Cimento* D **7** 447–68

[Se86b] Seke J 1986 Exact solutions of collective spontaneous emission from an assembly of N atoms in the case of a single-atom excitation *Phys. Rev.* A **33** 739–41

[Sl86] Steudel H 1986 Super-fluorescence from a system of atoms in front of a mirror *Ann. Phys., Lpz.* **7/43** 615–20

[T86] Trifonov E D 1986 Zero-phonon lines in super-radiance. *Trudy Inst. Fiz. Akad. Nauk ESSR* **59** (Tartu, Estonia) pp 205–15 (Engl. Transl. 1988 *Zero-phonon Lines and Spectral Hole Burning in Spectroscopy and Photochemistry* ed O Sild and K Haller (Berlin: Springer) pp 157–63)

[VGKMMBT86] Varnavsky O P, Golovlev V V, Kirkin A N, Malikov R F, Mozharovsky A M, Benedict M G and Trifonov E D 1986 Coherent propagation of small-area pulses in activated crystals *Zh. Eksp. Teor. Fiz.* **90** 1596–1609 (Engl. Transl. 1986 *Sov. Phys.–JETP* **63** 937–44)

[VGGS86] Vlasov R A, Gadomsky O N, Gadomskaya I V and Samartsev V V 1986 Non-linear reflection and refraction of ultra-short light pulses by surfaces of resonant media and phase 'memory' effects 1986 *Zh. Eksp. Teor. Fiz.* **90** 1938–51 (Engl. Transl. 1986 *Sov. Phys.–JETP* **63** 1134–41)

[Z86] Zaĭtsev A I 1986 Semi-classical theory of super-radiance of small Fresnel number systems *Cooperative Radiation and Photon Statistics* ed E D Trifonov and A S Troshin (Leningrad: State Pedagogical Institute Press) pp 103–17 (in Russian)

1987

[ASS87] Andrianov S N, Samartsev V V and Sheibut Yu E 1987 Coherent spontaneous radiation of Frenkel excitons *Teor. Mater. Fiz.* **72** 286–95 (Engl. Transl. 1987 *Theor. Math. Phys.* **72** 884–91)

[ATS87a] Andreev A V, Tikhomirov O Yu and Shaĭymkulov M O 1987 Dynamics of super-radiance of bulk media *Dokl. Acad. Nauk. SSSR* **296** 77–9 (Engl. Transl. 1987 *Sov.Phys.–Dokl.* **32** 712–3)

[ATS87b] Andreev A V, Tikhomirov O Yu and Shaĭymkulov M O 1987 Bragg diffraction of super-radiance under conditions governing the crystal lattice structure *Zh. Tekh. Fiz.* **57** 1782–90 (Engl. Transl. 1987 *Sov. Phys.: Tech. Phys.* **32** 1066]

[ASSZMRS87] Andrianov S N, Samartsev V V, Sheibut Yu E, Zinoviev P V, Malyukin Yu V, Rudenko E N and Silaeva N B 1987 Two-frequency super-radiance from pyrene guest centres in biphenyl *Fiz. Nizk. Temp.* **13** 957–66 (Engl. Transl. 1987 *Sov. J. Low Temp. Phys.* **13** 545)

[BMKS87] Bogoliubov N N, Moldoyarov A A, Fam Le Kien and Shumovskii A S 1987 Behaviour of a super-radiance pulse with allowance for inhomogeneous line broadening *Teor. Mat. Fiz.* **68** 449–60 (Engl. Transl. 1987 *Theor. Math. Phys.* **68** 940–8)

[C87] Crubellier A 1987 Super-radiance and sub-radiance. III. Small samples *J. Phys. B: At. Mol. Phys.* **20** 971–96

[CP87] Crubellier A and Pavolini D 1987 Super-radiance and sub-radiance. V. Atomic cascades between degenerate levels *J. Phys. B: At. Mol. Phys.* **20** 1451–70

[DS87] Duncan A and Stehle P 1987 Super-radiance: a numerical study *Phys. Rev.* A **35** 4181–5

[MMSB87] Malcuit M S, Maki J J, Simkin D J and Boyd R W 1987 Transition from super-fluorescence to amplified spontaneous emission *Phys. Rev. Lett.* **59** 1189–92

[RY87] Rupasov V I and Yudson V I 1987 Nonlinear resonant optics of thin films, the inverse method *Zh. Eksp. Teor. Fiz.* **93** 494–9 (Engl. Transl. 1987 *Sov. Phys.–JETP* **66** 282–4)

[SSS87]   Schiller A, Schwan L O and Schmid D 1987 Large-sample effects in super-fluorescence of O_2^- centres in KCl *J. Lumin.* **38** 243–6

[SR87]   Seke J and Rattay F 1987 Exact equations of motion for density matrix elements and numerical results for many-atom spontaneous emission in a damped cavity *J. Mod. Opt.* **34** 651–63

[ZM87]   Zakharov S M and Manykin E A 1987 Phase matching of the photon echo excited in a thin resonance layer at the interface of two media *Opt. Spektrosk.* **63** 1069–72 (Engl. Transl. 1987 *Sov. Phys.–Opt. Spectrosc.* **63** 630–2)

[ZKK87]   Zheleznyakov V V, Kocharovsky V V and Kocharovsky Vl V 1986 Cyclotron super-radiance—classical analogue of Dicke super-radiance *Izv. VUZ Radiofiz.* **29** 1095–116 (Engl. Transl. 1987 *Radiophys. Quantum Electron.* **29** 830–48)

1988

[ARS88]   Alicki R, Rudnicki S and Sadovski S 1988 Symmetry properties of product states for a system of N n-level atoms *J. Math. Phys.* **29** 1158–62

[AEI88]   Andreev A V, Emelyanov V I and Ilyinskii Yu A 1988 *Cooperative Phenomena in Optics: Super-radiance. Bistability. Phase Transitions* (Moscow: Nauka) (in Russian)

[ATF88]   Andreev A V, Tikhomirov O Yu and Fedotov M V 1988 Super-radiance, super-luminescence, and self-excitation in an optical cavity *Zh. Eksp. Teor. Fiz.* **94** 40–8 (Engl. Transl. 1988 *Sov. Phys.–JETP* **67** 1757–61)

[AHM88]   Auzel F, Hubert S and Meichenin D 1988 Very low threshold CW excitation of super-fluorescence at 2.72 μm in Er_3^+ *Europhys. Lett.* **7** 459–62

[BM88]   Basharov A M and Maimistov A I 1988 Polarized solitons in three-level media *Zh. Eksp. Teor. Fiz.* **94** 61–75 (Engl. Transl. 1988 *Sov. Phys.–JETP* **67** 1741–8)

[BKP88]   Bazhanov N A, Kovalev A I, Polyakov V V, Trautman V Yu and Svedchikov A V 1988 *Preprint* 1358 (Leningrad: Institute of Nuclear Physics) (in Russian)

[BT88]   Benedict M G and Trifonov E D 1988 Coherent reflection as super-radiation from the boundary of a resonant medium *Phys. Rev.* A **38** 2854–62

[BSKL88]   Bogoliubov N N, Shumovsky A S, Kudryavtsev I K and Lyagushin S F 1988 Super-fluorescence in a system with external sources *Physica* A **151** 293–302

[BMP88]   Bonifacio R, Maroli C and Piovella N 1988 Slippage and super-radiance in the high-gain FEL. Linear theory *Opt. Commun.* **68** 369–74

[BN88]   Bonifacio R and McNeil M J W 1988 Slippage and superradiance in the high gain FEL *Nucl. Instrum. Meth. Phys. Res.* A **272** 280–8

[GSK88]   Golenischev-Kutuzov V A, Samartsev V V and Khabibulin V M 1988 *Impulse Optical and Acoustical Coherent Spectroscopy* (Moscow: Nauka)

[KPSY88]   Kiselev Yu F, Prudkoglyad A F, Shumovskiĭ A S and Yukalov V I 1988 Detection of super-radiant emission from a system of nuclear magnetic moments *Zh. Eksp. Teor. Fiz.* **94** 344–9 (Engl. Transl. 1988 *Sov. Phys.–JETP* **67** 413–5)

[KK88]   Kocharovskaya O A and Khanin Ya I 1988 Coherent amplification of an ultrashort pulse in a three level medium without population inversion *Pis'ma. Zh. Eksp. Teor. Fiz.* **48** 581–4 (Engl. Transl. 1988 *Sov. Phys.–JETP Lett.* **48** 630–4)

[Me88]   Manogue C A 1988 The Klein paradox and superradiance *Ann. Phys., NY* **181** 261–83

[PT88]   Pirogov V Yu and Trifonov E D 1988 Quantum statistical properties of super-fluorescence *Opt. Spektrosk.* **64** 836–41 (Engl. Transl. 1988 *Sov. Phys.–Opt. Spectrosc.* **64** 498–501)

[P88]   Preparata G 1988 Quantum field theory of the free electron laser *Phys. Rev.* A **38** 233–7

[Ru88]   Rupasov V I 1988 Cooperative Raman scattering of light by a concentrated atomic system *Zh. Eksp. Teor. Fiz.* **94** 84–100 (Engl. Transl. 1988 *Sov. Phys.–JETP* **67** 2002–12)

[SSS88]   Schiller A, Schwan L O and Schmid D 1988 Spatial coherence in large sample super-fluorescence of O_2^- centres in KCl *J. Lumin.* **40&41** 541–2

[SH88]   Seke J and Herfort W 1988 Deviations from the exponential decay in the case of spontaneous emission from a two level atom *Phys. Rev.* A **38** 833–40

[SBA88]   Stroud C R, Bowden C M and Allen L 1988 Self-induced transparency in self-chirped media *Opt. Commun.* **67** 387–90

[Y88]   Yukalov V I 1988 Inversion polariton filamentation in laser media *J. Mod. Opt.* **35** 35–48

[YL88]   Yang Y and Luty F 1988 Observation of a possible IR super-fluorescence from *F*-centre/CN⁻ defect pairs in CsCl *J. Lumin.* (*Int. Conf. on Luminescence—Excited State Processes in Condensed Matter (Beijing, 1987)*) **40–41** 565–6

[ZMT88]   Zaĭtsev A I, Malyshev V A and Trifonov E D 1988 Effect of inhomogeneous broadening on super-fluorescence *Opt. Spektrosk.* **65** 1018–24 (Engl. Transl. 1988 *Sov. Phys.–Opt. Spectrosc.* **65** 599–602)

1989

[A89]   Andreev A V 1989 Collective super-radiance *Zh. Eksp. Teor. Fiz.* **95** 1562–70 (Engl. Transl. 1989 *Sov. Phys.–JETP* **68** 903–7)

[ATF89]   Andreev A V, Tikhomirov O Yu and Fedotov M V 1989 Dynamics of the super-radiance of cylindrically symmetric bulk media *Dokl. Akad. Nauk USSR* **305** 1347–9 (Engl. Transl. 1989 *Sov. Phys.–Dokl.* **34** 323)

[AZMT89]   Avetisyan Yu A, Zaĭtsev A I, Malyshev V A and Trifonov E D 1989
             Diffraction effects in super-fluorescence *Zh. Eksp. Teor. Fiz.* **95**
             1541–52 (Engl. Transl. 1989 *Sov. Phys.–JETP* **68** 1989 891–7)

[BBKH89]   Bausch R, Borgs P, Kree R and Haake F 1989 Forward–backward
             synchronization in solid state super-fluorescence *Europhys. Lett.* **10**
             445–9

[BBKPTTS89] Bazhanov N A, Bulyanitsa D S, Kovalev A I, Polyakov V V, Trautman
             V Yu, Trifonov E D and Shvedchikov A V 1989 Super-radiance at
             the NMR frequency in a system of proton spins in a solid state sample
             Fiz. Tverd. Tela **31** 206–8 (Engl. Transl. 1989 *Sov. Phys.–Solid State*
             31 291–2)

[BV89]   Belenov E M and Vasilev P P 1989 Coherent effects in ultra-short light
             pulses generated by a semiconductor injection laser *Zh. Eksp. Teor.
             Fiz.* **96** 1629–37 (Engl. Transl. 1989 *Sov. Phys.–JETP* **69** 922–6)

[BNP89]   Bonifacio R, McNeil B W J and Pierini P 1989 Superradiance in the
             high gain free electron laser *Phys. Rev.* A **40** 4467–75

[GLP89]   Goldobin I S, Lukyanov V N, Plyavenek A G, Sevegin V F and
             Yakubovich S G 1989 Observation of a sub-picosecond time
             structure in hetero-laser pulses with a non-linear absorber *Kvant.
             Elektron.* **16** 1305–7 (Engl. Transl. 1989 Sov. J. Quantum Electron.
             19 841–2)

[H89]   Harris S E 1989 Laser without inversion: interference of life-time-
             broadened resonances *Phys. Rev. Lett.* **62** 1033–6

[K89]   Kaneva E N 1989 Statistics of counter-propagating waves in a 1D super-
             radiation problem *Opt. Spektrosk.* **67** 127–31 (Engl. Transl. 1989
             Sov. Phys.–Opt. Spectrosc. **67** 71–4)

[KM89]   Knoester J and Mukamel S 1989 Intermolecular forces, spontaneous
             emission and superradiance in a dielectric medium—polariton
             mediated interactions *Phys. Rev.* A **40** 7065–80

[KS89]   Kuprionis Z and Svedas V 1989 Investigating the properties of super-
             radiance of Na atoms. One step excitation *Litov. Fiz. Sbor.* **29** 583–9
             (Engl. Transl. 1989 *Lith. Phys. J.* **29** 42–7)

[MBEZMS89]   Manykin E A, Basharov A M, Elyutin S O, Zakharov S M, Maimistov
             A I and Sklyarov Yu M 1989 Resonant non-linear optics of thin films
             Izv. Akad. Nauk SSSR: Ser. Fiz. **53** 2350–7 (Engl. Transl. 1989 *Bull.
             Acad. Sci. USSR Phys. Ser.* **53** 74–81)

[MMRB89]   Maki J J, Malcuit M S, Raymer M G and Boyd R W 1989 Influence of
             collisional de-phasing processes on super-fluorescence *Phys. Rev.* A
             40 5135–42

[PK89]   Palma G M and Knight P L 1989 Phase sensitive population decay:
             the two-atom Dicke model in a broad-band squeezed vacuum *Phys.
             Rev.* A **39** 1962–9

[RRB89]   Rzążewski K, Raymer M G and Boyd R W 1989 Delay time statistics
             of cooperative emission in the presence of homogeneous line
             broadening *Phys. Rev.* A **39** 5789

[SSS89a]   Schmid D, Schwan L O and Schiller A 1989 O_2^--defects: the first
             solid state model for super-fluorescence *Cryst. Latt. Defects Amorph.
             Mater.* **18** 27–42

[SSS89b]   Schwan L O, Schwendimann P and Sigmund E 1989 Correlations in extended high-density super-fluorescence: a self-organized distributed feedback laser *Phys. Rev.* A **40** 7093–6

[SH89]   Seke J and Herfort W 1989 Finite-time deviations from exponential decay in the case of spontaneous emission from a two-level hydrogenic atom *Phys. Rev.* A **40** 1926–40

[SR89]   Seke J and Rattay R 1989 Influence of the Fock state field on the many atom radiation process in a damped cavity *Phys. Rev.* A **39** 171–83

[SM89a]   Spano F C and Mukamel S 1989 Super-radiance in molecular aggregates *J. Chem. Phys.* **91** 683–700

[SM89b]   Spano F C and Mukamel S 1989 Nonlinear susceptibilities of molecular aggregates: enhancement of $\chi^{(3)}$ by size *Phys. Rev.* A **40** 5783–801

[TOZH89]   Thorn J R G, Osaka Y, Zeigler J M and Hochstrasser R M 1989 Two-photon spectroscopy in polysilanes *Chem. Phys. Lett.* **162** 455–60

[ZKK89a]   Zheleznyakov V V, Kocharovsky V V and Kocharovsky Vl V 1989 Polarization waves and super-radiance in active media *Usp. Fiz. Nauk* **159** 193–260 (Engl. Transl. 1989 *Sov. Phys.–Usp.* **32** 835–70)

[ZKK89b]   Zheleznyakov V V, Kocharovsky V V and Kocharovsky Vl V 1989 Super-radiance: the approach of electrodynamics of continuous active media *Rev. Rep. Phys. Non-Linear Waves* **2** 136

1990

[AP90]   Agarwal G S and Puri R R 1990 Cooperative behavior of atoms irradiated by broad-band squeezed light *Phys. Rev.* A **41** 3782–91

[AM90]   Agranovich V M and Mukamel S 1990 Note on super-radiance of excitonic molecules *Phys. Lett.* A **147** 155–60

[A90]   Andreev A V 1990 Optical super-radiance: new ideas and new experiments *Usp. Fiz. Nauk* **160** 1–46 (Engl. Transl. 1990 *Sov. Phys.–Usp.* **33** 997–1020)

[ABST90]   Arutyunyan R V, Bol'shov L A, Strizhov V F and Tkalya E V 1990 Excitation and super-radiance of isomeric nuclei in a laser plasma in a high intensity laser radiation field *Kvant. Elektron.* **17** 496–500 (Engl. Transl. 1990 *Sov. J. Quantum Electron.* **20** 430–4)

[AAL90]   Azarenkov A N, Al'tshuler G B and Lebed'ko V E 1990 High intensity super-radiance in a two-dimensional system of active centres *Kvant. Elektron.* **17** 200–2 (Engl. Transl. 1990 *Sov. J. Quantum Electron.* **20** 155)

[BBZKMT90]   Bazhanov N A, Bulyanitsa D S, Zaĭtsev A I, Kovalev A I, Malyshev V A and Trifonov E D 1990 Super-radiance in a system of proton spins *Zh. Eksp. Teor. Fiz.* **97** 1995–2004 (Engl. Transl. 1990 *Sov. Phys.–JETP* **70** 1128–33)

[BMTZ90]   Benedict M G, Malyshev V A, Trifonov E D and Zaĭtsev A I 1990 *Opt. Spektrosk.* Mirrorless bistability and ultrashort light pulses through a thin layer with resonant two-level atoms **68** 812–7 (Engl. Transl. 1990 *Sov. Phys.–Opt. Spectrosc.* **68** 473–6

[BT90]   Benedict M G and Trifonov E D 1989 On boundary value problems in coherent resonant optics *Potsdamer Forsch.* B **64** 9–38

[BCCSPP90] Bonifacio R, Casagrande F, Cerchioni G, de Salvo Souza L, Pierini P and Piorella N 1990 Physics of the high gain FEL and super-radiance *Riv. Nuovo Cimento* **13** 1–69

[CCB90] Cai S Y, Cao J and Bhattacharjee A 1990 Linear theory of super-radiance in a free-electron laser *Phys. Rev.* A **42** 4120–26

[DLF90] Delfyett P J, Lee C-H, Florez L Tl, Stoffel N G, Guritter T J, Andreadakis N C, Alphonse G A and Connolly J C 1990 Generation of subpicosecond high-power optical pulses from a hybrid mode-locked semiconductor laser *Opt. Lett.* **15** 1371–3

[KT90] Kaneva E N and Trifonov E D 1990 The self-organisation of polarisation in super-radiant systems *Potsdamer Forschungen* B **64** 51–63

[KTQ90] Knight P L and Trang Quang 1990 Sub-Poissonian statistics and squeezing in fluorescence from N atoms in a cavity *Phys. Rev.* A **41** 6255–60

[ODW90] Overwijk M H F, Dijkhuis J I and de Wijn H W 1990 Super-fluorescence and amplified spontaneous emission of 29 cm^{-1} phonons in ruby *Phys. Rev. Lett.* **65** 2015–8

[PVLOKL90] Palma G M, Vaglica A, Leonardi C, De Oliveira F A M and Knight P L 1990 Effects of broad-band squeezing on the quantum onset of super-radiance *Opt. Commun.* **79** 377–80

[Pa90] Preparata G 1990 Superradiance effects in a gravitational antenna *Mod. Phys. Lett.* A **5** 1–5

[SLT90a] Samson A M, Logvin Yu A and Turovets S I 1990 Interaction of short optical pulses with an inverted thin film consisting of two-level atoms *Kvant. Elektron.* **17** 1223–6 (Engl. Transl. 1990 *Sov. J. Quantum Electron.* **20** 1133–6)

[SLT90b] Samson A M, Logvin Yu A and Turovets S I 1990 Induced super-radiance in a thin film of two-level atoms *Opt. Commun.* **78** 208–12

[Sv90] Shirokov M I 1990 Para-Fermi operators in the theory of super-radiance *J. Phys. B: At. Mol. Phys.* **23** 1923–31

[Sl90] Steudel H 1990 Inverse scattering theory of super-fluorescence *Quantum Opt.* **2** 387–407

[TTZH90] Tilger A, Trommsdorf H P, Zeigler J M and Hochstrasser R M 1990 Excitation dynamics in polysilanes *J. Lumin.* **45** 373–6

[T90] Trache M 1990 Statistical fluctuations in equilibrium super-radiance *Europhys. Lett.* **12** 501–6

[XRW90] Xie H, Ram-Mohan L R and Wolff P A 1990 Electron–hole recombination in narrow band gap $Hg_{1-x}Cd_xTe$ and stimulated emission of LO phonons *Phys. Rev.* B **42** 3620–7

1991

[AZMT91] Avetisyan Yu A, Zaïtsev A I, Malyshev V A and Trifonov E D 1991 Diffraction angular structure of super-fluorescence *Opt. Spektrosk.* **70** 1345–8 (Engl. Transl. 1991 *Sov. Phys.–Opt. Spectrosc.* **70** 786)

[BKK91]   Belyanin A A, Kocharovsky V V and Kocharovsky Vl V 1991 Super-radiance phenomenon in semiconductor magneto-optics *Solid State Commun.* **870** 243–6

[BMTZ91]   Benedict M G, Malyshev V A, Trifonov E D and Zaĭtsev A I 1991 Reflection and transmission of ultrashort light pulses through a thin resonant medium: local-field effects *Phys. Rev.* A **43** 3845–53

[GS91]   Ginzburg N S and Sergeev A S 1991 Super-radiance in layers of excited classical and quantum oscillators *Zh. Eksp. Teor. Fiz.* **99** 438–46 (Engl. Transl. 1991 *Sov. Phys.–JETP* **72** 243)

[HF91]   Hongen G and Fanxin C 1991 Observation of super-radiance from F_2^- and F_3^+ centres in LiF crystals *Chinese J. Lasers* **18** 72–4 (Engl. Transl. 1991 *Chinese Phys.–Lasers* **18** 72)

[KK91]   Kalafati Yu D and Kokun V A 1991 Picosecond relaxation processes in a semiconductors laser excited by a powerful ultra-short light pulse *Zh. Eksp. Teor. Fiz.* **99** 1793–1803 (Engl. Transl. 1991 *Sov. Phys.–JETP* **72** 1003–8)

[K91]   Kaneva E N 1991 Self-organisation in super-radation *Opt. Spektrosk.* **70** 69–73 (Engl. Transl. 1991 *Sov. Phys.–Opt. Spectrosc.* **70** 94–7)

[KIN91]   Kawaguchi H, Iwata H and Tan-No N 1991 Picosecond optical pulse generation from mutually coupled laser diodes by mode-locking *Conf. Lasers and Electro-Optics* (Washington, DC: Optical Society of America) pp 122–4

[L91]   Lambruschini C L P 1991 A semi-classical approach for two-colour four-mode solid state super-fluorescence *Opt. Commun.* **85** 291–8

[LOD91]   Levi A F J, O'Gorman J, Dykaar D *et al* 1991 Ultrashort pulse generation using intracavity loss-modulated quantum well lasers *Conf. Lasers and Electro-Optics* (Washington, DC: Optical Society of America) p 120

[LT91]   Liu W S and Tombesi P 1991 Squeezing in a super-radiant state *Quantum Opt.* **3** 93–104

[LST91]   Logvin Yu A, Samson A M and Turovets S I 1991 Instabilities and chaos in bistable thin films of two-level atoms *Opt. Commun.* **84** 99–103

[Ls91]   Los V F 1991 Collective relaxation of spins and superradiance of magnons *J. Phys.: Condens. Matter* **3** 7027–38

[M91]   Malyshev V A 1991 Localisation length of a 1-D exciton and temperature dependence of the radiative lifetime in frozen dye solutions with J aggregates *Opt. Spektrosk.* **71** 873–5 (Engl. Transl. 1991 *Sov. Phys.–Opt. Spectrosc.* **71** 505–6)

[MTZB91]   Malyshev V A, Trifonov E D, Zaĭtsev A I and Benedict M G 1991 Mirrorless optical bistability of thin layer with resonant two-level atoms in ultra short light pulse field *Non-Linear Dynamics in Optical Systems* vol 7, ed N B Abraham, E Garmire and P Mandel (Washington, DC: Optical Society of America) pp 231–5

[MYHK91]   Misawa K, Yao H, Hayashi T and Kobayashi T 1991 Super-radiance quenching by confined acoustic phonons in chemically prepared microcrystallites *J. Chem Phys.* **94** 4131–40

[MYK91]   Misawa K, Yao H and Kobayashi T 1991 Quenching of the super-radiance decay by coupled acoustico-phonons in CdS microcrystals *J. Lumin.* **48&49** 269–72

[NUY91]   Nishikawa T, Uesugi N and Yumoto J 1991 Parametric super-fluorescence in $KTiOPO_4$ crystals pumped by 1 ps pulses *Appl. Phys. Lett.* **58** 1943–5

[P91]   Piovella N 1991 A hyperbolic secant solution for the super-radiance in free electron lasers *Opt. Commun.* **83** 92–6

[PSM91]   Potemski M, Stepniewski R, Maan J C, Martinez G, Wyder P and Etienne B 1991 Auger recombination within Landau levels in a two-dimensional electron gas *Phys. Rev. Lett.* **66** 2239–42

[SKKK91]   Scherrer D P, Kälin A W, Kesselring R and Kneubühl F K 1991 Ultrashort far-infrared super-radiant emissions optically pumped by truncated hybrid 10 μm CO_2 laser pulses *Appl. Phys. B* (*Photophys. Laser Chem.*) **53** 250–2

[Sn91]   Schwan L O 1991 Spontaneous emission of coherent light in highly excited media *J. Lumin.* **48&49** 289–94

[Sy91]   Scully M O 1991 Enhancement of the index of refraction via quantum coherence *Phys. Rev. Lett.* **67** 1855–8

[Se91a]   Seke J 1991 Analysis of the projection-operator methods: inconsistencies and their removal *J. Phys. A: Math. Gen.* **24** 2121–9

[Se91b]   Seke J 1991 Many-atom Jaynes–Cummings model without rotating wave approximation *Quantum Opt.* **3** 127–36

[Te91]   Trache M 1991 Quantum fluctuations in equlibrium superradiance *J. Mod. Opt* **38** 2361–70

[YCT91]   You L, Cooper J and Tippenbach M 1991 Alternative treatment for the initiation of super-fluorescence *J. Opt. Soc. Am. B* **8** 1139–48

1992

[BSK92]   Bakos J S, Scherrer D P and Kneubühl F K 1992 Observation of two-photon noise-initiated fluctuations in far-infrared Dicke's super-radiance *Phys. Rev. A* **46** 410–3

[BKK92]   Belyanin A A, Kocharovsky V V and Kocharovsky Vl V 1992 Recombination super-radiance in semiconductors *Laser Phys.* **2** 952–64

[BCSPP92]   Bonifacio R, Corsini R, De Salvo L, Pierini P and Piovella N 1992 New effects in the physics of high-gain free-electron lasers; a proposed experiment and possible applications *Riv. Nuovo Cimento* **15** 1–52

[BGJLP92]   Bouchiat M A, Guena J, Jacquier P, Lintz M and Pottier L 1992 From linear amplification to triggered super-radiance: illustrative examples of stimulated emission and polarization spectroscopy for sensitive detection of a pulsed excited forbidden transition *J. Physique* II **2** 727–47

[DGP92]   Delguidice E, Giunta B and Preparata G 1992 Superradiance and ferromagnetic behavior *Nuovo Cimento* D **14** 1145–55

[HKS92]   Haake F, Kolobov M I and Steudel H 1992 Dynamical models for forward–backward coupling in super-fluorescence *Opt. Commun.* **92** 385–92

[JS92]   Jansen D and Stahl A 1992 Correlation between counter-propagating pulses in super-fluorescence *Europhys. Lett.* **18** 33–8

[KSS92]   Keitel C H, Scully M O and Süssman G 1992 Triggered super-radiance *Phys. Rev.* A **45** 3242–9

[Kr92]   Knoester J 1992 Exciton superradiance in molecular-crystal slabs *J. Lumin.* **53** 101–4

[MT92]   Michalska-Trautman R 1992 Conservation laws in super-fluorescence *Phys. Rev.* A **46** 7270–6

[RB92]   Rai I and Bowden C M Quantum statistical analysis of super-fluorescence and amplified spontaneous emission in dense media *Phys. Rev.* A **46** 1522–9

[Se92]   Seke J 1992 Effect of the counter-rotating terms in the many-atom Jaynes–Cummings model with cavity losses *Opt. Lett.* **17** 355–7

[WD92]   Wu Ding 1991/92 Super-radiance in an optical klystron *Nucl. Instrum. Meth. Phys. Res.* A **318** 588–91

[ZKK92]   Zheleznyakov V V, Kocharovsky V V and Kocharovsky Vl V 1991 Generation of ultra-short, high-power pulses in super-radiating systems with complex energy spectrum *Izv. Ross. Akad. Nauk: Ser. Fiz.* **56** 140–57 (Engl. Transl. 1992 *Bull. Russ. Acad. Sci. Phys.* **56** 1395–410)

1993

[AEI93]   Andreev A V, Emel'yanov V I and Il'inskii Yu A 1993 *Cooperative Effects in Optics* (Bristol: Institute of Physics)

[AP93a]   Andreev A V and Polevoy P V 1993 Generation of ultrashort pulses by 2-component super-radiating media *Kvant. Elektronika* **20** 991–8 (Engl. Transl. 1993 *Sov. J.–Quantum Electron.* **23** 863–9)

[AP93b]   Andreev A V and Polevoy P V 1993 Superradiance of 2-component media *JETP Lett.* **57** 107–10

[DEMP93]   Delgiudice E, Enz C P, Mele R and Preparata G 1993 Solid He-4 as an ensemble of superradiating nuclei *Nuovo Cimento* D **15** 1415–20

[HKFGR93]   Haake F, Kolobov M, Fabré C, Giacobino E and Reynaud S 1993 Super-radiant laser *Phys. Rev. Lett.* **71** 995–8

[KLG93]   Kweon G I, Lawandy N M and Gomes A S L 1993 Superradiance effects on the infrared absorption and vibrational temperature of optically pumped molecules *Infrared Phys.* **34** 629–34

[LP93]   Lyakhov G A and Popyrin S L 1993 Effect of multimode dissipation and nonuniform population in two-level systems on the temporal structure of the superradiation pulse *Opt. Spektrosk.* **75** 442–4 (Engl. Transl. 1993 *Sov. Phys.–Opt. Spectrosc.* **75** 261–2)

[MMV93]   Malikov R F, Malyshev V A and Varnavsky O P 1993 Coherent amplification and compression of light pulses with a narrow spectrum in an inhomogeneously broadened two-level medium with population inversion *Opt. Spektrosk.* **74** 331–41 (Engl. Transl. 1993 *Sov. Phys.–Opt. Spectrosc.* **74** 203–9)

[M93]   Malyshev V A 1993 Localisation length of a one-dimensional exciton and low temperature behaviour of radiative lifetime of J aggregated dye solutions *J. Lumin.* **55** 225–30

[MTH93]   Manabe Y, Tokihiro T and Hanamura E 1993 Superradiance of interacting Frenkel excitons in a linear system *Phys. Rev.* B **47** 2019–30; 1993 *Phys. Rev.* B **48** 2773–6

[Se93]   Seke J 1993 The effect of the counter-rotating terms in the Dicke model *Phys. Rev.* A **193** 587–602

[TM93]   Tachiazawa T and Maeda K 1993 Superradiation in the Kerr–De Sitter space time *Phys. Lett.* A **172** 325–30

1994

[AP94]   Andreev A V and Polevoy 1994 Superradiance of 2-component media *Quantum Opt.* **6** 57–72

[A94]   Auerbach N 1994 'Super-radiant' states in intermediate energy nuclear physics *Phys. Rev.* C **50** 1606–10

[BR94]   Bartholdtsen D and Rinkleff R H 1994 Superfluorescent transitions in an external magnetic field *Z. Phys.* D **30** 265–73

[BPJY94]   Bjork G, Pau S, Jacobson J and Yamamoto Y 1994 Wannier exciton superradiance in a quantum-well microcavity *Phys. Rev.* B **50** 17 336–48

[BDPPP94a] Bonifacio R, Desalvo L, Pierini P, Piovella N and Pellegrini C 1994 A study of linewidth, noise and fluctuations in a FEL operating in SASE *Nucl. Instrum. Meth. Phys. Res.* A **341** 181–5

[BDPPP94b] Bonifacio R, Desalvo L, Pierini P, Piovella N and Pellegrini C 1994 Spectrum, temporal structure, and fluctuations in a high-gain free-electron laser starting from noise *Phys. Rev. Lett.* **73** 70–3

[BPPRT94] Bonifacio R, Pierini P, Pellegrini C, Rosenzweig J and Travish G 1994 Slippage, noise and superradiant effects in the UCLA FEL experiment *Nucl. Instrum. Meth. Phys. Res.* A **341** 285–8

[BIM94]   Boursey E, Itji H and Meziane M 1994 Coherent ringing in Te-130(2) superfluorescent emission *IEEE J. Quantum Electron.* **30** 2653–6

[DGJ94]   Drobny G, Gantsog T and Jex I 1994 Phase properties of a field mode interacting with N two-level atoms *Phys. Rev.* A **49** 622–5

[GGL94]   Gagel R, Gadonas R and Laubereau A 1994 Evidence for biexcitons and dynamic Stark effect in J-aggregates from femtosecond spectroscopy *Chem. Phys. Lett.* **217** 228–33

[GD94]   Genack A Z and Drake J M 1994 Laser physics—scattering for super-radiation *Nature* **368** 400–1

[GNS94a]   Ginzburg N S, Novozhilova Y V and Sergeev A S 1994 Superradiance of ensembles of classical electron-oscillators as a method for generation of ultrashort electromagnetic pulses *Nucl. Instrum. Meth. Phys. Res.* A **314** 230–3

[GNS94b]   Ginzburg N S, Novozhilova Y V and Sergeev A S 1994 Cyclotron superemission of electron clusters as the method for generation of ultrashort electromagnetic pulses *Zh. Tekh. Fiz.* **64** 83–95

[GKK94]   Golubyatnikova E R, Kocharovsky V V and Kocharovsky Vl V 1994 Non-linear Bragg scattering and mode super-radiance in an open Fabry–Perot resonator *Kvant. Elektron.* **24** 849–54 (Engl. Transl. 1994 *Sov. J. Quantum Electron.* **24** 791–6)

[GHLLZP94]   Gover A, Hartemann F V, Lesage G P, Luhmann N C, Zhang R S and Pellegrini C 1994 Time and frequency domain analysis of superradiant coherent synchrotron radiation in a waveguide free-electron laser *Phys. Rev. Lett.* **72** 1192–5

[KSTY94]   Kamalov V F, Struganova I A, Tani T and Yoshihara K 1994 Temperature dependence of superradiant emission of PIC J-aggregates *Chem. Phys. Lett.* **220** 257–61

[KS94a]   Kobayashi S and Sasaki F 1994 Dynamical properties of large coherence length excitons in PIC J-aggregates *Japan. J. Appl. Phys. Part 2 Lett.* **34** Suppl. 34-1 279–81

[KS94b]   Kobayashi S and Sasaki F 1994 Excitation wavelength-dependent superradiant decay in PIC J aggregates *J. Lumin.* **60** 824–6

[KM94]   Koga J and Maeda K 1994 Superradiance around rotating dilatonic black holes *Phys. Lett.* B **340** 29–34

[KH94]   Kumarakrishnan A and Han X L 1994 Investigations of superfluorescent cascades *Opt. Commun.* **109** 348–60

[MTS94]   Malyshev V A, Trifonov E D and Schwan L O 1994 Self-locking of counter-directional pulses of SF at high density of active centres *Opt. Spekrosk.* **76** 524–8 (Engl. Transl. 1994 *Sov. Phys.–Opt. Spectrosc.* **76** 470–4)

[MG94]   Manassah J T and Gross B 1994 Effects of different broadening mechanisms on pulse amplification is the superradiant regime *Opt. Commun.* **113** 213–25

[MBR94]   McPherson A, Boyer K and Rhodes C K 1994 X-ray superradiance from multiphoton excited clusters *J. Physique* B **27** L637–41

[MTB94]   Meziane J, Itji H, Topouzkhanian A and Boursey E 1994 Superradiance spectroscopy—the Te2 isotopomers *J. Physique IV* **4** 705–8

[RDG94]   Reichertz L A, Dutz H, Goertz S, Kramer D, Meyer W, Reicherz G, Thiel W and Thomas A 1994 Polarization reversal of proton spins in a solid-state target by superradiance *Nucl. Instrum. Meth. Phys. Res.* A **340** 278–82

[RY94]   Reshetov V A and Yevseyev I V 1994 Polarization properties of superradiance in the case of pumping pulses with small areas *Laser Phys.* **4** 109–11

[Sy94]   Scully M O 1994 Resolving conundrums in lasing without inversion via exact solutions to simple models *Quantum Opt.* **6** 203–15

[Se94]   Seke J 1994 Spontaneous decay of an atomic state in non-relativistic QED: a complete treatment including gauge invariance, renormalisation and non-Markovian time behaviour *J. Phys. A: Math. Gen.* **27** 263–74

[SL94]   Steudel H and Leonhardt I 1994 Superfluorescence with pumping self-similar solution *Opt. Commun.* **107** 88–92

[T94]   Trifonov E D 1994 Internal reflection as the reason of the correlation of counter-propagating pulses in SR *Opt. Spektrosk.* **77** 61–4 (Engl. Transl. 1994 *Opt. Spectrosc.* **77** 51–3)

[VEFC94]   Vasilev V V, Egorov V S, Fedorov A N and Chekhonin I A 1994 Lasers and laser systems based on cooperative effects in optically dense resonance media without population inversion *Opt. Spektr.* **76** 146–60 (Engl. Transl. 1994 *Opt. Spectrosc.* **76** 134–49)

1995

[AP95]   Andreev A V and Polevoy P V 1995 Superradiance in IR and optical transitions of molecules *Infrared Phys. Technol.* **36** 15–23

[BS95]   Baldwin G C and Solem J C 1995 Recent proposals for gamma-ray lasers *Laser Phys.* **5** 231–9

[BKN95]   Balko B, Kay I W and Neuberger J W 1995 Nuclear super-fluorescence: A feasibility study based on the generalized Haake–Reibold theory *Phys. Rev.* B **52** 858–69

[BKSS95]   Balko B, Kay I W, Silk J D and Sparrow D A 1995 Pumping requirements for achieving nuclear superfluorescence *Laser Phys.* **5** 355–61

[BN95]   Benedict M G and Németh I 1995 An inverse scattering method in resonant optical problems with external triggering *J. Mod. Opt.* **42** 2265–73

[Br95]   Brewer G 1995 Two-ion super-radiance theory *Phys. Rev.* A **52** 2965–70

[GK95]   Gadomsky O N and Krutitsky K V 1995 Annihilation superradiation *Laser Phys.* **5** 379–95

[IMOB95]   Itji H, Meziane J, Oullemine S and Boursey E 1995 Optical feedback effects in Te-2 superfluorescence *Opt. Commun.* **117** 251–5

[JQ95]   John S and Quang T 1995 Localization of superradiance near a photonic band gap *Phys. Rev. Lett.* **74** 3419–22

[JSV95]   Jutte M, Stolz H and Vonderosten W 1995 Propagation of coherent small-area light pulses in CdS at the I–2 bound exciton. *Europhys. Lett.* **32** 161–6

[KB95]   Kalman P and Brabec T 1995 Generation of coherent hard-x-ray radiation in crystalline solids by high-intensity femtosecond laser pulses *Phys. Rev.* A **52** R21–4

[KT95]   Kaneva E N and Trifonov E D 1995 Reflection effect on the correlation of superradiant pulses *Opt. Spektrosk.* **79** 293–8 (Engl. Transl. 1995 *Opt. Spectrosc.* **79** 370–4)

[KP95]   Kinrot O and Prior Y 1995 Nonlinear interaction of propagating short pulses in optically dense media *Phys. Rev.* A **51** 4996–5007

[KK95]   Kocharovsky V V and Kocharovsky Vl V 1995 Infrared neoclassical super-radiance in a system of molecules with quasiequidistant spectrum of vibrational levels *Infrared Phys. Technol.* **36** 1003–6

[LKL95]   Lee S J, Khurgin J B and Li S 1995 Switching of superradiance in semiconductor superlattices *Appl. Phys. Lett.* **66** 3316–8

[LBH95]   Lu X, Brownell J H and Hartmann S R 1995 Coherence inhibition in cascade superfluorescence *Laser Phys.* **5** 522–5

[MC95]   Malyshev V and Conjero Jarque E 1995 Optical hysteresis and instabilities inside polariton band gap *J. Opt. Soc. Am.* B **12**

[MST95]   Malyukin Y V, Seminozhenko V P and Tovmachenko O G 1995 Manifestation of autolocalization of exciton excitation in J-aggregates *Zh. Eksp. Teor. Fiz.* **107** 812–23

[MG95a]   Manassah J T and Gross B 1995 The effects of atomic coherence on a three-level homogeneously broadened amplifier in the super-radiant regime *Opt. Commun.* **119** 663–72

[MG95b]   Manassah J T and Gross B 1995 Amplifiers in the superradiant regime 1. Two-level system *Laser Phys.* **5** 509–16

[MG95c]   Manassah J T and Gross B 1995 Amplifiers in the superradiant regime 2. Three-level homogeneously broadened system *Laser Phys.* **5** 517–21

[Mi95]   Miklaszewski W 1995 Near resonant propagation of the light pulse in a homogeneously broadened two level medium *J. Opt. Soc. Am.* **12** 1909–17

[PS95]   Pike E R and Sarkar S 1995 *The Quantum Theory of Radiation* (Oxford: Clarendon)

[PG95]   Pinhasi Y and Gover A 1995 Mode locked super radiant free electron laser oscillator *Nucl. Instrum. Meth. Phys. Res.* A **358** 86–9

[P95]   Piovella N 1995 Transient regime and superradiance in a short-pulse free-electron-laser oscillator *Phys. Rev.* E **51** 5147–50

[PRSZS95]   Plekhanov A I, Rautian S G, Safonov V P, Zhyravlev F A and Shelkovnikov V V 1995 4-photon light scattering by J-aggregates of pseudoisocyanine in a polymer matrix *Opt. Spektrosk.* **78** 92–9 (Engl. Transl. 1995 *Opt. Spectrosc.* **78** 81–7)

[Rv95]   Reshetov V A 1995 Polarization properties of superradiance from levels with hyperfine structure *J. Phys. B: At. Mol. Phys.* **28** 1899–904

[SASZS95]   Samartsev V V, Andrianov S N, Sheibut Y E, Zinoviev P V and Silaeva N B 1995 Optical superradiance in a crystal of biphenyl with pyrene *Laser Phys.* **5** 534–41

[Sr95]   Schiffer M 1995 Black-hole physics and the universalities of superradiance and of grey-body radiation *Gen. Relat. Grav.* **27** 1–8

[S95a]   Seke J 1995 Squeezing and Rabi oscillations in the Dicke model within and without the rotating-wave approximation *Physica A* **213** 587–96

[S95b]   Seke J 1995 Squeezing in the many atom Jaynes–Cummings model within and without the rotating wave approximation *Quantum Semiclass. Opt.* **7** 161–7

[SD95]   Skorobogatov G A and Dzevitskii B E 1995 New observations of collective nuclear superradiation in the isomeric transitions Te-125m2 $\rightarrow$ Te- 125m1+$h\nu$(109.3 keV) and Te-123m2 $\rightarrow$ Te-123m1+$h\nu$(88.46 keV) *Laser Phys.* **5** 258–67

[TMH95]   Tokihiro T, Manabe Y and Hanamura E 1995 Superradiance of Frenkel excitons with any degree of excitation prepared by a short-pulse laser *Phys. Rev.* B **51** 7655–68

[Wg95]   Wang J 1995 Two-atom super-radiance in the photonic band gap *Phys. Lett.* A **204** 54–8

[WR95]   Wang N J and Rabitz H 1995 Optimal control of optical pulse propagation in a medium of three-level systems *Phys. Rev.* A **52** R17–20

[WZZGY95]   Wang H Z, Zheng X G, Zhao F L, Gao Z L and Yu Z X 1995 Superradiance of high density Frenkel excitons at room temperature *Phys. Rev. Lett.* **74** 4079–82

[YC95]   You L and Cooper J 1995 Semiclassical study of spectral evolution under propagation in an active medium *Phys. Rev.* A **51** 4194–202

[Yv95a]   Yukalov V I 1995 Transient coherent phenomena in radiofrequency region *Laser Phys.* **5** 526–33

[Yv95b]   Yukalov V I 1995 Theory of coherent radiation by spin maser *Laser Phys.* **5** 970–92

[Yv95c]   Yukalov V I 1995 Origin of pure spin super-radiance *Phys. Rev. Lett.* **75** 3000–3

1996

[DVB96]   DeVoe R G and Brewer R G 1996 Observation of superradiant and subradiant spontaneous emission of two trapped ions *Phys. Rev. Lett.* **76** 2049–52

Other references

[1] Abragam A 1970 *The Principles of Nuclear Magnetism* (Oxford: Clarendon)

[2] Abramowitz M and Stegun I A 1965 *Handbook of Mathematical Functions* (New York: Dover)

[3] Agranovich V M 1968 *Theory of Excitons* (Moscow: Nauka) (in Russian)

[4] Al'tshuler S A and Kozyrev B M 1983 *Electron Paramagnetic Resonance of Compounds of the Transition Group Elements* (New York: Academic)

[5] Arecchi F T 1969 Photocount distributions and field statistics *Quantum Optics, Proc. Int. School of Phys. E. Fermi XLII* ed R J Glauber (London: Academic) pp 57–110

[6] Bellman R, Birnbaum G and Wagner W G 1963 Transmission of monochromatic radiation in a two-level material *J. Appl. Phys.* **34** 780–2

[7] Ben Aryeh Y, Bowden C M and Englund J C 1986 Intrinsic optical bistability in collections of spatially distributed two-level atoms *Phys. Rev.* A **34** 3917–26

[8] Białynicki-Birula I and Białynicka-Birula Z 1990 Angular correlation of photons *Phys. Rev.* A **42** 2829–38

[9] Born M and Wolf E 1989 *Principles of Optics* 6th edn (Oxford: Pergamon)

[10] Bree A and Vilkos V V 1971 A study of some singlet and triplet electronic states of pyrene *Spectrochim. Acta* A **27** 2333–54

[11] Chesnut D W and Suna A 1963 Fermion behaviour of one-dimensional excitons *J. Chem. Phys* **39** 146–9

[12] Cohen-Tannoudji C, Dupont-Roc J and Grynberg G 1989 *Photons and Atoms* (New York: Wiley)

[13] Courant R and Hilbert D 1962 *Methods of Mathematical Physics* vol II *Partial Differential Equations* (New York: Interscience)

[14] Davydov A S 1971 *Theory of Molecular Excitons* (New York: Plenum)

[15] Demtröder W 1995 *Laser Spectroscopy* (Berlin: Springer)

[16] Eilbeck J C 1972 Reflection of short pulses in linear optics *J. Phys. A: Math. Gen.* **5** 1335–63

[17] Elsaesser T, Shat J, Rota L and Lugle P 1991 Initial thermalization of photoexcited carriers in GaAs studied by femtosecond luminescence spectroscopy *Phys. Rev. Lett.* **66** 1757–60

[18] Fidder H, Knoester J and Wiersma D A 1991 Optical properties of disordererd molecular aggregates: a numerical study *J. Chem. Phys* **95** 7880–90

[19] Fidder H, Terpstra J and Wiersma D A 1991 Dynamics of Frenkel excitons in disordered molecular aggregates *J. Chem. Phys* **94** 6895–907

319

[20] Frantz L M and Nordvik J S 1963 Theory of pulse propagation in a laser amplifier *J. Appl. Phys.* **34** 2346–9

[21] Friedberg R, Hartmann S R and Manassah J T 1989 Mirrorless optical bistability condition *Phys. Rev.* A **39** 3444–6

[22] Galitskii V M and Yelesin V F 1986 *Resonance Interaction between Electromagnetic Fields and a Semiconductor* (Moscow: Nauka)

[23] Gardiner C W 1985 Inhibition of the atomic phase decay by squeezed light: a direct effect of squeezing *Phys. Rev. Lett.* **56** 1917–20

[24] Gardiner C W 1991 *Quantum Noise* (Berlin: Springer)

[25] Glauber R J 1965 Optical coherence and photon statistics *Quantum Optics and Electronics* ed C DeWitt, A Blandin and C Cohen-Tannoudji (New York: Gordon and Breach) pp 63–185

[26] Goldman M 1970 *Spin Temperature Magnetic Resonance in Solids* (Oxford: Clarendon)

[27] Gradshteyn I S and Ryzhik I M 1980 *Tables of Integrals Series and Products* (London: Academic)

[28] Haken H 1977 *Synergetics* (Berlin: Springer)

[29] Heitler W 1960 *The Quantum Theory of Radiation* 3rd edn (Oxford: Oxford University Press)

[30] Hochstrasser R M and Small G J 1968 Spectra and structure of mixed organic crystals *J. Chem. Phys* **48** 3612–24

[31] Hopf F A, Bowden C M and Louisell W H 1984 Mirrorless optical bistability with the use of the local-field correction *Phys. Rev.* A **29** 2591–6

[32] Hopf F A and Bowden C M 1985 Heuristic stochastic model of mirrorless optical bistability *Phys. Rev.* A **32** 268–75

[33] Inguva R and Bowden C M 1990 Spatial and temporal evolution of the first order phase transition in intrinsic optical bistability *Phys. Rev.* A **41** 1670–6

[34] Jackson J D 1975 *Classical Electrodynamics* 2nd edn (New York: Wiley)

[35] Jaynes E T and Cummings F W 1963 Comparison of quantum and semi-classical radiation theories with application to the beam maser *Proc. IEEE* **51** 89

[36] Jeffries C 1963 *Dynamic Nuclear Orientation* (New York: Interscience)

[37] Jelley E E 1936 Spectral absorption and fluorescence of dyes in the molecular state *Nature* **138** 1009–10

[38] Jordan P and Wigner E 1928 Über das Paulische Äquivalenzverbot *Z. Phys.* **47** 631–51

[39] Khalfin L A 1956 *Dokl. Akad. Nauk SSSR* On the theory of the decay of a quasi–stationary state **115** 277–80 (in Russian)

[40] Kitaigorodskii A I 1955 *Organic Crystal Chemistry* (Moscow: Academic) (in Russian)

[41] Konstantinov O V and Pèrel V I 1960 A graphical technique for computation of kinetic quantities *Zh. Eksp. Teor. Fiz.* **39** 197–208 (Engl. Trans. 1961 *Sov. Phys.–JETP* **12** 142)

[42] van Kranendonk J and Sipe J E 1977 Foundations of the macroscopic electromagnetic theory of dielectric media *Progress in Optics XV* ed E Wolf (Amsterdam: North-Holland) pp 245–350

[43] Lamb G L 1980 *Elements of Soliton Theory* (New York: Wiley)

[44] Lamb W E, Schlicher R R and Scully M O 1987 Matter–field interaction in atomic physics and quantum optics *Phys. Rev.* A **36** 2763–72

[45] Landau L D and Lifshitz E M 1976 *Classical Mechanics* (Oxford: Pergamon)

[46] Landau L D and Lifshitz E M 1968 *The Classical Theory of Fields* (Oxford: Pergamon)

[47] Landau L D and Lifshitz E M 1977 *Quantum Mechanics* (Oxford: Pergamon)

[48] Lieb E, Schultz T and Mattis D 1961 Two soluble models of an antiferromagnetic chain *Ann. Phys., NY* **16** 407–66

[49] Lorentz H A 1952 *Theory of Electrons* 2nd edn (New York: Dover)

[50] Loudon R 1983 *The Quantum Theory of Light* (Oxford: Clarendon)

[51] Madey J M J 1971 Stimulated emission of bremsstrahlung in a periodic magnetic field *J. Appl. Phys.* **42** 1906–13

[52] Mandel L 1976 The case for and against semiclassical radiation theory *Progress in Optics XIII* ed E Wolf (Amsterdam: Elsevier North-Holland) pp 27–68

[53] Mandel L 1982 Squeezing and antibunching in coherent generation *Opt. Commun.* **42** 437–39

[54] Pike E R 1969 Some problems in the statisics of optical fields *Quantum Optics, Proc. Int. School of Phys. E. Fermi XLII* ed R J Glauber pp 160–75 (London: Academic)

[55] Pons M L and Roso-Franco L 1990 Feasibility of nonlinear reflection of light by a collisionless molecular beam *Opt. Lett.* **15** 1230–2

[56] Pons M L and Roso-Franco L 1990 Space singularities in the penetration of a plane wave inside a very dense saturable absorber *Europhys. Lett.* **12** 507–12

[57] Power E and Zienau S 1959 Coulomb gauge in non-relativistic quantum electrodynamics and the shape of spectral lines *Phil. Trans. R. Soc.* A **251** 427–54

[58] Provotorov B N 1961 On magnetic resonance saturation in crystals *Zh. Eksp. Teor. Fiz.* **41** 1582–91 (Engl. Transl. 1962 *Sov. Phys.–JETP* **14** 1126–31)

[59] Provotorov B N 1962 Quantum statistical theory of cross-relaxation *Zh. Eksp. Teor. Fiz.* **42** 882–8 (Engl. Transl. 1962 *Sov. Phys.–JETP* **15** 611–4)

[60] Rebane K K and Rebane L A 1974 *Pure Appl. Chem.* **37** 161

[61] Ritsch H, Marte M A M and Zoller P 1992 Quantum noise reduction in Raman lasers *Europhys. Lett.* **19** 7–12

[62] Roso-Franco L 1985 Self-reflected wave inside a very dense saturable absorber *Phys. Rev. Lett.* **55** 2149–51

[63] Roso-Franco L 1987 Propagation of light in a nonlinear absorber *J. Opt. Soc. Am.* B **4** 1878–84

[64] Roso-Franco L and M L Pons 1990 Reflection of a plane wave at the boundary of a saturable absorber: normal incidence *J. Mod. Opt.* **37** 1645–53

[65] Scheibe G 1936 Transformation of absorption spectra of some sensitised dyes and its origin *Angew. Chem.* **49** 563

[66] Spano F C 1991 Fermion excited states in one-dimensional molecular aggregates with site disorder: nonlinear optical response *Phys. Rev. Lett.* **67** 3424–7

[67] Talanov V I 1964 Propagation of ultra-short pulses in an active medium *Izv. Vyssh. Uchebn. Zav. Radiofiz.* **7** 491–6

[68] Tanabe Y and Aoyagi K 1982 Excitons in magnetic insulators *Excitons: Modern Problems in Condensed Matter Science* vol 2, ed E I Rashba and M D Sturge (Amsterdam: North-Holland) p 603

[69] Teich M C and Saleh B E A 1989 Squeezed states of light *Quantum Opt.* **1** 153–91

[70] Trifonov E D 1971 Theory of excitation transfer in crystals with impurity centres *Izv. Akad. Nauk SSSR: Ser. Fiz.* **35** 1330–1335 (in Russian)

[71] Walls D F and Milburn G J 1994 *Quantum Optics* (Berlin: Springer)

[72] *J. Mod. Opt.* 1987 **34** Nos 6,7 (Topical issue on squeezed states of light)

[73] *J. Opt. Soc. Am.* B 1987 **4** No 10 (Topical issue on squeezed states of the electromagnetic fields)

[74] *Appl. Phys.* B 1992 **55** No 3 (Topical issue on quantum noise reduction in optical systems)

[75] *Quantum Optics* 1994 **6** No 8 (The Crested Butte Conference on atomic coherence effects)

Index